Learn Azure Synapse Data Explorer

A guide to building real-time analytics solutions to unlock log and telemetry data

Pericles (Peri) Rocha

BIRMINGHAM—MUMBAI

Learn Azure Synapse Data Explorer

Publishing Product Managers: Birjees Patel and Arindam Majumder
Content Development Editor: Shreya Moharir
Technical Editor: Devanshi Ayare
Copy Editor: Safis Editing
Project Coordinator: Farheen Fathima
Proofreader: Safis Editing
Indexer: Tejal Daruwale Soni
Production Designer: Shankar Kalbhor
Marketing Coordinator: Nivedita Singh

First published: February 2023

Production reference: 1200123

Published by Packt Publishing Ltd.
Livery Place
35 Livery Street
Birmingham
B3 2PB, UK.

ISBN 978-1-80323-395-6

www.packtpub.com

To my daughter, Isabella, I love you to the moon and all the way back. To my wife, Cecilia, my partner, and the love of my life, thank you for your patience, love, friendship, and partnership in life. I love you. To my brother, Plinio, my best friend, and my favorite companion in the things we do together. And last but not least, in loving memory of my mother, Yara, and my father, Jose. This work is dedicated to all of you.

Contributors

About the author

Pericles (Peri) Rocha is a technical product manager, architect, and data scientist with more than 25 years of experience. He has worked with diverse challenges from building highly available database environments to data science projects. He holds an MSc degree in data science from UIUC and is a member of Tau Beta Pi. He currently works at Microsoft as a product manager in the Azure Synapse engineering team. Originally from São Paulo, Brazil, Peri worked in Europe for three years before relocating to the USA in 2016. In his spare time, he enjoys playing music, studying karate, and reading. He lives near Redmond, WA, with his wife, daughter, two dogs, and nine guitars.

I'd like to thank everyone who crossed my path through 25 years of professional experience. All of you helped me shape my own story and I am deeply thankful for it.

About the reviewer

Felipe Andrade is a client technical lead at Microsoft Canada. He has been at Microsoft for 9 years and has been working with data analytics for over 10 years. He has spent most of his career at Microsoft in analytics technical roles working with Power BI, SQL, Synapse, Databricks, and machine learning. He also worked in a couple of startups as a software engineer running social network analytics.

I'd like to thank Peri Rocha for inviting me to be a technical reviewer for his book. Thanks to my family, Leticia, Luisa, and Alice, for their patience and kindness.

Table of Contents

2

Creating Your First Data Explorer Pool 35

3

Exploring Azure Synapse Studio 61

4

Real-World Usage Scenarios 85

7

Data Visualization with Power BI 177

8

Building Machine Learning Experiments 195

9

Exporting Data from Data Explorer Pools 219

Part 3: Managing Azure Synapse Data Explorer

10

System Monitoring and Diagnostics 237

11

Tuning and Resource Management 253

12

Securing Your Environment 265

Preface

Large volumes of data are generated daily from applications, websites, internet of things devices, and other free-text, semi-structured data sources. Azure Synapse Data Explorer helps you collect, store, and analyze such data, and enables you to work with other analytical engines, such as Apache Spark, to develop advanced data science projects and maximize the value you get from your log and telemetry data.

This book offers a comprehensive view of Azure Synapse Data Explorer, covering not only the core scenarios of Data Explorer but also how it integrates into the whole picture within Azure Synapse. From data ingestion, through data visualization and advanced analytics, you will learn an end-to-end approach to maximizing the value of unstructured data and driving powerful insights using data science capabilities. With real-world usage scenarios, you'll learn how to identify key projects where Azure Synapse Data Explorer can help you achieve your business goals. You will also learn how to manage big data as part of a platform as a service offering, tune, secure, and serve data at scale to end users.

By the end of this book, you will have mastered the big data life cycle and be able to implement advanced analytical scenarios from raw telemetry and log data.

Who this book is for

If you are a data engineer, data analyst, or business analyst working with unstructured data and want to learn how to maximize the value of such data, this book is for you. To maximize your learning experience from this book, you should be familiar with working with data and performing simple queries using SQL or KQL. Even though it is not a requirement, familiarity with Python will help you get more from the examples. This book is also excellent for professionals already working with Azure Synapse who want to incorporate unstructured data into their data science projects.

What this book covers

Chapter 1, *Introducing Azure Synapse Data Explorer*, is the first of four chapters in *Part 1*, *Introduction to Azure Synapse Data Explorer*, where you will be introduced to the product and learn the basics that you need before you start to work with data. It welcomes you to Azure Synapse Data Explorer and elaborates on the need for a fast and highly scalable data exploration service for telemetry and log data. It introduces Azure Synapse and explains how the Data Explorer service fits under the Azure Synapse umbrella. Finally, it discusses the architecture and infrastructure of Data Explorer pools, and the scale of the service today.

Chapter 2, Creating Your First Data Explorer Pool, gets your hands busy by walking you through the creation of your first Azure Synapse workspace and a Data Explorer pool using the Azure portal, Azure Synapse Studio, or the Azure **Command-Line Interface** (**CLI**). If you are not familiar with Azure yet, don't worry; this chapter guides you through the steps to create your first free Azure account, allowing you to follow the examples in the book.

Chapter 3, Exploring Azure Synapse Studio, introduces the development and management environment of Azure Synapse. You will learn about the user interface elements of Azure Synapse Studio, and where to find what you are looking for by navigating through the hubs. In addition to that, in this chapter, you will load some data into a database and run your first query to help you familiarize yourself with the query editor. This chapter closes with an overview of where to manage and monitor your environment using Azure Synapse Studio.

Chapter 4, Real-World Usage Scenarios, describes some example solution architectures you can use in common log and telemetry data analytics scenarios. It looks at five real-world use cases that integrate Azure Synapse Data Explorer with other Azure services and helps you understand the blueprints so that you can build your own.

Chapter 5, Ingesting Data into Data Explorer Pools, kicks off *Part 2, Working with Data*. It walks you through the data loading process, choosing your own data loading strategy, and walks you through different ways to load data into Data Explorer pools. This chapter builds the data assets that you will use in most chapters of the book.

Chapter 6, Data Exploration and Analysis with KQL and Python, is all about learning how to query, transform, and get insights from your data using **Kusto Query Language** (**KQL**) and Python. You will learn how to use KQL to explore the data you have at hand and familiarize yourself with the schema, plot simple charts in the query editor, obtain percentiles, and even use native KQL commands to look at trends in your data using linear regression. In the second half of this chapter, you will create an Azure Synapse notebook to explore and transform data using Python and create a lake database.

Chapter 7, Data Visualization with Power BI, complements the previous chapter by helping you configure Power BI integration with Azure Synapse and author new Power BI reports directly from Azure Synapse Studio. It walks you through the creation of reports that connect to data in Data Explorer pools, as well as to your new lake database.

Chapter 8, Building Machine Learning Experiments, provides an overview of applied machine learning, and how to introduce advanced analytics to your Azure Synapse projects using **automated machine learning** (**AutoML**). You will use Python to prepare your data for machine learning experiments, train a series of models, and find the best model to help you predict values.

Chapter 9, Exporting Data from Data Explorer Pools, closes *Part 2, Working with Data*, by walking you through data export scenarios. It explains scenarios where data exports are needed and walks you through different options you have available to perform data exports, including continuous data exports.

Chapter 10, System Monitoring and Diagnostics, is the first of four chapters in *Part 3, Managing Azure Synapse Data Explorer*. In this chapter, you will learn about managing a platform-as-a-service service such as Azure Synapse, and which parts of the service you should be concerned with. Through code examples and guidance through the user interface, you will learn how to stay on top of your Data Explorer pools and proactively monitor them. By setting up alerts, you'll learn how to get notified on your phone if an event of interest happens in your environment.

Chapter 11, Tuning and Resource Management, introduces resources to help you provide predictable performance to end users and using cache policies to speed up queries. It walks you through the implementation of resource management to help you categorize user requests to prioritize the execution of critical workloads while queueing requests that can wait.

Chapter 12, Securing Your Environment, provides you with the information you need to make sure your data is secure at rest and in transit, and that only people who are intended to access your data have access to it. It walks you through an overview of the security issues you need to consider for your own implementations, how to double-encrypt your data for an added layer of security, how to authenticate and authorize users, and how to protect the network environment that transits your data.

Chapter 13, Advanced Data Management, covers how to adhere to governmental regulations for data handling, including how to permanently purge personal data. You will learn how to use extents, or data shards, in Azure Synapse Data Explorer to move large volumes of data quickly for archival.

To get the most out of this book

To maximize your learning experience, you should have a basic understanding of concepts around data integration, data retrieval, and building basic data visualizations. Previous experience with SQL, KQL, and Python is not required, but it will help you understand the concepts in the code examples more quickly.

Software/hardware covered in the book	Operating system requirements
Azure Synapse Studio	Windows, macOS, or Linux
The Azure portal	Windows, macOS, or Linux
Power BI Desktop	Windows
Microsoft Azure App	iOS or Android

The Azure portal and Azure Synapse Studio are web-based tools that are used to manage, develop, and build solutions for Azure Synapse Data Explorer. Microsoft supports the latest versions of the following browsers: Microsoft Edge, Safari (Mac only), Chrome, and Firefox.

To install Power BI Desktop, visit `https://learn.microsoft.com/power-bi/fundamentals/desktop-get-the-desktop`.

To install the Microsoft Azure App, visit `http://aka.ms/getazureapp` on your mobile device, or look for the Microsoft Azure App in your device's app store.

If you are using the digital version of this book, we advise you to type the code yourself or access the code from the book's GitHub repository (a link is available in the next section). Doing so will help you avoid any potential errors related to the copying and pasting of code.

Download the example code files

You can download the example code files for this book from GitHub at `https://github.com/PacktPublishing/Learn-Azure-Synapse-Data-Explorer`. If there's an update to the code, it will be updated in the GitHub repository.

We also have other code bundles from our rich catalog of books and videos available at `https://github.com/PacktPublishing/`. Check them out!

Download the color images

We also provide a PDF file that has color images of the screenshots and diagrams used in this book. You can download it here: `https://packt.link/DQQ7A`.

Conventions used

There are a number of text conventions used throughout this book.

`Code in text`: Indicates code words in text, database table names, folder names, filenames, file extensions, pathnames, dummy URLs, user input, and Twitter handles. Here is an example: "To create or alter a new workload group, use the `.create-or-alter workload_group` command."

A block of code is set as follows:

````
.alter-merge workload_group ['Engineering Department WG'] ```
{
  "RequestQueuingPolicy": {
      "IsEnabled": true
  }
} ```
````

Any command-line input or output is written as follows:

```
az synapse kusto pool create --name "droneanalyticsadx"
--resource-group "rg-AzureSynapse" --sku name="Compute
optimized" size="Small" --workspace-name "drone-analytics"
```

Bold: Indicates a new term, an important word, or words that you see onscreen. For instance, words in menus or dialog boxes appear in **bold**. Here is an example: "To enable it, you must select the **Enable** option next to **Double encryption using a customer-managed key**, in the **Security** tab of the **Create Synapse workspace** wizard."

> **Tips or important notes**
> Appear like this.

Get in touch

Feedback from our readers is always welcome.

General feedback: If you have questions about any aspect of this book, email us at `customercare@packtpub.com` and mention the book title in the subject of your message.

Errata: Although we have taken every care to ensure the accuracy of our content, mistakes do happen. If you have found a mistake in this book, we would be grateful if you would report this to us. Please visit `www.packtpub.com/support/errata` and fill in the form.

Piracy: If you come across any illegal copies of our works in any form on the internet, we would be grateful if you would provide us with the location address or website name. Please contact us at `copyright@packt.com` with a link to the material.

If you are interested in becoming an author: If there is a topic that you have expertise in and you are interested in either writing or contributing to a book, please visit `authors.packtpub.com`.

Share Your Thoughts

Once you've read *Learn Azure Synapse Data Explorer*, we'd love to hear your thoughts! Scan the QR code below to go straight to the Amazon review page for this book and share your feedback.

https://packt.link/r/1-803-23395-8

Your review is important to us and the tech community and will help us make sure we're delivering excellent quality content

Download a free PDF copy of this book

Thanks for purchasing this book!

Do you like to read on the go but are unable to carry your print books everywhere? Is your eBook purchase not compatible with the device of your choice?

Don't worry, now with every Packt book you get a DRM-free PDF version of that book at no cost.

Read anywhere, any place, on any device. Search, copy, and paste code from your favorite technical books directly into your application.

The perks don't stop there, you can get exclusive access to discounts, newsletters, and great free content in your inbox daily

Follow these simple steps to get the benefits:

1. Scan the QR code or visit the link below

`https://packt.link/free-ebook/9781803233956`

2. Submit your proof of purchase
3. That's it! We'll send your free PDF and other benefits to your email directly

Part 1

Introduction to Azure Synapse Data Explorer

To maximize your learning experience, you should quickly become familiar with the core concepts and tools you will work with when reproducing the examples and learning new concepts, and how these concepts can help you in real-life projects. The first part of the book focuses on introducing Azure Synapse Data Explorer and all of its layers. You will learn about the service architecture, all of the platform elements within Azure Synapse, and how to create your own lab environment to run through the book examples. You will also become familiar with Azure Synapse Studio, and the development and management interface of Azure Synapse. Finally, you will learn about solution templates from real-world usage scenarios that will help you speed up your own Azure Synapse Data Explorer implementations.

This part comprises the following chapters:

- *Chapter 1, Introducing Azure Synapse Data Explorer*
- *Chapter 2, Creating Your First Data Explorer Pool*
- *Chapter 3, Exploring Azure Synapse Studio*
- *Chapter 4, Real-World Usage Scenarios*

1
Introducing Azure Synapse Data Explorer

Every day, applications and devices connected to the internet generate massive amounts of data. To give some perspective, we expect to have 50 billion connected devices by 2030 generating data, and up to 175 **zettabytes** (**ZB**) of data generated by 2025 (from every possible source). As more and more new connected devices reach the market every year, and as companies make greater use of unstructured data from application logs, the amount of data generated daily will become difficult to measure. In fact, some companies are keeping certain types of data, such as telemetry and application logs, for no longer than a certain period (such as 90 to 120 days) because even with the fact that storage has never been cheaper, storing and managing large volumes of data can quickly become cost-prohibitive.

Being able to store, manage, and quickly analyze unstructured data has become a critical business need for most companies. From application logs, you can predict the behavior of users and respond quickly to user demand. By analyzing device telemetry, you can anticipate hardware failures, reduce downtime in factories, predict the weather, and detect patterns that help optimize your operation. Most importantly, the ability to correlate application and device data, apply **machine learning** (**ML**) algorithms, and visualize data in real time allows you to respond quickly to operational challenges, as well as customer and market demands.

Azure Synapse Data Explorer complements the **Synapse Structured Query Language** (**Synapse SQL**) engine and Apache Spark engine already present in Azure Synapse to offer a big data service that helps acquire, store, and manage big data to unlock insights from device telemetry and application logs. It works just like the **Azure Data Explorer** standalone service, but with the benefit of tightly integrating with the other services offered by Azure Synapse, allowing you to build **end-to-end** (**E2E**) advanced analytics projects from data ingestion to rich visualizations using Power BI.

By the end of this chapter, you should have a thorough understanding of where Azure Synapse Data Explorer fits in the data lifecycle, how to describe the service and differentiate it from the standalone service, and when to use Data Explorer pools in Azure Synapse.

In this chapter, we will go through the following topics:

- Understanding the lifecycle of data
- Introducing the Team Data Science Process
- The need for a fast and highly scalable data exploration service
- What is Azure Synapse?
- What is Azure Synapse Data Explorer?
- Integrating Data Explorer pools with other Azure Synapse services
- Exploring the Data Explorer pool infrastructure and scalability
- What makes Azure Synapse Data Explorer unique?
- When to use Azure Synapse Data Explorer

Technical requirements

To build your own environment and experiment with the tools shown in this chapter (and throughout the book), you will need an Azure account and a subscription. If you don't have an Azure account, you can create one for free at `https://azure.microsoft.com/free/`. Microsoft offers $200 in Azure credit for 30 days, as well as some popular services for free for 1 year. Azure Synapse is not one of the free services, but you should be able to use your free credit to run most examples in this book as long as you adhere to the following practices:

- **Using the smallest pool sizes**: Azure Synapse Data Explorer offers pool sizes ranging from extra small (2 cores per instance) to large (16 cores per instance). Picking the smallest pool size options will help you save money and still learn about Azure Synapse Data Explorer without any constraints.
- **Keeping your scale to a minimum**: As with pool sizes, you don't need several instances running on your cluster to learn about Azure Synapse Data Explorer. Avoid using autoscale (discussed in *Chapter 2*), and keep your instance count to a minimum of two.
- **Manage your storage**: Azure Synapse Data Explorer also charges you by storage usage, so if you're trying to save costs in your learning journey, make sure you only have the data you need for your testing.
- **Stop your pools when not in use**: You are charged for the time your cluster is running, even if you are not using it. Make sure you stop your Data Explorer pools when you are done with your experiments so that you are not charged. You can resume your pools next time you need them!

One or more examples in this chapter make use of the *New York Yellow Taxi* open dataset available at `https://docs.microsoft.com/en-us/azure/open-datasets/dataset-taxi-yellow?tabs=azureml-opendatasets`.

> **Note**
>
> The Azure free account offer may not be available in your country. Please check the conditions before you apply.

Understanding the lifecycle of data

The typical data lifecycle in the world of analytics begins with data generation and ends with data analysis, or visualization through reports or dashboards. In between these steps, data gets ingested into an analytical store. Data may or may not be transformed in this process, depending on how the data will be used. In some cases, data can be updated after it has been loaded into an analytical store, even though this is not optimal. Appending new data is quite common.

Big data is normally defined as very large datasets (volume) that can be structured, semi-structured, or unstructured, without necessarily having a pre-defined format (variety), and data that changes or is produced fast (velocity). *Volume*, *variety*, and *velocity* are known as *the three Vs* of big data.

> **Note**
>
> While most literature defines the three Vs of big data as volume, variety, and velocity, you may also see literature that defines them as five Vs: the previously mentioned volume, variety, velocity, but also veracity (consistency, or lack of) and value (how useful the data is). It is important to understand that a big data solution needs to accommodate loading large volumes of data at low latency, regardless of the structure of the data.

For data warehousing and analytics scenarios in general, you will typically go through the following workflow:

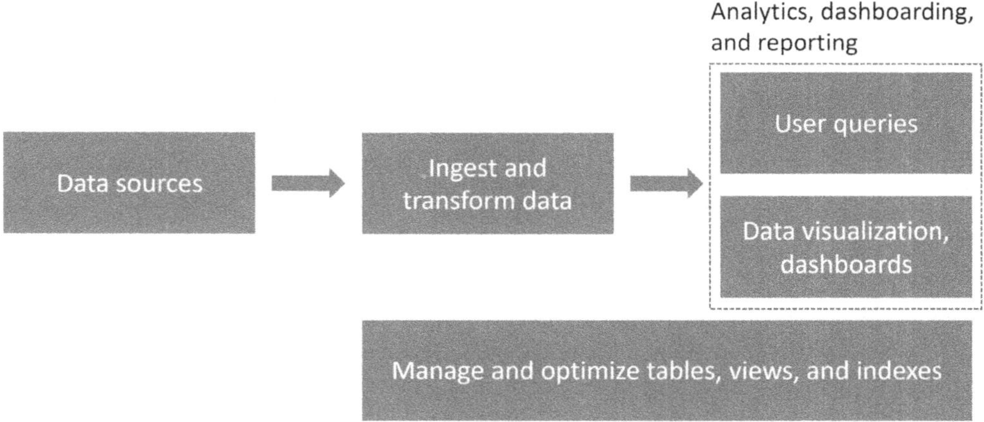

Figure 1.1 – A typical workflow in analytics

Let us break down the steps in this process, as follows:

- **Data sources**: This is where data originates from. Some examples of data sources may include a sales application that stores transactions on a database (in which case, the database in question would be the source), telemetry data from **internet of things** (**IoT**) devices, application log data, and much more.
- **Create database objects**: The first step is to create the database itself, and any objects you will need to start loading data. Creating tables at this stage is common, but not required—in many cases, you will create destination tables as part of the data ingestion phase.
- **Ingest and transform data**: The second step is to bring data to your analytical store. This step involves acquiring data, copying it to your destination storage location, transforming data as needed, and loading it to a final table (not necessarily in this order—sometimes, you will load data and transform it in the destination location) that will be retrieved by user queries and dashboards. This can be a complex process that may involve moving data from a source location to a data lake (a data repository where data is stored and analyzed in its raw form), creating intermediary tables to transform data (sort, enrich, clean data), creating indexes and views, and other steps.
- **User queries, data visualization, and dashboards**: In this step, data is ready to be served to end users. But this does not mean you are done—you need to make sure queries are executed at the expected performance level, and dashboards can refresh data without user interaction while reducing overall system overhead (we do not want a dashboard refreshing several times per day if that's not needed).
- **Manage and optimize tables, views, and indexes**: Once the system is in production and serving end users, you will start to find system bottlenecks and opportunities to optimize your analytical environment. This will involve creating new indexes (and maintaining the ones you have created before!), views, and materialized views, and tuning your servers.

The lifecycle of big data can be similar to that of a normal data warehouse (a robust database system used for reporting and analytics), but it can also be very specific. For the purpose of this book, we'll look at big data from the eyes of a data scientist, or someone who will deliver advanced analytics scenarios from big data. Building a pipeline and the processes to ensure data travels quickly from when it is produced to unlock insights without compromising quality or productivity is a challenge for companies of all sizes.

The lifecycle of data described here is widely implemented and well proven as a pattern. With the growth of the data science profession, we have observed a proliferation of new tools and requirements for projects that went well beyond this pattern. With that, came the need for a methodology that helps govern ML projects from gathering requirements up to model deployment, and everything in between, allowing data scientists to focus on the outcomes of their projects as opposed to building a new approach for every new project. Let's look at how the TDSP helps achieve that.

Introducing the Team Data Science Process

In 2016, Microsoft introduced the **Team Data Science Process** (**TDSP**) as an agile, iterative methodology to build data science solutions at scale efficiently. It includes best practices, role definitions, guidelines for collaborative development, and project planning to help data scientists and analysts build E2E data science projects without having to worry about building their own operational model.

Figure 1.2 illustrates the stages in this process:

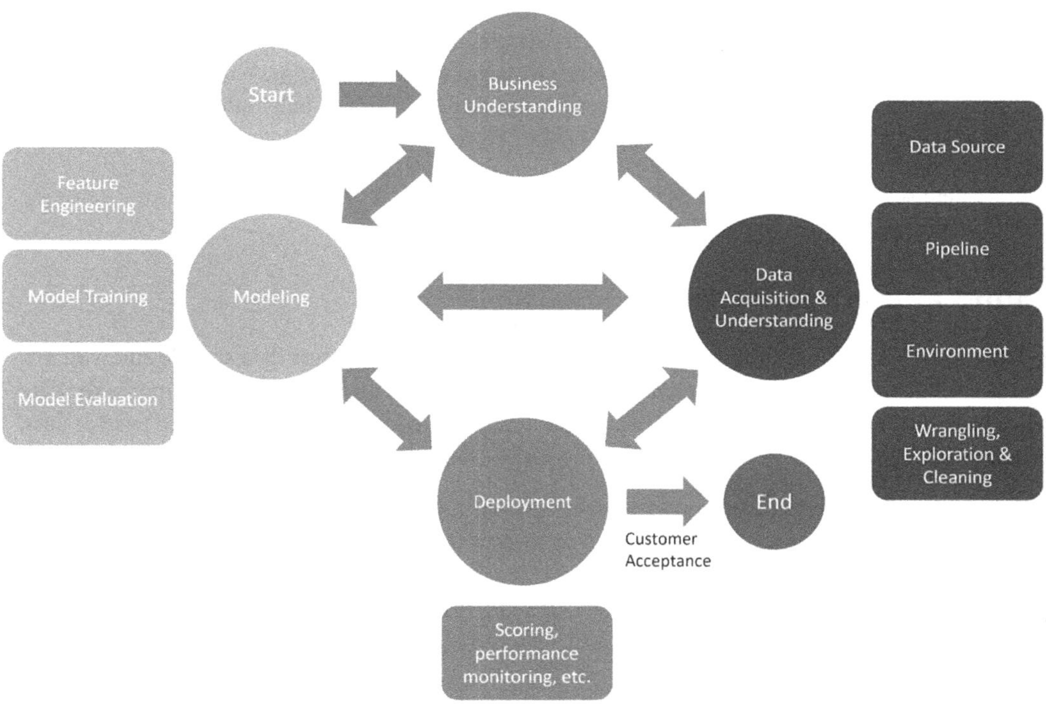

Figure 1.2 – The TDSP lifecycle

At a high level, the TDSP lifecycle outlines the following stages of data science projects:

1. **Business Understanding**: This stage involves working with project stakeholders to assess and identify the business problems that are being addressed by the project, as well as to define the project objectives. It also involves identifying the source data that will be used to answer the business problems that were identified.

2. **Data Acquisition & Understanding**: At this stage, the actual ingestion of data begins, ensuring a clean, high-quality dataset that has a clear relationship with the business problems identified in the **Business Understanding** stage. After having performed initial data ingestion, in this stage, we explore the data to determine whether thedata quality is, in fact, adequate.
3. **Modeling**: After ensuring we have the right data that help address the business problems, we now perform **feature engineering** (**FE**) and model training. By creating the right features from your source data and finding the model that best answers the problem specified in the **Business Understanding** stage, in this stage we determine the model that is best suited for production use.
4. **Deployment**: This is where we operationalize the model that was identified in the *Modeling* stage. We build a data pipeline, deploy the model to production, and prepare the interfaces that allow model consumption from external applications.
5. **Customer Acceptance**: By now, we have a data pipeline in place and a model that helps address the business challenges identified at the beginning of our project. At the **Customer Acceptance** stage, we get agreement from the customer that this project in fact helps address our challenges and identify an entity to whom we hand off the project for ongoing management and operations.

For more details about the TDSP, refer to `https://docs.microsoft.com/en-us/azure/architecture/data-science-process/overview`.

Tooling and infrastructure

Big data projects will require specialized tools and infrastructure to process data at scale and with low latency. The TDSP provides recommendations for infrastructure and tooling requirements for data science projects. These recommendations will include the underlying storage systems used to store data, the analytical engines (such as SQL and Apache Spark), cloud services to host ML models, and more.

Azure Synapse offers the infrastructure and development tools needed in big data projects from data ingestion through data storage, with the option of analytical engines for data exploration and to serve data to users at scale, as well as modeling, and data visualization. In the next sections, we will explore the full data lifecycle and how Azure Synapse helps individuals deliver E2E advanced analytics and data science projects.

The need for a fast and highly scalable data exploration service

Data warehouses, and SQL-based databases, have reached a level of maturity where the technologies are stable, widely available from a variety of vendors, and popularly adopted by enterprises. *Structured* databases are efficiently stored, and queries are resolved by using techniques such as indexing and materialized views (among other techniques) to quickly retrieve the data requested by the user.

Unstructured data, however, does not have a pre-defined schema, or structure. Storing unstructured data optimally is challenging, as data pages cannot be calculated in advance the way they are in typical SQL databases. The same challenges apply to the processing and querying of unstructured data.

Application logs and IoT device data are good examples of unstructured data that is produced at low latency. They are text-heavy but without pre-defined text sizes. An application log can not only contain clickstreams, user feedback, and error messages, but also dates and device **identifiers (IDs)**. IoT device data may include facts such as a count of objects scanned and measures, but also barcode numbers, descriptive text, coordinates, and more.

This is all high-value data that companies now realize can be useful to improve products and respond quickly to market changes and user feedback. Therefore, being able to efficiently store, process, query, and maintain unstructured data is a real requirement for companies of all sizes. But managing big data by itself is not enough—we need the means to efficiently acquire, manage, explore, model, and serve data to end users. In short, we need to realize the full data lifecycle to unlock insights and maximize the value of data. On top of that, we need to make sure that your company's data, being such a valuable asset, is well protected from unauthorized access, and that the analytical environment adheres to mission-critical requirements imposed by enterprises. Let us now look at how Azure Synapse helps address these needs.

What is Azure Synapse?

Azure Synapse is a unified analytics platform that brings together several cloud services to help you manage your data science projects from data ingestion all the way to serving data to end users. *Figure 1.3* illustrates the service architecture for Azure Synapse Analytics:

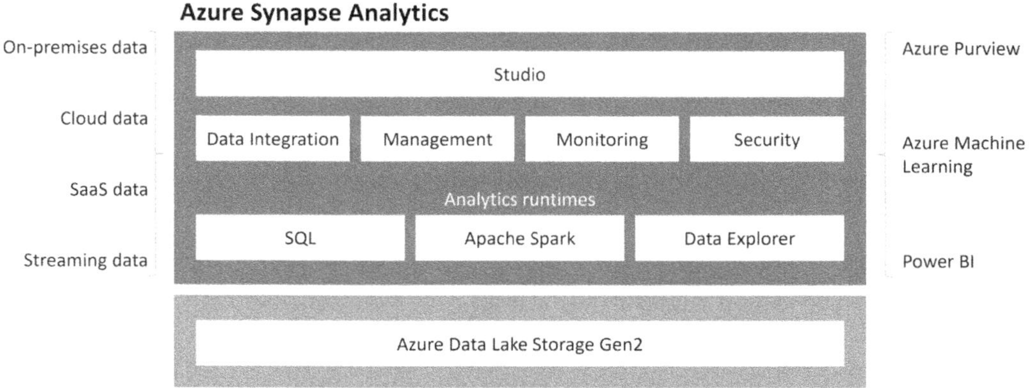

Figure 1.3 – Azure Synapse Analytics service architecture

All these capabilities are managed by the *umbrella* Azure Synapse service in the form of what is called an Azure Synapse *workspace* (the shaded area on top in *Figure 1.3*). When you provision a new Azure Synapse workspace, you are offered a single point of entry and single point of management for all services included in Azure Synapse. You don't need to go to different places to create data ingestion pipelines, explore data using Apache Spark, or author Power BI reports—instead, all the work is done through one development and management environment called **Azure Synapse Studio**, reachable through `https://web.azuresynapse.net`.

Before Azure Synapse, E2E advanced analytics and data science projects were built by putting together several different services that could be hosted on the cloud or on-premises. The promise of Azure Synapse is to offer one platform where all advanced analytics tasks can be performed.

Let's look in detail at the capabilities offered by Azure Synapse.

Data integration

Leveraging the **Azure Data Factory** (**ADF**) code base, Azure Synapse pipelines offer a code-free experience to build data integration jobs that enable data ingestion from data sources in the cloud, on-premises, **Software-as-a-Service** (**SaaS**) sources, data streaming, and more. It includes native connectors to more than 95 data sources.

With Azure Synapse pipelines, data engineers can build E2E workflows for data moving and processing. Azure Synapse supports nested activities, linked services, and execution triggers, and offers common data transformation and data wrangling (transforming data, or mapping data to other columns) activities. In Azure Synapse, you can even add a notebook activity that contains complex logic to process data using an **Azure Synapse notebook**, a code-rich experience, or use flowcharts with a rich user-friendly interface that implements complex pipelines using a code-free experience. This is illustrated in *Figure 1.4*.

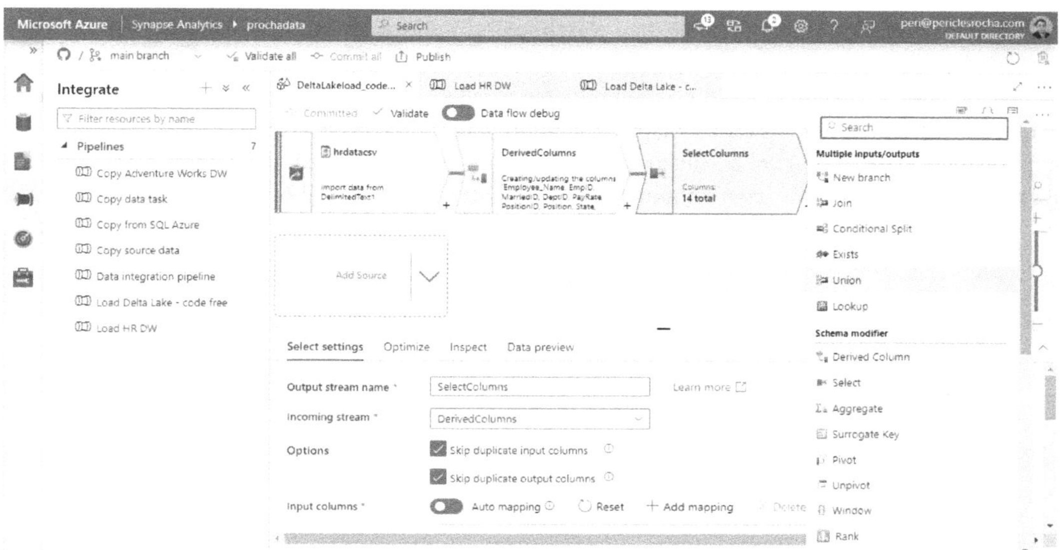

Figure 1.4 – Code-free pipeline and data flow authoring experience

> **Note**
>
> Not all features offered in ADF are available in Azure Synapse Analytics. To explore the differences between ADF and Azure Synapse Analytics, visit `https://docs.microsoft.com/en-us/azure/synapse-analytics/data-integration/concepts-data-factory-differences`.

You should avoid loading data into Azure Synapse SQL using traditional methods that are normal practice in Microsoft SQL Server. For example, issuing batch `INSERT`, `UPDATE`, and `DELETE` statements to load or update data is not an optimal process in Azure Synapse SQL, because the **massively parallel processing** (**MPP**) engine in Azure Synapse SQL was not designed for singleton operations. To help load data efficiently, besides using pipelines, Azure Synapse offers a convenient `COPY` **Transact-SQL** (**T-SQL**) command that helps move data from **Azure Data Lake Storage Gen2** (**ADLS Gen2**) to Synapse SQL tables in an optimal fashion.

Enterprise data warehousing

An enterprise data warehouse—or, simply put, a data warehouse—is a centralized system that integrates data from disparate data sources to enable reporting and analytics in organizations. It stores the data efficiently and is configured to serve data through reporting or user queries, without inflicting overhead on transactional systems.

Azure Synapse offers a highly scalable data warehousing solution through Synapse SQL pools—a distributed query engine to process SQL queries at **petabyte** (**PB**) volume. The SQL analytical engine in Azure Synapse is an evolution of a product previously called Azure SQL Data Warehouse. An Azure Synapse workspace can have several Synapse SQL pools, and the user can run queries using any of the compute pools available. *Figure 1.5* illustrates the ability to pick the desired compute pool for a given query. Pools that have a gray icon without a checkmark are either stopped or not available for use:

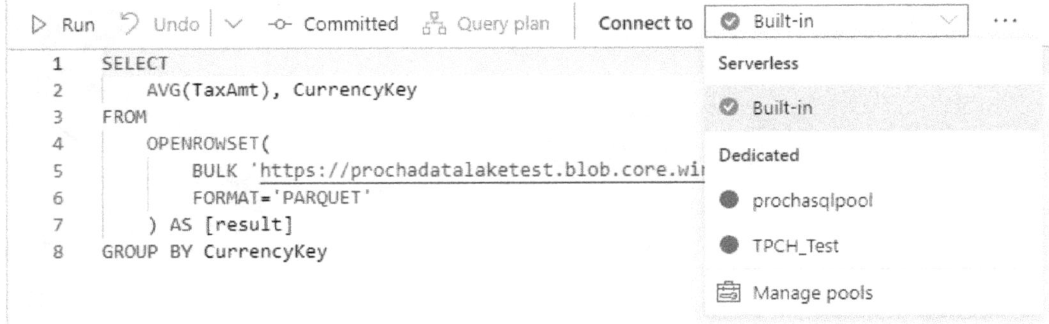

Figure 1.5 – Picking a SQL pool in the query editor

Synapse SQL is offered in two flavors, as detailed here:

- **Dedicated SQL pool:** This is a pre-provisioned compute cluster that offers predictable performance and cost. When a dedicated SQL pool is provisioned, the cluster capacity is reserved and kept online to respond to user queries, unless you choose to pause the SQL pool—a strategy to save money when the cluster is not in use. Dedicated SQL pools run an MPP engine to distribute data across nodes on a cluster and retrieve data. A central **control node** receives user queries and distributes them across the cluster nodes, resolving the user queries in parallel. When you provision a new dedicated SQL pool, you specify its desired cluster size based on your **service-level objective**. Dedicated SQL pool sizes range from one cluster node to up to 60 nodes processing user queries.
- **Serverless SQL pools**: A query engine that is always available to use when needed, mostly applicable for unplanned use, or *bursty* workloads. You do not need to pre-provision serverless SQL pools. You are charged based on the data volume processed in queries. Every Azure Synapse workspace includes a serverless SQL pool. The distributed query processing engine that runs serverless SQL pools is more robust and more complex than the engine that runs dedicated SQL pools and assigns resources to the cluster as needed. You do not control the size of your compute pool or how many resources are allocated to user queries.

Figure 1.6 illustrates the service architecture for dedicated and serverless SQL pools:

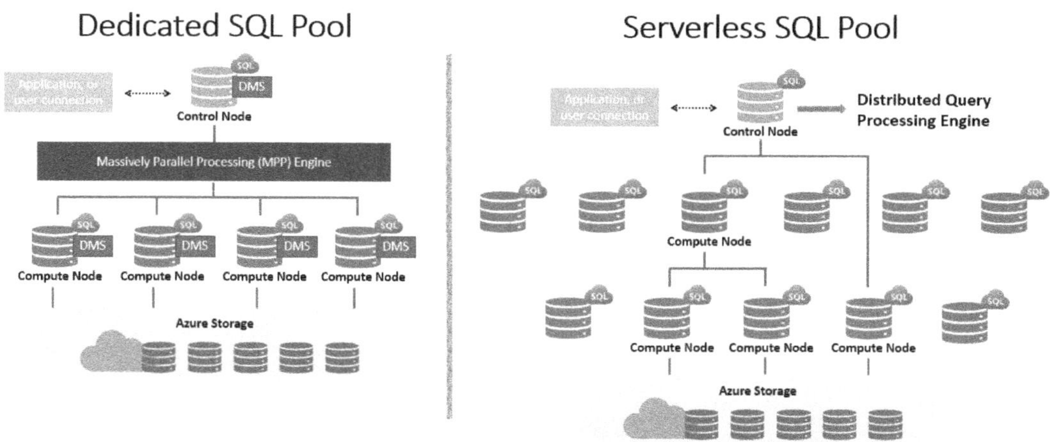

Figure 1.6 – Service architecture for dedicated and serverless SQL pools (adapted from https://docs.microsoft.com/azure/synapse-analytics/sql/overview-architecture)

To learn in depth how serverless pools work in Azure Synapse, I recommend the *POLARIS: The Distributed SQL Engine in Azure Synapse* white paper, which can be found at `https://www.vldb.org/pvldb/vol13/p3204-saborit.pdf`.

Exploration on the data lake

Through native integration with ADLS Gen2, serverless SQL pools allow you to query data directly from files residing on Azure Storage. You can store data in a variety of file formats, such as Parquet or **comma-separated files** (**CSV**), and query it using the familiar T-SQL language.

Exploration on the data lake offers a quick alternative for users who want to explore and experiment with the existing data before it gets loaded into Synapse SQL tables for high-performance querying. In *Figure 1.7*, you can see a T-SQL query that uses the `OPENROWSET` operator to reference data from a Parquet file stored on ADLS:

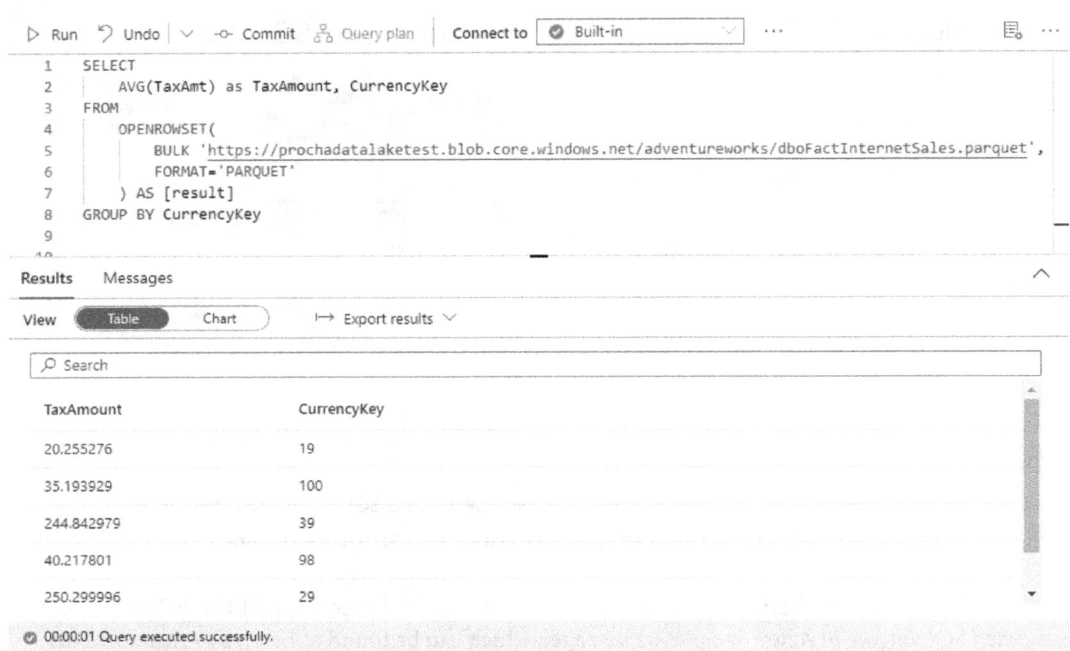

Figure 1.7 – Running T-SQL queries to query data stored on the data lake

This is a powerful capability for users who want to explore data before they decide how to store it to enable the processing of queries at scale. To learn more about exploring data on the data lake using serverless SQL pools, visit `https://learn.microsoft.com/azure/synapse-analytics/get-started-analyze-sql-on-demand`.

Apache Spark

Apache Spark is an open source, highly scalable big data processing engine. It achieves high performance by supporting in-memory data processing and automatically distributing jobs across nodes in a cluster of servers. Apache Spark is widely popular in the data science community not only for its performance benefits (achieved due to its support for in-memory processing and scalability, as described), but also due to the fact that it has built-in support for popular languages such as Python, R, and Scala. Some Apache Spark distributors add additional support for third-party languages as well, such as SQL or C#.

Azure Synapse includes a fully managed Spark service that can be used for data exploration, data engineering, data preparation, and creating ML models and applications. You can choose from a range of programming languages for your data exploration needs, including C#, Scala, R, PySpark, and Spark SQL. The Apache Spark service offered by Azure Synapse is automatically provisioned based on your workload size, so you do not need to worry about managing the actual instances in the cluster.

Apache Spark in Azure Synapse comes with a rich set of libraries, including some of the most used by data engineers and data scientists, such as NumPy, Pandas, Scikit-learn, Matplotlib, and many others. You can also install any packages that are compatible with the Spark distribution used in your Apache Spark pool.

> **Note**
>
> To see a full list of libraries that are pre-installed in Apache Spark pools in Azure Synapse, visit `https://docs.microsoft.com/en-us/azure/synapse-analytics/spark/apache-spark-version-support` and select your desired Spark runtime version.

To explore data using Apache Spark, Azure Synapse offers a notebook experience that allows users to use markdown cells and code cells, as with other popular notebook experiences that are available on the market. This experience is illustrated in *Figure 1.8*.

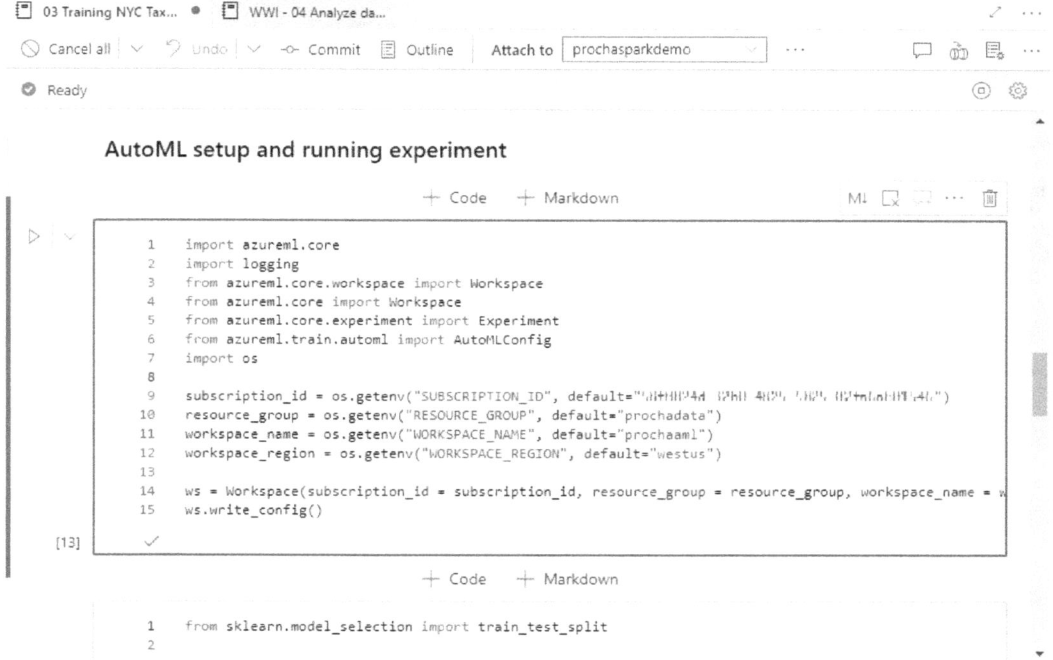

Figure 1.8 – Synapse notebooks authoring experience (subscription ID obfuscated)

Notebooks are saved in your Synapse workspace (or in your source control system if you configured Git integration) just like other workspace artifacts, so anyone connecting to the same workspace will be able to collaborate on your notebooks.

Log and telemetry analytics

Azure Synapse includes native integration with Azure Data Explorer to bring log and telemetry data to E2E advanced analytics and data science projects. You can pre-provision Data Explorer pools in Azure Synapse and have reserved compute capacity for your analytical needs.

Through Azure Synapse Data Explorer (in preview at the time of writing), Data Explorer pools in Azure Synapse enable interesting new scenarios for analysts, data scientists, and data engineers. For example, they offer integration with notebooks in Azure Synapse, allowing you to explore data using your language of choice, in a fully collaborative environment. As you can see in *Figure 1.9*, by right-clicking a table on a Data Explorer pool database and selecting **New notebook**, Azure Synapse Studio can automatically generate a notebook with code to load that table to a Spark data frame:

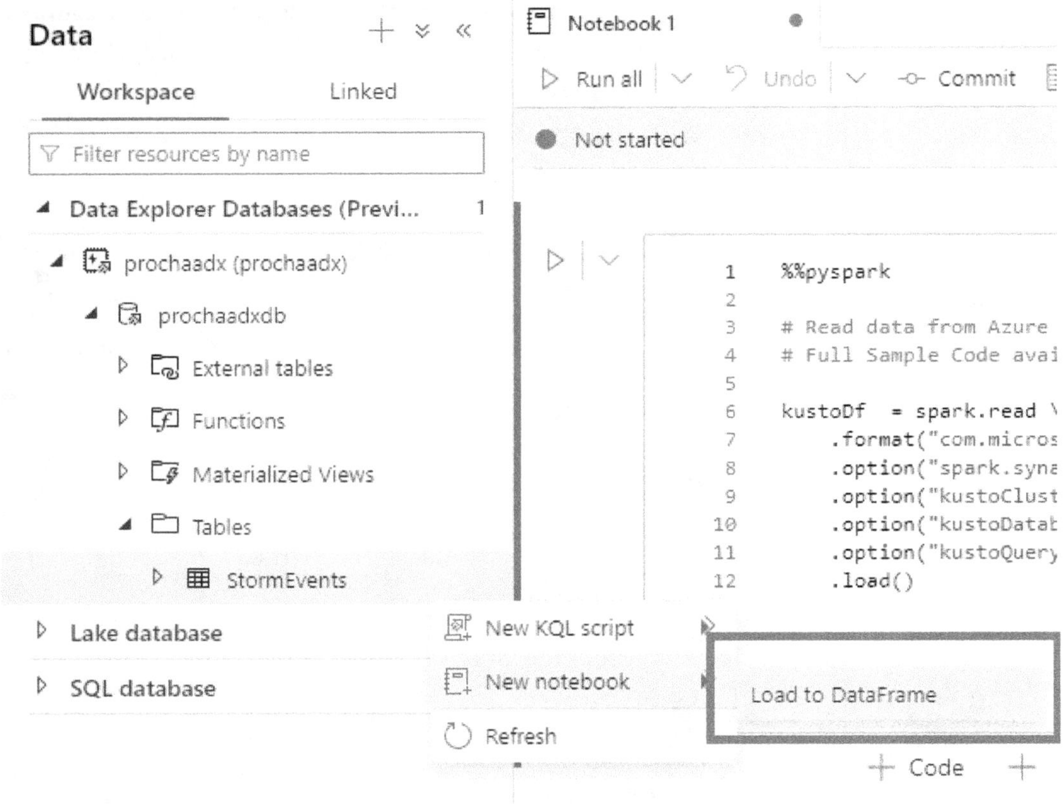

Figure 1.9 – Notebooks reading data from Data Explorer pools

The notebook experience is the same as with Apache Spark reading data from the data lake, except in this case, Azure Synapse generates code for you to load data from Data Explorer into a Spark DataFrame (a table-like data structure that allows you to work with data).

Integrated business intelligence

Having all these data capabilities at your fingertips, it would make sense to be able to richly visualize data. Azure Synapse offers integration with Microsoft Power BI, allowing you to add Power BI datasets, edit reports directly in Azure Synapse Studio, and automatically see those changes reflected in the Power BI workspace that hosts that report.

> **Note**
>
> Azure Synapse does not provision new Power BI workspaces for you. Instead, you add your existing Power BI workspaces to the Azure Synapse workspace using a linked service connection. A separate license to Power BI may be required.

Thanks to the **Azure Active Directory** (**AAD**) integration, connecting to the Power BI service is a simple process. Synapse Studio uses your credentials to look for Power BI workspaces in your AAD tenant and allows you to select the desired one from a combobox, as illustrated in *Figure 1.10*.

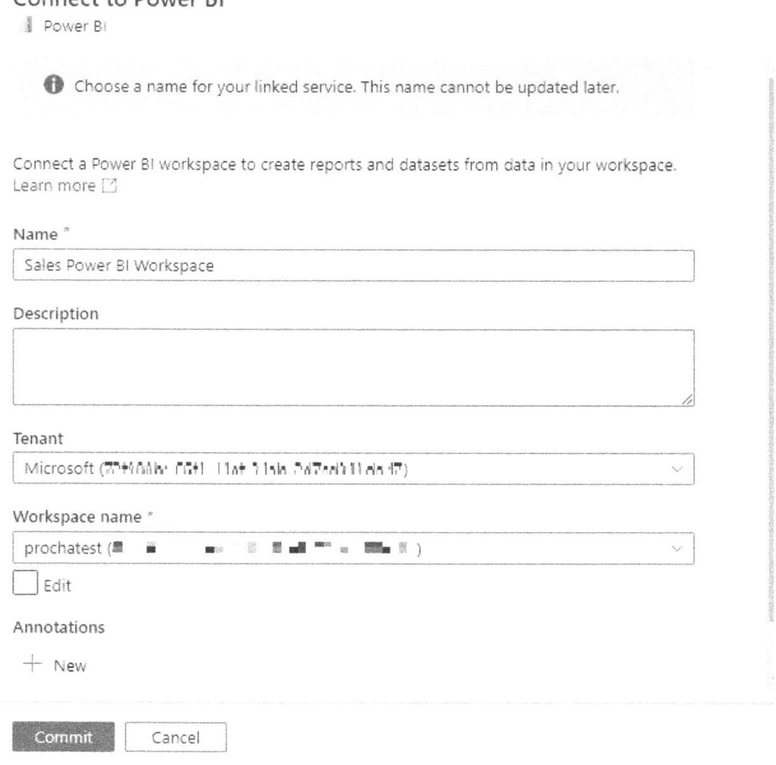

Figure 1.10 – Adding a Power BI workspace as a linked service in Azure Synapse (tenant and workspace globally unique IDs (GUIDs) obfuscated)

This is a powerful and quite useful capability. Not only does it allow you to be more productive and avoid switching between apps to do different work, but it also allows analysts to see the shape and form of their data while they are still exploring it and experimenting with transformations.

Data governance

With the growth of a data culture in corporations, an explosion happened in the number of data marts, data sources, and amount of data that can be used for analytical needs. In Azure Synapse alone, projects normally consume data residing on several data sources. This data can then be copied to a data lake on Azure, transformed, and eventually copied to SQL tables. That is a lot of data and metadata to maintain! How do you get a global view of all the data assets in your organization, and how do you govern this data and classify sensitive data so that it is not misused?

Microsoft Purview is Microsoft's data governance solution for enterprises. It connects to data on-premises, on the cloud, and even to SaaS sources, giving companies a unified view of their data estate. It has advanced data governance features such as data catalogs, data classification, data lineage, data sharing, and more. You can learn more about Microsoft Purview at `https://azure.microsoft.com/en-us/services/purview/`.

> **Note**
>
> Just as with the Power BI integration, Microsoft Purview requires you to have a Purview account, with the appropriate rights, configured separately from your Synapse workspace. To learn how to connect a Synapse workspace to a Purview account, visit `https://docs.microsoft.com/en-us/azure/synapse-analytics/catalog-and-governance/quickstart-connect-azure-purview`.

Configuring your integration with Purview is a simple process: Azure Synapse Studio allows you to pick the Purview account from a list of subscriptions, or to provide the details of your Purview account manually. Once you have configured the integration, you can manage it from Synapse Studio, as illustrated in *Figure 1.11:*

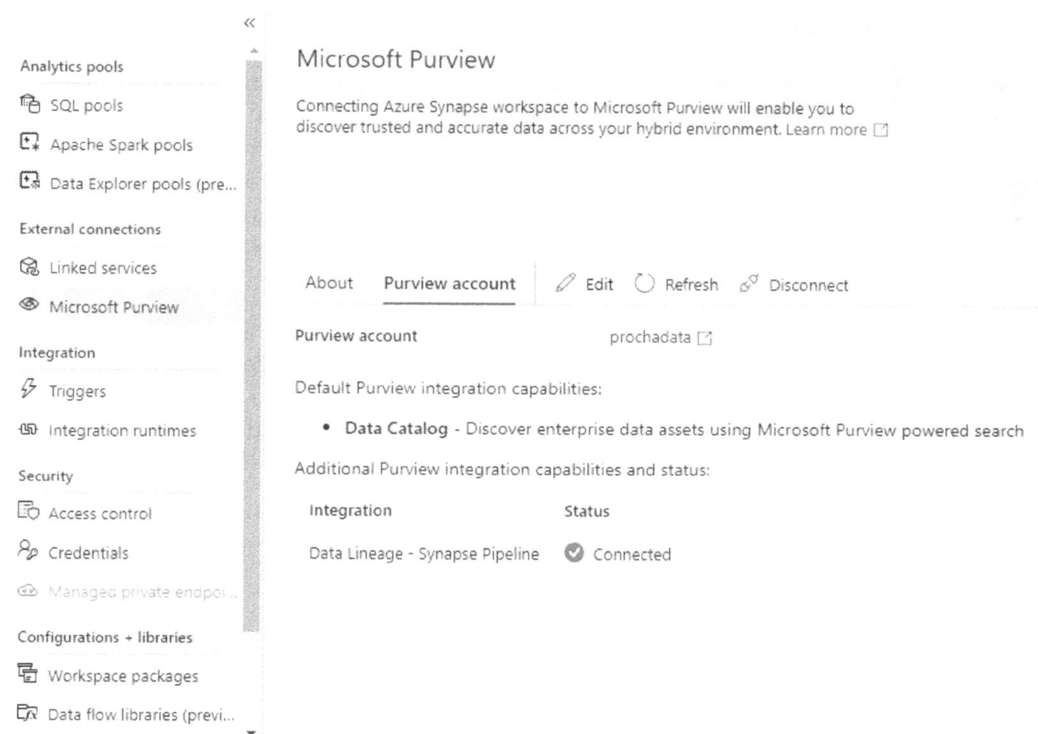

Figure 1.11 – Managing the Microsoft Purview integration

After configuring the Purview integration on a Synapse workspace, you benefit from the following features:

- **Discovery**: Search for any data assets cataloged by Purview by using the global search box.
- **Data lineage**: Understand how the data traveled through the organization and how it was transformed before it landed in the current shape and location. It also allows you to see the raw form of data before it was transformed.
- **Connect to new data**: Having discovered new data assets, instantly connect to them using linked services. From here, you can leverage any service on Azure Synapse to work with the data, from experimentation on Apache Spark to moving data using pipelines.
- **Push lineage back to Microsoft Purview**: After you apply transformations to data and create new datasets on Azure Synapse, you can push metadata that describes your new datasets to Microsoft Purview's central repository for discovery from future users.

While Purview integration is outside of the scope of this book, make sure you understand how to make governance a first-class citizen in your projects—it is a critical aspect of analytical projects and is quickly becoming a hard prerequisite for enterprises.

Broad support for ML

ML is a first-class citizen in Azure Synapse. Models can be trained on Apache Spark, as discussed previously, using a variety of algorithms and libraries, such as Spark MLlib or Scikit-learn. Another option is to connect to the Azure Machine Learning service from within an Azure Synapse notebook and train models using Azure Machine Learning's compute engine. Because Azure Machine Learning offers **automated ML** (**AutoML**), you do not even need to know the best algorithm and features to achieve your objective: AutoML tests a series of parameters and algorithms with the given data and offers you the best algorithm based on a series of results.

Besides model training, Azure Synapse can consume models that were previously trained to run batch scoring with data residing on Azure Synapse. The SQL analytical engine in Azure Synapse can import **Open Neural Network Exchange** (**ONNX**) models (which can be generated from Azure Machine Learning) into its model registry and allow you to use the `PREDICT` function in T-SQL to score columns in real time, as part of regular SQL queries. This is quite powerful!

For example, given a `dbo.Trips` SQL table that contains New York taxi trip data, the query shown in *Figure 1.12* uses the model scored with `Id = 60` to predict taxi fares, given other columns such as passenger count, trip distance, and the date and time of the trip:

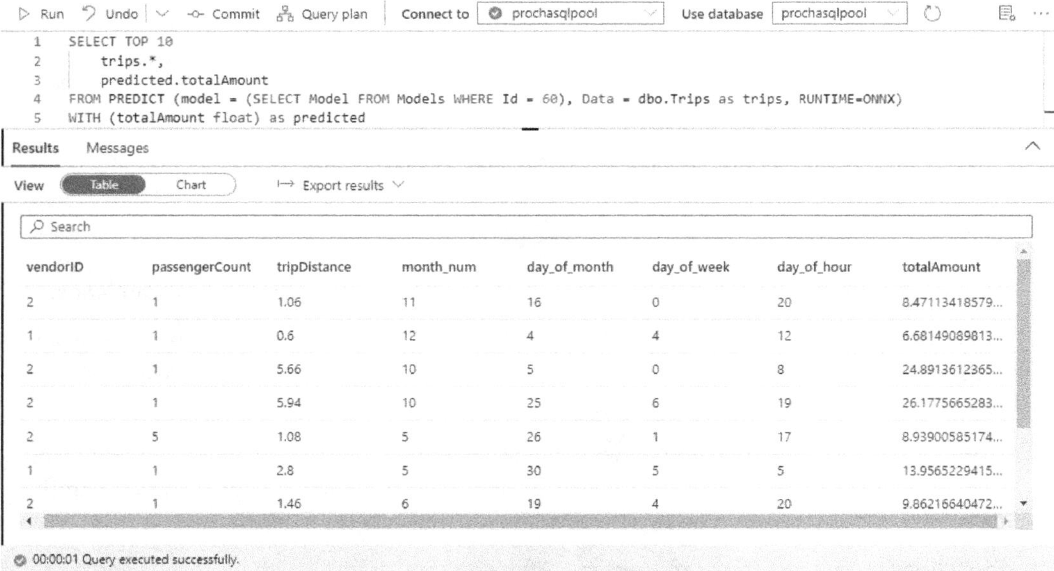

Figure 1.12 – Using PREDICT to score columns in a SQL query

Note that in this example, the model is stored in Azure Synapse SQL's model registry and scoring is done locally, which produces a quick query response time with negligible impact on the query plan. No external services are called.

For consumption of more complex models, or in scenarios where an ML application needs more complex logic that can be better achieved by using a language other than T-SQL, Apache Spark is a perfect alternative. By leveraging Spark pools in Azure Synapse, the regular Notebook experience can also be used for batch scoring.

Security and Managed Virtual Network

The fact that Azure Synapse offers all these cloud services on a single platform may give the impression that it is hard to protect your data. In reality, Azure Synapse workspaces can be created so that they are fully isolated from other workspaces, at the network layer. This is achieved by using **Managed Virtual Network** (or **Managed VNet**) in Azure Synapse.

Managed VNet manages the network isolation, and you do not need to configure **network security group** (**NSG**) rules to allow traffic to and from your virtual network. When associated with a managed private endpoint, workspaces configured on a Managed VNet are fully protected against data exfiltration.

Besides network isolation, Azure Synapse offers an industry-leading set of security features for data protection, **role-based access control** (**RBAC**), different authentication mechanisms, threat protection, and more.

> **Note**
>
> For a detailed view of the security capabilities across all Azure Synapse services, refer to the *Azure Synapse Analytics security* white paper at `https://docs.microsoft.com/en-us/azure/synapse-analytics/guidance/security-white-paper-introduction`.

Management interface

As you can tell by now, Azure Synapse offers several different services that allow you to have a unified platform to build your advanced analytics and data science projects—from data ingestion, all the way to serving data to end users using Power BI. To manage all these services and to build your projects, the primary (and almost only) tool we will use is Azure Synapse Studio.

If you have used ADF before, you will find the **user experience** (**UX**) in Azure Synapse Studio a familiar one. It organizes your work and resources you are using by implementing the concept of hubs (seen on the left-hand side of the **user interface** (**UI**)), as follows:

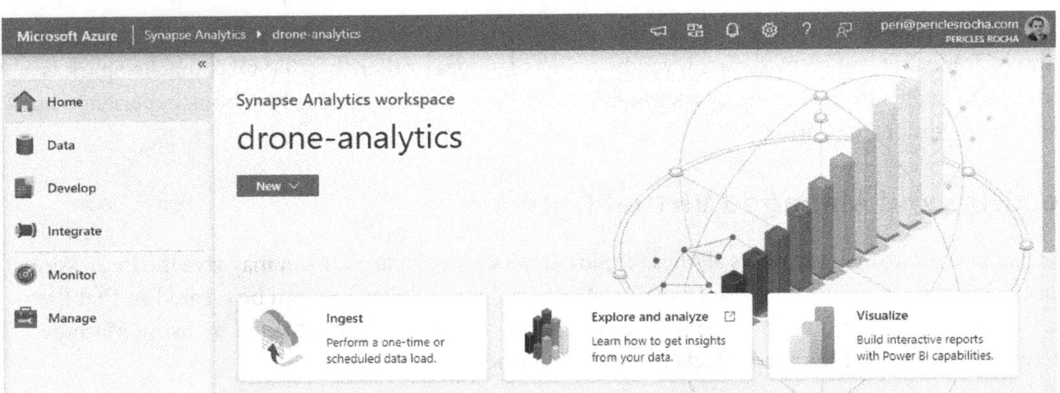

Figure 1.13 – The home page of Azure Synapse Studio

The hubs in Azure Synapse Studio are set out here:

- **Home**: This is the landing page of Azure Synapse Studio. It offers quick links to recently used resources, pinned resources, links to the Knowledge center, and other links to useful content.
- **Data**: The **Data** hub lets you navigate through any SQL, Apache Spark, or Data Explorer pools that you provisioned for your environment, as well as any linked data sources, such as ADLS. It provides a tree view where you can navigate through your compute pools and databases, and even glance at data by right-clicking tables and generating preview scripts.
- **Develop**: Go to the **Develop** hub to find all the SQL/**Kusto Query Language** (**KQL**) scripts you are working on, as well as Synapse notebooks, Power BI reports, and even data flows. All the work you do and *save* (the actual word used in Synapse Studio is *publish*) is stored with the Synapse service, in the context of your workspace. However, you can configure source control and collaboration using Git, and save your work in a Git repository, on Azure DevOps, or on GitHub.
- **Integrate**: This is where you manage all the data integration pipelines created in your Synapse workspace. Azure Synapse Studio provides a code-free experience to build pipelines in a workflow-like environment with a rich UX that mirrors the same experience on ADF.
- **Monitor**: The **Monitor** hub is your **single pane of glass** (**SPOG**) to monitor the status of your compute pools and pipeline runs, Apache Spark jobs, and more. You can also browse through a history of recent activities and verify their execution.
- **Manage**: Finally, the **Manage** hub is where you configure connections to linked services and integration runtimes, scale your SQL, Data Explorer, or Apache Spark pools, and even create pipeline triggers.

Other tools can be used to connect to Azure Synapse services to perform queries and some basic management tasks. Tools such as **SQL Server Management Studio** (**SSMS**) and ADS can be used to run queries on SQL pools. For overall service management and provisioning, while some tasks can be accomplished via T-SQL statements, Azure Synapse can be managed using PowerShell, as well as through the Azure portal.

As you can see, Azure Synapse brings together several different services from data ingestion, through data processing (using your choice of analytical engine) and data visualization to deliver an E2E approach to analytics. It is an industry-defining service offering that brings pieces of the analytical puzzle together for a 360-degree view of your data estate.

This book will focus on Data Explorer in Synapse and how it integrates with these services. So, let's look into it in detail.

What is Azure Synapse Data Explorer?

Before we talk about how Data Explorer is used in Azure Synapse, you may be asking, *what is Azure Data Explorer anyways?* Azure Data Explorer is a cloud-based big data platform that enables analytics on large volumes of data, on unstructured, semi-structured, and structured data, with high performance.

Azure Data Explorer comes from a tool built internally at Microsoft for the exploration of telemetry data, which was named *Kusto*. The French explorer Jacques Cousteau inspired the name. The query language it uses is called KQL. Microsoft still extensively uses Azure Data Explorer for telemetry data across its product teams.

At a high level, Azure Data Explorer has the following key features:

- **Data ingestion**: Supports a series of diverse ways to ingest data, from managed pipelines (for example, Event Grid or IoT Hub), connectors and plugins (for example, Kafka Connect or Apache Spark connector), programmatic ingestion through **software development kits** (**SDKs**) or external data loading tools. It supports ingesting up to 200 **MB** of data per second, per cluster node, and load performance responds linearly as you scale the service in and out.
- **Time-series analysis**: Azure Data Explorer is optimized for time-series analysis and processes thousands of time series in a few seconds.
- **Cost-effective queries and storage**: Usage of Azure Data Explorer is charged by compute hours, not by queries, so you can stop your cluster when not in use. It is also charged by storage used. To save on compute hours, Azure Data Explorer supports auto-stop, to automatically stop your cluster after a certain time of inactivity—or you can stop it manually and start again when needed. On storage, Azure Data Explorer offers retention policies, so you can control how long you want to keep your data, also to optimize costs. For long-term storage or cold data, you can always store your data on Azure Storage.
- **Fast read-only query with high concurrency**: Azure Data Explorer is a columnar store and offers fast text indexing. It allows you to retrieve data from a billion records in less than a second.

- **Fully managed and globally available**: You do not need to worry about provisioning hardware, managing operating systems, patching, backup, or even the service infrastructure. Azure Data Explorer is a fully managed **Platform-as-a-Service** (**PaaS**) offering, so you only need to worry about your data. Also, it is globally available, allowing you to provision services closer to where your data is, reducing network latency and respecting data residency.
- **Enables custom solutions**: Azure services such as Azure Monitor, Microsoft Sentinel, and others are built with Azure Data Explorer in their backend. You can leverage the service's REST API and client libraries to build your custom solutions on top of Azure Data Explorer.

> **Note**
>
> This book explores Azure Synapse Data Explorer, and how it integrates with other Azure Synapse services. To learn more about the standalone service Azure Data Explorer and KQL, a good resource is *Scalable Data Analytics with Azure Data Explorer*, available at `https://www.packtpub.com/product/scalable-data-analytics-with-azure-data-explorer/9781801078542`.

Azure Synapse brings the standalone service Azure Data Explorer to Synapse workspaces, enabling you to complement SQL and Apache Spark pools with an interactive query experience optimized for log and telemetry data. As with dedicated SQL pools, Data Explorer pools are provisioned by you, and compute capacity is reserved while the pool is running. You select your desired cluster size based on your service-level requirements.

As expected, you can use Azure Synapse Studio to run queries on Data Explorer, resume and pause pools, manage the size of your pools by scaling up or down, and view details of your pool such as the instance count, **CPU** utilization, cache utilization, and more.

In Azure Synapse workspaces, when you navigate to the **Develop** hub, you create KQL scripts to explore data on Data Explorer pools. KQL has grown in popularity in recent years due to its adoption by other Azure services, such as Azure Monitor, Microsoft Sentinel, and others.

Integrating Data Explorer pools with other Azure Synapse services

As mentioned previously, before Azure Synapse, data science and advanced analytics projects required engineers to put together several pieces of a puzzle to deliver an E2E solution to users. By bringing Azure Data Explorer natively to Azure Synapse through Data Explorer pools, you no longer need to maintain external connectors and manage services separately. Furthermore, you benefit from the productivity gains of Azure Synapse workspaces, building everything they need on Azure Synapse Studio.

Data Explorer pools on Synapse workspaces offer several benefits, as detailed next.

Query experience integrated into Azure Synapse Studio's query editor

You can query Data Explorer pools using the same tools and the same look and feel you experience with dedicated or serverless SQL pools. Additionally, you can go back and forth between a KQL query on a Data Explorer pool and a T-SQL query on a dedicated SQL pool to get the full context of your data, without having to switch browser tabs or different applications, enabling data correlation across all data sources. Finally, all your KQL scripts can be saved with your SQL scripts and Synapse notebooks into your workspace for future use (or merged into the Git source control mechanism of your choice). In *Figure 1.14*, you can see the **Develop** hub bringing together all your scripts, notebooks, data flows, and Power BI reports:

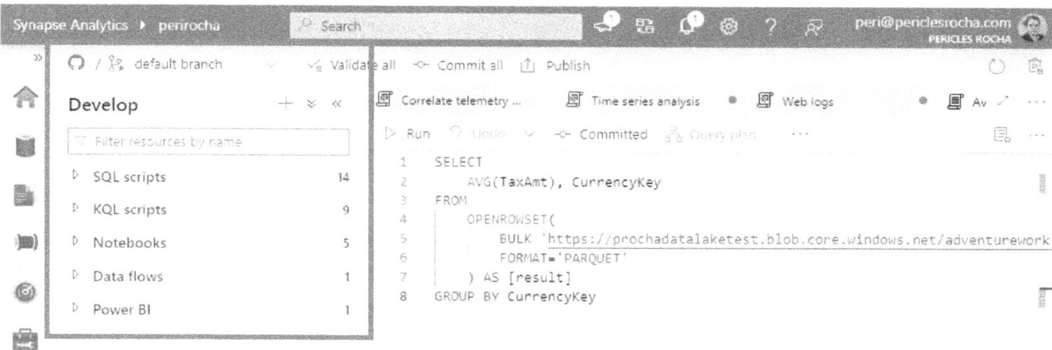

Figure 1.14 – Integrated authoring experience for all your Azure Synapse assets, with source control

> **Note**
>
> Azure Synapse exposes an endpoint for Data Explorer pools the same way as the standalone service Azure Data Explorer. You can still use Azure Data Explorer query tools such as Kusto.Explorer, the Azure Data Explorer web UI, and even the Kusto **command-line interface (CLI)** to perform queries if you wish to use them.

Exploring, preparing, and modeling data with Apache Spark

As discussed previously, you can simply right-click a table on a Data Explorer pool and quickly start a new Synapse notebook to use your programming language of choice for data exploration and preparation and to train (and consume!) ML models leveraging Apache Spark. Therefore, you can leverage other benefits of Apache Spark in Synapse, such as Azure Machine Learning integration, and use services such as AutoML.

Data ingestion made easy with pipelines

Among the diverse ways you can load data into Data Explorer pools, as you would expect, Synapse pipelines offer full, native support for the service. If you have existing pipelines and data flows, incorporating Data Explorer pools into your workflows is a simple task.

Unified management experience

Having a SPOG to manage and monitor your services is a huge productivity gain. From Azure Synapse Studio, you can create, delete, pause, resume, and scale Data Explorer pools up or down. You can also monitor the health of pools. Finally, you can control security and access-control rules. When managing settings for your Synapse workspace in the Azure portal, you will also find a central location under **Analytics pools** to create, pause, or delete your Data Explorer pools the same way you do it for SQL and Apache Spark pools. This is illustrated in *Figure 1.15*.

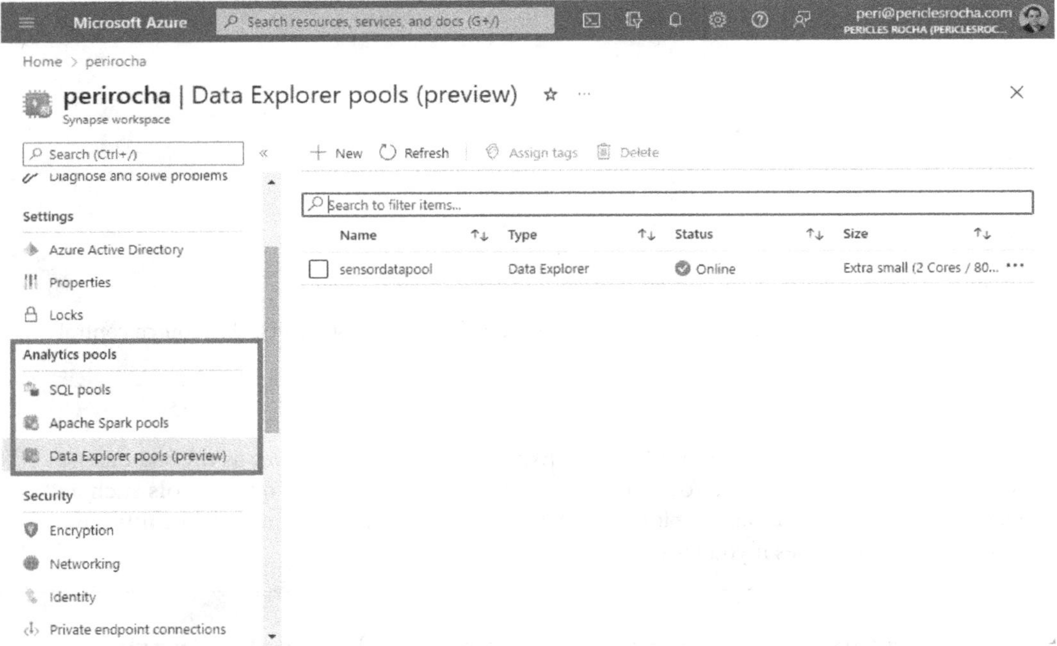

Figure 1.15 – Seamless experience across all analytics pools in the Azure portal

As you can see, Data Explorer is a native service in Azure Synapse and benefits from all the aspects mentioned. It's different from Power BI and Purview in the sense that you don't need to configure it as an external service—instead, Data Explorer pools are like natural cousins of SQL pools and Apache Spark pools, and they share the same experience.

Exploring the Data Explorer pool infrastructure and scalability

Let us look at how Data Explorer pools work behind the curtains.

Any typical deployment of Data Explorer, regardless of being the standalone service or Data Explorer pools in Azure Synapse, will almost always consist of two major services working together, as follows:

- **The Engine service**: Serves user queries, processes data ingestion, and accepts control commands that create or change databases, tables, or other metadata objects (a.k.a. **data definition language (DDL)** for seasoned SQL users).
- **The Data Management service**: Connects the Engine service with data pipelines, orchestrates and maintains data ingestion processes, and manages data purging tasks (a.k.a. data grooming) that run on the Engine nodes of the cluster.

These services are deployed through **virtual machines (VMs)** in Microsoft Azure, building a cluster of Data Explorer **compute nodes**. These nodes perform different tasks in the architecture of the Data Explorer pool, which we will discuss next.

Data Explorer pool architecture

The Engine service is the most important component in the architecture of Data Explorer pools. There are four types of cluster nodes defined by their respective roles supporting the Engine service, as follows:

- **Admin node**: This node maintains and performs all metadata transactions across the cluster.
- **Query Head node**: When users submit queries, the Query Head node builds a distributed query plan and orchestrates query execution across the Data nodes in the cluster. It holds a read-only copy of the cluster metadata to make decisions for optimal query performance.
- **Data node**: As the *worker bee* in the cluster, it receives part of the distributed query from the Query Head node and executes that portion of the query to retrieve the data that it holds. Data shards are cached in the Data nodes. These nodes also create new data shards when new data is ingested into the database.
- **Gateway node**: Acts as a broker for the Data Explorer REST API. It receives control commands and dispatches them to the Admin node, and sends any user queries it receives to a Query Head node. It is also responsible for authenticating clients that connect to the service via external API calls.

You do not need to worry about how many nodes of which types your cluster contains. The actual implementation of the cluster is transparent to the end user, and you don't have control over the individual nodes.

Scalability of compute resources

Data Explorer was designed to scale vertically and horizontally to achieve companies' requirements, and to accommodate periodical changes in demand. By scaling vertically, you are adding or removing CPU, cache, or **RAM size** for each node in the cluster. By scaling horizontally, you are adding more instances of the specified node size to the cluster. For example, you can configure your Data Explorer pool to start with two instances with eight cores each, and then scale your environment horizontally or vertically as needed.

> **Note**
> You cannot control the specific number of CPUs, amount of RAM, or cache size for the VMs used in your clusters. Azure Synapse Data Explorer has a pre-defined set of VM sizes from **Extra Small** (two cores) to **Large** (16 cores) to choose from. These VM sizes have a balanced amount of each compute resource.

Sometimes, it is hard to anticipate how much of a compute resource you will need for a given task throughout the day. Furthermore, if you have high usage of your analytics environment at one point in time during the day but less usage at separate times, you would want to adjust the service to scale automatically as users demand more and fewer resources.

Data Explorer allows you to do just that through **Optimized autoscale**: just set the minimum number of instances you want to have running at any given time of the day, and a maximum number of instances the service can provision in case there's more user demand than the currently allocated resources can support, and Data Explorer pools will scale in and out automatically. So, if your cluster is underutilized, Data Explorer will scale in to lower your cost (while scaling out if the cluster is overutilized). This can be configured in the Azure portal or in Azure Synapse Studio, as seen in *Figure 1.16*.

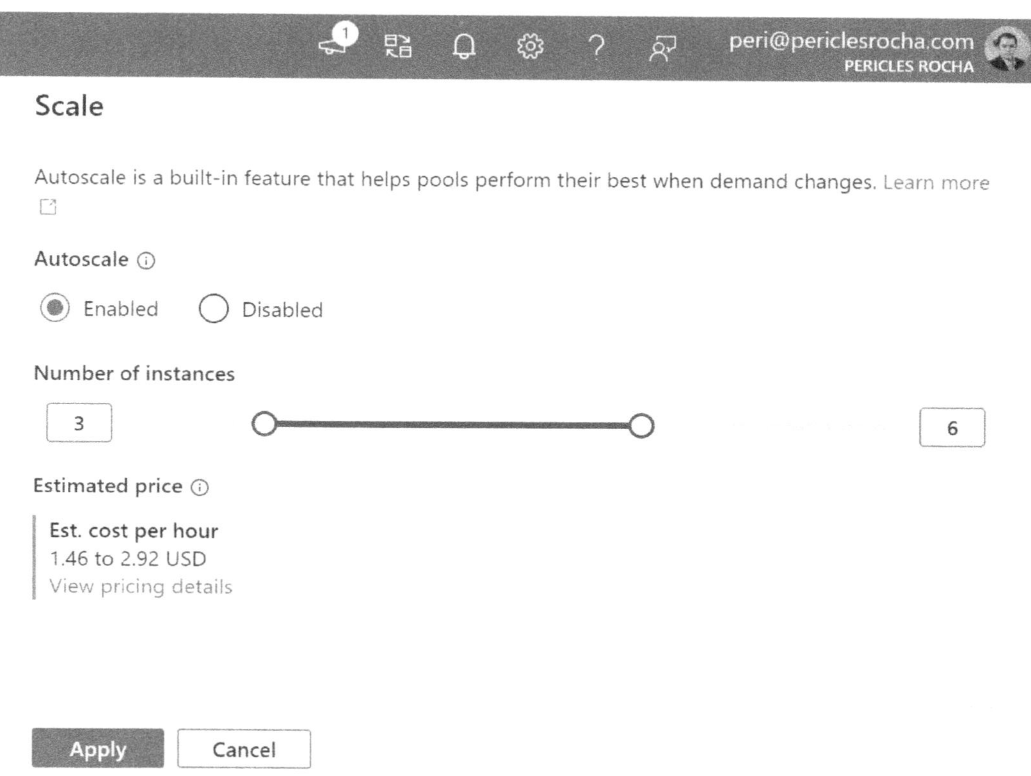

Figure 1.16 – Specifying a minimum and maximum number of instances on Autoscale

The maximum number of instances seen on the slider in *Figure 1.16* scales with the workload size that you selected. With a large compute cluster, you can scale to up to 1,000 instances.

Managing data on distributed clusters

Scaling in and out is great, but you must be thinking: how about my data? The architecture of Data Explorer decouples the storage layer from the compute layer, meaning these layers can scale independently. If more storage is needed, then more resources are allocated to the storage layer. If more compute is needed, your compute VMs will increase in size, or you may have more instances of them.

The Data Explorer service implements database sharding to distribute data across its storage. The Engine service has awareness of each data shard and distributes queries across them. For almost all cases, you don't need to know details about the physical data shards themselves, as Data Explorer exposes data simply through logical tables.

Data is physically persisted in storage, but to deliver a fast query experience, Data Explorer pools cache data in **solid-state drives** (**SSDs**). We will look at how you can define caching policies to balance costs and performance in *Chapter 10, System Monitoring and Diagnostics*.

Data shards are distributed across Data nodes using a hash function, which makes the process deterministic—using this hash function, the cluster can determine at any time which Data node is the preferred one for a certain shard. When you scale a Data Explorer pool in or out, the cluster then redistributes the data shards equally across the Data nodes available.

Mission-critical infrastructure

For enterprises, it is not enough to be able to store large amounts of data and retrieve them quickly. Data is a critical asset for companies, and the infrastructure that holds their data needs to be bulletproof to protect it from security and availability challenges and needs to offer developer productivity and sophisticated tooling for monitoring.

Data Explorer pools inherit several of the mission-critical features in Azure Synapse Analytics (and some of these were described in the *What is Azure Synapse?* section of this chapter). Let us look at other features that it offers that are relevant to building mission-critical environments, as follows:

- **AAD integration**: AAD is Microsoft's cloud-based **identity and access management** (**IAM**) service for the enterprise. It helps users sign in to a corporate network and access resources in thousands of SaaS applications such as Microsoft Office 365, Azure services, and third-party applications built with support for AAD.
- **Azure Policy support**: This allows companies to enforce standards and evaluate compliance with services provisioned by users. For Data Explorer, you can use policies such as forcing Data Explorer encryption at rest using a customer-managed key, or force-enabling double encryption, among other policies.
- **Purging of personal data**: Companies have a responsibility to protect customer data, and the ability to delete personal data from the service is a strong asset to help them satisfy the **General Data Protection Regulation's** (**GDPR's**) obligation. Data Explorer supports purging individual records, the purging of an entire table, or the purging of records in materialized views. This operation permanently deletes data and is irreversible.
- **Azure Availability Zones**: Built for **business continuity and disaster recovery** (**BCDR**), Azure Availability Zones replicate your data and services to at least three different data centers in an Azure region. Your data residency is still respected, but in the case of a local failure on a region's data center, your application will fail over to one of the copies in a different data center, but on the same Azure region.
- **Integrated with Azure Monitor**: Collect and analyze telemetry data from your Data Explorer pools to understand cluster metrics and track query, data ingestion, and data export operations performance.

- **Globally available**: At the time of this writing, Azure was available in more than 60 regions worldwide, and the list of regions continues to grow every year. This allows organizations to deploy applications closer to their users to reduce latency and offer more resiliency and recovery options, but also respect data residency rules. For an updated list of Azure regions, visit `https://azure.microsoft.com/explore/global-infrastructure/`.

> **Note**
>
> Not every Azure service is available in every Azure region. For a detailed view of Azure service availability per Azure region, use the *Products available by region* tool at `https://azure.microsoft.com/en-us/global-infrastructure/services/`.

How much scale can Data Explorer handle?

As of July 2022, Microsoft claimed the following usage statistics for Azure Data Explorer globally:

- 115 PB of data ingested daily
- 2.5 billion queries daily
- 8.1 **exabytes** (**EB**) in total data size
- 2.4 million VM cores running at any given time
- More than 350,000 KQL developers

These are important numbers for a managed service. What is even more impressive is that Microsoft claims those numbers are growing close to 100% year over year.

All the details mentioned here about the service architecture and scalability are characteristics of the standalone Azure Data Explorer service too. There are a few special things about Data Explorer in Azure Synapse, so let's explore that next (no pun intended).

What makes Azure Synapse Data Explorer unique?

Even though the underlying service of Data Explorer pools in Azure Synapse is the same as Azure Data Explorer, some capabilities are available exclusively in Azure Synapse. Let us investigate those differences, as follows:

- **Firewall**: Azure Synapse workspaces include a firewall and allow you to configure **IP** firewall rules to grant or deny access to a workspace. This is not available in the standalone service.
- **Availability Zones**: Enabled by default for Azure Synapse workspaces where Availability Zones are available. This can optionally be enabled when using Azure Data Explorer alone.

- **VM sizes for compute**: Azure Data Explorer offers more than 20 different VM configurations to choose from. For Azure Synapse Data Explorer, a simplified subset of the VM configurations is offered, ranging from extra small (two cores) to large (16 cores).
- **Code control**: As previously mentioned, in Azure Synapse you can connect your workspace with a Git repository, Azure DevOps, or GitHub. This option is not available when using the standalone service.
- **Pricing**: For a Azure Synapse workspace, Data Explorer pools pricing is simplified to two meters: **VCore** and **Storage**. When using Azure Data Explorer as a standalone service, you may be charged by using multiple meters such as **Compute**, **Storage**, **Networking**, and the Azure Data Explorer IP markup, which is applied when you make use of fast data ingestion, caching, querying, and management capabilities. Additionally, **Reserved Instances**, which offer discounted prices when you make a commitment to use an Azure service for a certain period (typically 1 or 3 years), are only available for the standalone service Azure Data Explorer, and not for Azure Synapse Data Explorer.

As seen from the preceding points, there is no significant loss of functionality by using Azure Synapse Data Explorer when compared to the standalone service Azure Data Explorer. Azure Synapse Data Explorer includes the benefits seen on the standalone service, and it also incorporates the enterprise features offered with Synapse workspaces. However, is Azure Synapse Data Explorer the solution to every analytical problem? In the next section, you will find out how to decide whether you need Data Explorer pools or not.

When to use Azure Synapse Data Explorer

By now, you should already understand that Azure Synapse Data Explorer is an analytical engine to process queries on unstructured, semi-structured, and structured data, with exceptionally large data volumes, low-latency ingestion, and blazing-fast queries. Data Explorer is not, however, the solution to every data problem. In some cases, you will be better off with a different solution. Let us look at some of the most common analytics scenarios and the most appropriate analytical store in each case, as follows:

- **Scenario**: *I need a classic data warehouse.*

  Recommendation: Do not use Azure Synapse Data Explorer. Use dedicated SQL pools in Azure Synapse, which are optimized for user queries in a typical star schema, even at large data volumes.

- **Scenario**: *My solution requires frequent updates on individual records, and singleton INSERT, UPDATE, and DELETE operations.*

  Recommendation: Do not use Azure Synapse Data Explorer. In such cases, a transactional, operational database will be a better solution. Consider options such as Azure SQL, SQL Server (on-premises, or in an Azure VM), MySQL, or even Cosmos DB for NoSQL scenarios.

- **Scenario**: *My solution needs to run on a cloud other than Microsoft Azure, or on-premises.*

  Recommendation: Do not use Azure Synapse Data Explorer, as it runs exclusively on Azure.

- **Scenario**: *My data demands constant transformation and long-running extract, transform, load (ETL)/extract, load, transform (ELT) processes.*

  Recommendation: Do not use Azure Synapse Data Explorer. Even though you have Synapse pipelines in your Synapse workspace, and you can constantly ingest data into Data Explorer pools, the core scenario for Data Explorer is to offer interactive analytics on big data. You are better off running your ETL/ELT pipelines on Azure Synapse pipelines, ADF, Apache Spark, or even Azure Batch.

- **Scenario**: *I need to train large ML models several times throughout the day.*

  Recommendation: This may be a good scenario for Azure Synapse Data Explorer. In this case, you can prepare data or train models on Apache Spark for Azure Synapse, but note that you will miss out on the real-time characteristic of data analysis that Data Explorer offers. Ideally, you want to use Data Explorer with data streaming from devices and applications in real time, but this still can be a valuable scenario for Azure Synapse Data Explorer. This may be less valuable when using the standalone service Azure Data Explorer, as it will not benefit from the native, in-product integration with Apache Spark (even though a connector for Spark is available for Azure Data Explorer uses).

- **Scenario**: *I have a very small amount of data to analyze.*

  Recommendation: It depends. If your analysis requires a full-text search or **JSON** documents, you may benefit from the indexing capabilities of Azure Synapse Data Explorer. It can also be a suitable alternative if you need to correlate this data with other data stored on Synapse SQL or in the data lake. If you are on a low budget and don't need the added benefit of Azure Synapse, you may be better served with SQL Server, Azure Cognitive Search, or even Cosmos DB.

- **Scenario**: *I need to perform time-series analysis on metric data from sensors, social media, websites, financial transactions, or other fast streaming data.*

  Recommendation: You should use Azure Synapse Data Explorer. Data Explorer pools are optimized for application log and IoT device data and can ingest data at high volumes offering insights in near real time.

- **Scenario**: *I have data in a diverse schema, and with high volumes of data in near real time.*

  Recommendation: You should use Azure Synapse Data Explorer. Data Explorer pools are optimized for unstructured, semi-structured, and structured data and allow you to run interactive analytics on data of any shape.

- **Scenario:** I need to correlate application logs or telemetry data from IoT devices with data sitting in a data warehouse and the data lake.

  Recommendation: You should use Azure Synapse Data Explorer. By leveraging the SQL analytical pools in Azure Synapse (dedicated and serverless), you can use one tool to query all your data, regardless of the analytical store that holds it.

The rule of thumb is to think about Data Explorer pools when you are managing telemetry or log analytics data at scale. You should use it with Azure Synapse when you need to combine your analysis with data from other sources or use the added benefits of Azure Synapse in your project.

Summary

Azure Synapse Data Explorer brings all the innovation that was built into the standalone service Azure Data Explorer into Azure Synapse. By using Data Explorer pools in Azure Synapse, you can correlate data from several different sources, residing on different analytical engines, to get a 360-degree view of all your data and unlock insights.

In this chapter, you learned about the lifecycle of data, the TDSP, and how Data Explorer fits into the analytics landscape. We explored the key components of Azure Synapse and Azure Synapse Data Explorer, and how Data Explorer pools benefit from integration with other Azure Synapse services such as Apache Spark.

Next, you learned about the infrastructure of Data Explorer pools and how they deliver massive scalability. We looked at the service architecture and how Data Explorer pools manage data in a distributed cluster. We also explored the mission-critical features of Azure Synapse Data Explorer that give the trust enterprises need to adopt the solution.

Finally, we discussed what makes Azure Synapse Data Explorer unique when compared to the standalone service Azure Data Explorer, and how to determine whether Azure Synapse Data Explorer is the right solution for you.

In the next chapter, we will explore how to create an Azure Synapse workspace and a Data Explorer pool.

2

Creating Your First Data Explorer Pool

In the previous chapter, we introduced Azure Synapse Data Explorer pools, its architecture and main components, and key scenarios where you should consider using the service.

In this chapter, we'll walk you through the creation of your first Azure Synapse Data Explorer pool and everything that's needed to get it created.

For starters, maybe you are not currently an Azure user. If that's the case, we'll walk you through creating your first Azure account and Azure subscription. If you're not familiar with Azure terminology, do not worry – we'll give you the details you need.

Next, we'll create an Azure Synapse workspace and log in for the first time. You'll learn how to navigate to Azure Synapse Studio and create your first Data Explorer pool.

Finally, we'll show you how to create the exact same service using the Azure portal and the Azure **Command Line Interface** (**CLI**). This can be useful if you're automating the creation of Azure services and can help you familiarize yourself with the different options available to manage your cloud infrastructure.

If you are a seasoned Azure professional, feel free to skip this chapter and move on to *Chapter 3*, *Exploring Azure Synapse Studio*, where you will learn how to use Azure Synapse Studio from the point of view of someone working with Data Explorer. Still, before you skip this chapter, I recommend exploring the creation of a Data Explorer pool on Azure Synapse Studio so that you understand what the settings are, and then skip the sections about the Azure portal and the CLI.

Here are the topics that we will cover in this chapter:

- Creating a free Azure account
- Creating an Azure Synapse workspace
- Creating a Data Explorer pool using Azure Synapse Studio

- Creating a Data Explorer pool using the Azure portal
- Creating a Data Explorer pool using the Azure CLI

Technical requirements

To create a free Azure account, you will need a Microsoft or GitHub account. If you need to create a Microsoft account, head over to `https://account.microsoft.com/account` and create one. If you don't want to create a Microsoft account, you can log in to Azure using your GitHub credentials.

Besides the Microsoft (or GitHub) account, you will also need to provide a credit card and a phone number to create your Azure account. These details are needed, even if you are not charged, to verify that you are a real person, so make sure you have that at hand when you create your Azure account. Debit cards are also accepted in most countries, except for Brazil and Hong Kong SAR. Keep in mind that Azure Synapse is not one of the free services included with the Azure free account, so you will be charged for the experiments you make – except if you're eligible for the $200 offer (which will be detailed in this chapter), but make sure to keep your expenses within that budget if you're not willing to pay for your tests!

Additionally, make sure you download the book materials from our code repository at `https://github.com/PacktPublishing/Learn-Azure-Synapse-Data-Explorer`. You can download the full repository by selecting **Code** and then **Download ZIP**, or by cloning the repository by using your Git client of choice. Starting with this chapter, you will find code samples and datasets that we will use throughout the book in this code repository. In this chapter, you will learn how to use the Azure CLI to create a Data Explorer pool. We will not, however, discuss the Azure CLI in detail. If you would like to learn more about it, refer to the documentation at `https://learn.microsoft.com/cli/azure/`. To run the Azure CLI, you will need access to a terminal such as Windows Terminal, Command Prompt, or Bash on Linux or macOS, and have the rights to install the Azure CLI on your computer. If that's not an option for you – maybe due to software policy requirements in your organization – you can still use Azure Cloud Shell at `https://shell.azure.com/`, a browser-based shell experience to manage Azure.

Creating a free Azure account

If you don't have an Azure account and have never created one, the Azure free account is the best way to get started with Azure. This offer gives you $200 in Azure credits to spend over the first 30 days after you create your account, and free monthly access up to a certain amount for some services for the first 12 months. It also allows you to use several other services that are always free, such as Azure App Service (10 web, mobile, or API apps with 1 GB storage), Azure Cosmos DB (limited amounts), and Azure Functions.

> **Note**
>
> Usage of some free services is unlimited, while usage of others is limited by service tier or for a limited period. For a complete and up-to-date list of all free services available with the Azure free account offering, refer to `https://azure.microsoft.com/free/free-account-faq/#free-services`. Additionally, if you have used Azure before and have already taken benefit of the Azure free account offering, you are no longer eligible for the free account offer and should create a pay-as-you-go subscription (or use an existing one).

Sadly, Azure Synapse is not one of the free services included in the Azure free account offering, but you can use your USD 200 credit to offset any costs you may have. This credit should be enough to run the examples in this book as long as you use the smallest analytical pool sizes and stop your pools when not in use.

To create a free Azure account, proceed with the following steps:

1. Go to `https://azure.microsoft.com/en-us/free/` and click **Start free**. If you are no longer eligible for the Azure free account offering, then select the **Pay as you go** link, as shown in *Figure 2.1*:

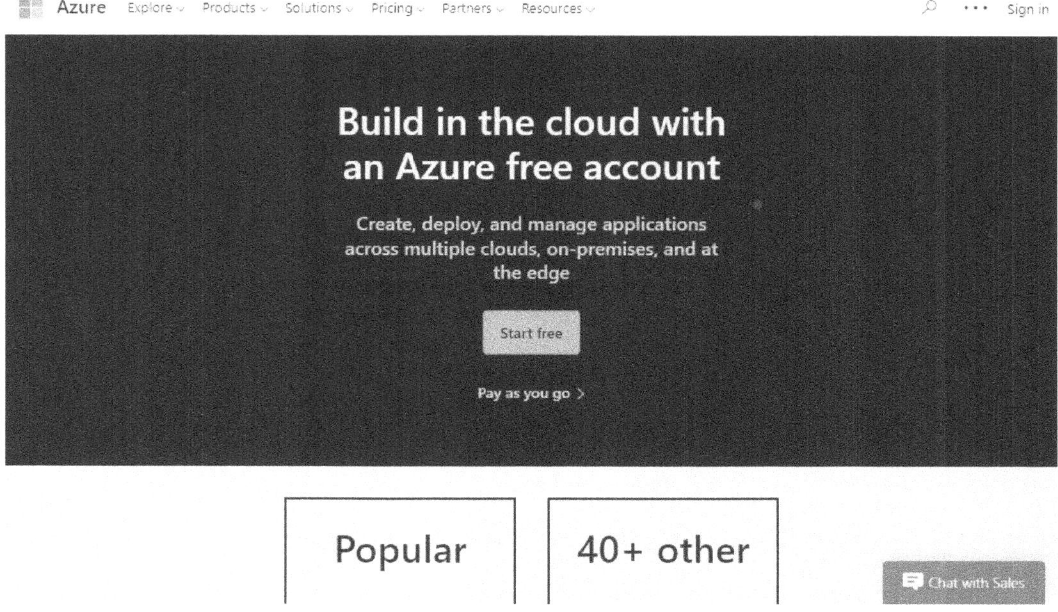

Figure 2.1 – Landing page of the Azure free account offering

2. When prompted, sign in using your Microsoft or GitHub account. If you do not have a Microsoft or a GitHub account, this page allows you to create one.

3. Depending on how much information is present on your Microsoft account, you may see a form asking you to provide more details about yourself, such as your full name, country, email address, and phone number. Provide the required details and click **Next**.
4. You will now be asked to perform identity verification by phone. Provide your phone number here to receive a text message or a phone call with a verification code that you will need to provide back to the sign-up form. Type the code you receive and click **Next**.
5. On the **Identity verification by card** page, provide your credit or debit card details and click **Next**. You will not be charged for the use of Azure services unless you upgrade to a **Pay as you go** subscription – this step is required to authenticate you as a real person signing up for the service, protecting Microsoft from fraud attempts.
6. Finally, on the **Agreement** page, review the subscription agreement, offer details, and privacy statement. When you are ready to proceed, check the **I agree** checkbox and click **Sign Up**.
7. The sign-up process may take a few seconds to a few minutes to complete. Once it processes your account, you'll be forwarded to a page that confirms you are ready to start with Azure and displays a blue button that says **Go to Azure portal**. Click this button to be forwarded to the Azure management portal.

The Azure free account sign-up process creates what's called an **Azure subscription** for you. The Azure subscription is your billing unit in Azure – every service you create in Azure is associated with a subscription. An Azure account can have several subscriptions, a strategy often used to charge departments in organizations based on their cloud consumption. This means that even when your Azure free services expire, you can create other subscriptions using your Azure account.

You did it! You now have what you need to log in to the Azure portal and create your first Azure Synapse workspace. In fact, that's exactly what we're going to do next.

Creating an Azure Synapse workspace

To create any service in Azure, you can use either the management portal, also known as the Azure portal, or the Azure CLI. To start your journey in Azure Synapse, we will show you how to create your very own Azure Synapse workspace using the Azure portal. We will detail all the parameters needed to create the service, which will help you understand the implementation details when you need to create new Synapse workspaces using the Azure portal or the Azure CLI.

To create your Azure Synapse workspace, proceed with the following steps:

1. Navigate to `https://portal.azure.com` and log in using your Azure account credentials.
2. Welcome to the Azure portal! If this is your first time here, you might as well start getting used to it. This is Azure's unified console to build, manage, and monitor all Azure services. To create an Azure service, you can click the **Create a resource** button and navigate to the service you are looking for. However, my preferred way of doing that, when you know what service you are

looking for, is to type the name of the service you want to create in the global search box and select it. So, you can simply type `Azure Synapse` in the search box and then select **Azure Synapse Analytics** in the **Services** list, as shown in *Figure 2.2*:

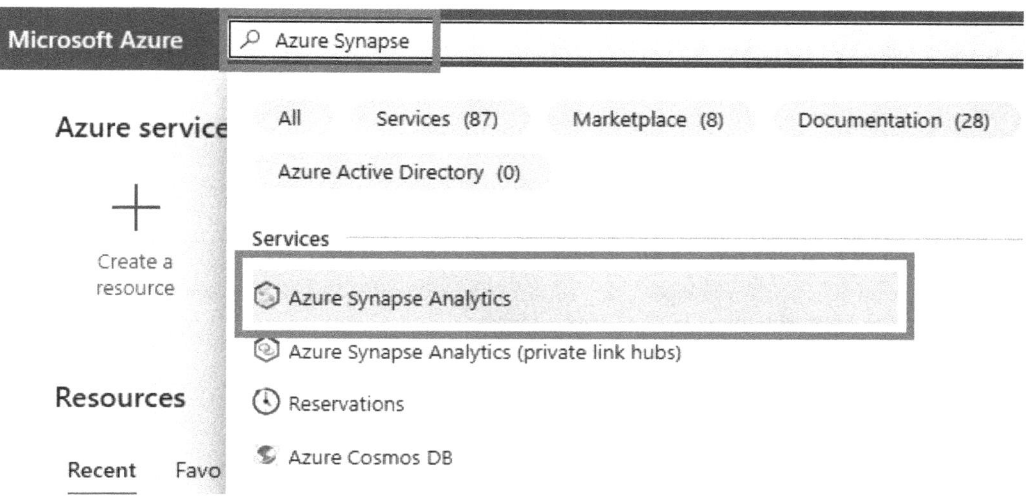

Figure 2.2 – Navigating to the Azure Synapse page in the Azure portal

3. This takes you to the Azure Synapse Analytics page of the Azure portal, which allows you to browse any previously created workspaces and create new ones. To start creating your first workspace, select the **Create** button, as highlighted in the following screenshot:

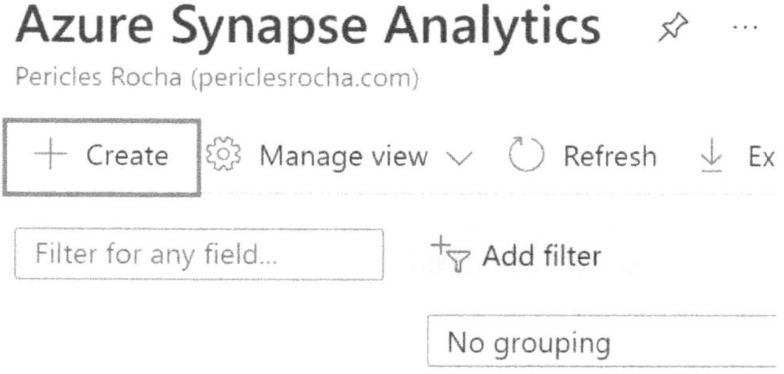

Figure 2.3 – Creating your Azure Synapse workspace

You will be directed to the **Create Synapse workspace** page. On this page, we will provide all the information needed to create a new Azure Synapse workspace via the Azure portal. Let's look at the settings that we need to provide.

Basics tab

First, we will provide the details for the **Basics** section of the **Create Synapse workspace** form, as shown in the following screenshot:

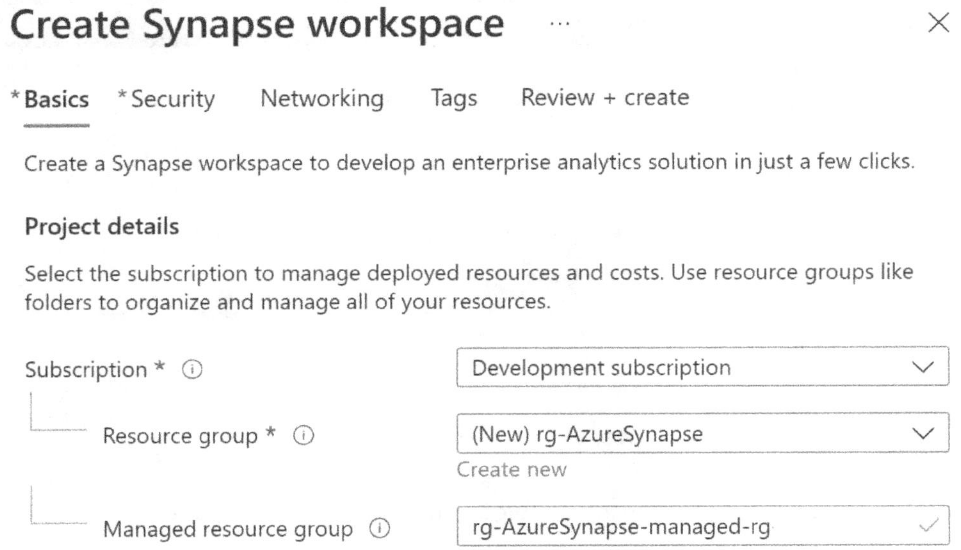

Figure 2.4 – The Project details section of the Basics tab

In the **Project details** section, we need to specify the subscription and resource group where the new workspace will be created. Let's look into those parameters in detail:

- **Subscription**: We'll start by choosing the subscription where this service will be created. Remember that the subscription is your billing unit in Azure, so any services that you create under this subscription will be charged using the payment information you provided for it. Note that your subscription may have a different name from the one in *Figure 2.4*, such as `Azure subscription 1` or a custom name you may have used if you chose to rename your subscription.
- **Resource group**: Next, you need to provide the name of the resource group where your services will be contained. When you create cloud applications, these applications will usually require more than one Azure service. A resource group is a logical container where you place your Azure services together – like a folder where you store your files – so that you can manage the resources your application needs together. For example, when you create an Azure Synapse workspace, Azure provisions the workspace itself, but it can also provision a new storage account, which will be placed in the same resource group. In this example, I'll call the resource group `rg-AzureSynapse`, but you can use any name that works for you or even use an existing resource group.

- **Managed resource group**: Finally, you may optionally provide the name for a managed resource group. This is where Azure Synapse may store some ancillary resources that may be needed by the workspace. If you don't provide a name for the managed resource group, the Azure portal will create one for you using a random name that starts with `synapseworkspace-managedrg`. Because I like to know what the resources in Azure were created for, I'll use an easy-to-identify name, `rg-AzureSynapse-managed-rg`, so that every time I see it, I know what this resource group is used for.

Under the **Project details** section (you may need to scroll down), you will find the **Workspace details** section, as illustrated in *Figure 2.5*:

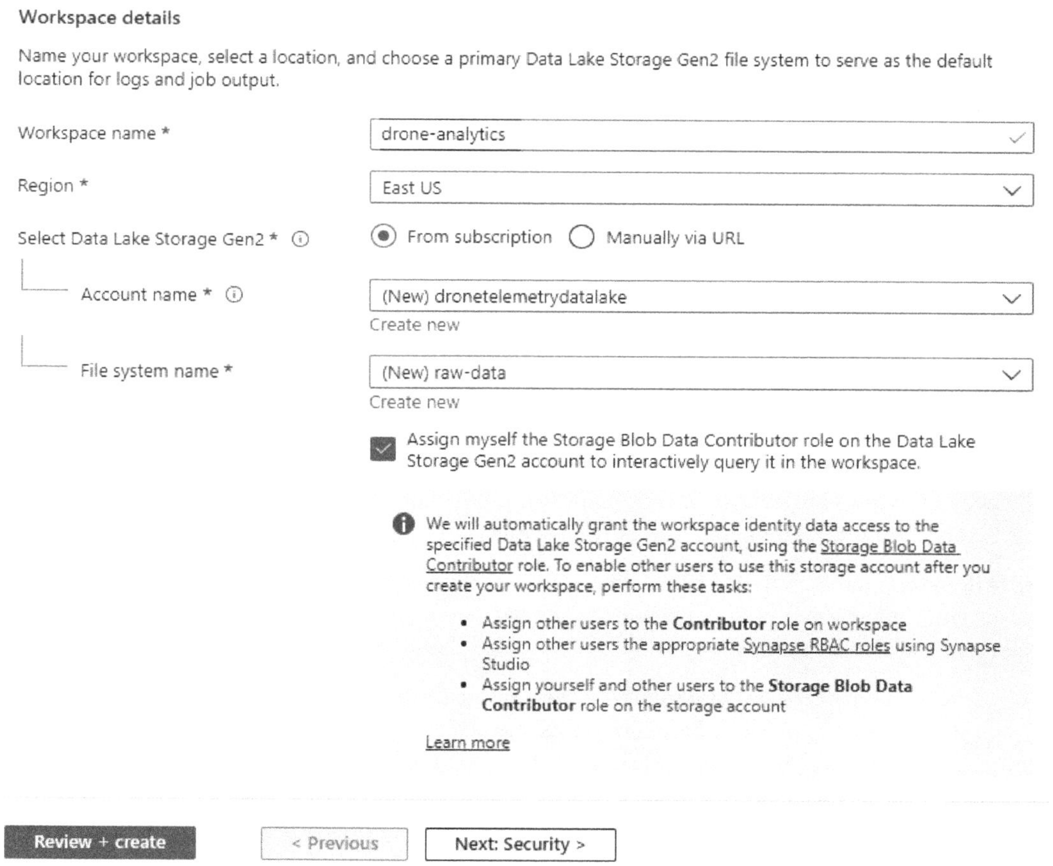

Workspace details

Name your workspace, select a location, and choose a primary Data Lake Storage Gen2 file system to serve as the default location for logs and job output.

Workspace name * drone-analytics

Region * East US

Select Data Lake Storage Gen2 * From subscription Manually via URL

Account name * (New) dronetelemetrydatalake

Create new

File system name * (New) raw-data

Create new

Assign myself the Storage Blob Data Contributor role on the Data Lake Storage Gen2 account to interactively query it in the workspace.

We will automatically grant the workspace identity data access to the specified Data Lake Storage Gen2 account, using the Storage Blob Data Contributor role. To enable other users to use this storage account after you create your workspace, perform these tasks:

- Assign other users to the **Contributor** role on workspace
- Assign other users the appropriate Synapse RBAC roles using Synapse Studio
- Assign yourself and other users to the **Storage Blob Data Contributor** role on the storage account

Learn more

Review + create < Previous Next: Security >

Figure 2.5 – The Workspace details section

In the **Workspace details** section, you provide the name of your workspace, the Azure region where it will be deployed, and details about your **Azure Data Lake Storage** (**ADLS**) Gen2 account. This is what you need to know about these parameters:

- **Workspace name**: This is where you choose the name of your Azure Synapse workspace. I recommend using a name that helps people identify the purpose of this workspace, as your co-workers may be working on other Synapse workspaces as well. For our example, I'll call it `drone-analytics`.
- **Region**: This is the Azure region where your workspace will be deployed to. This is an important choice and will determine the level of latency you may experience while using the service. Ideally, you should pick either the region that's the closest to your physical location, the region where data will be written from, or the location where you need your data to reside for compliance reasons. For our example, we'll pick the **East US** region.

> **Note**
>
> Each Azure region contains specific data recovery, high availability, and service availability characteristics. To see what each Azure region offers and choose the one that meets your business needs, refer to `https://azure.microsoft.com/global-infrastructure/geographies/`.

- **Select Data Lake Storage Gen2**: Azure Synapse workspaces need an ADLS Gen2 account. Here, you define whether you will use an ADLS Gen2 account from your subscription, create a new one, or manually provide the details of an existing account. For this example, I'll create a new storage account that I named `dronetelemetrydatalake`. Note that Azure storage account names are unique across Azure, so you need to choose a different name for yours. Azure storage account names can only have numbers and lowercase letters.
- **File system name**: This will be the name of the container that will be created with your storage account (or you can pick an existing one if you are using an existing storage account). I will name ours `raw-data`.
- Make sure you check the **Assign myself the Storage Blob Data Contributor role…** checkbox, as your account will need to be assigned this role to allow querying files in your storage account using the serverless SQL pools. In fact, any accounts that need access to the same storage account via serverless SQL pools will need to be assigned the same role. Azure Synapse also assigns this role to the workspace identity, as described in the blue box in *Figure 2.5*.

When you are ready, click the **Next: Security** > button, or simply select the **Security** tab at the top to move to the next section.

Security tab

The next step is to choose the security settings for our new workspace. As illustrated in *Figure 2.6*, the **Security** tab is the place to provide these settings.

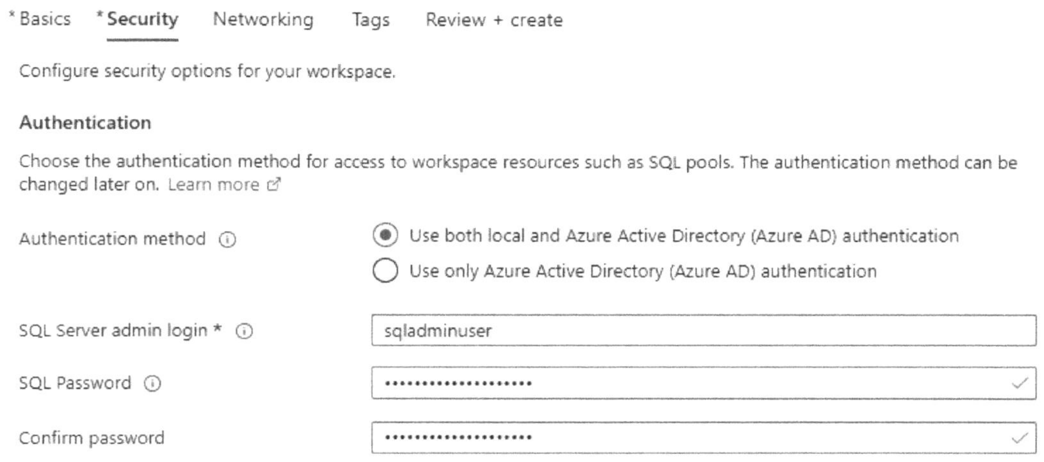

* Basics * **Security** Networking Tags Review + create

Configure security options for your workspace.

Authentication

Choose the authentication method for access to workspace resources such as SQL pools. The authentication method can be changed later on. Learn more

Authentication method — (•) Use both local and Azure Active Directory (Azure AD) authentication / ( ) Use only Azure Active Directory (Azure AD) authentication

SQL Server admin login * — sqladminuser

SQL Password — ••••••••••••••••••••

Confirm password — •••••••••••••••••••

Figure 2.6 – Choosing authentication options

Azure Synapse allows you to authenticate and access workspace resources (such as SQL pools) by using your **Azure Active Directory** (**AAD**) account, by using local authentication through SQL logins, or by using **shared access signatures** (known as **SAS tokens**). Authentication via SQL logins is referred to as local authentication.

There are two configuration options for your workspace: you can configure it to exclusively use AAD, or to allow a combination of AAD and local authentication. Microsoft SQL Server professionals will be quite familiar with local authentication, as there is a server-side configuration option called **Mixed Mode Authentication**.

You should use the **AAD-only** authentication option if your company prohibits the use of local authentication or if you are trying to simplify your environment. By using AAD-only, you don't need to manage additional users, passwords, and password policies and provide a login every time you connect to a workspace resource. AAD simplifies authentication by providing single sign-on, central account and password management, more robust authentication protocols, and many other advantages.

You should use the local and AAD authentication options if you have legacy applications that do not support AAD or clients that can't authenticate to a Windows domain or are from untrusted domains, or in specialized scenarios. When you choose this option, you will be required to provide a name for the SQL Server admin login (which you can change from the default name, `sqladminuser`) and the password for this account.

> **Note**
>
> Data scientists, data engineers, and analysts may use a mixed set of data science and business intelligence tools to connect to their analytical environment. Therefore, it is very common to use local authentication in Azure Synapse. Make sure you understand your user requirements before you decide which authentication option to use.

Next, we need to make important decisions about allowing network access to the ADLS Gen2 account, and about using double encryption.

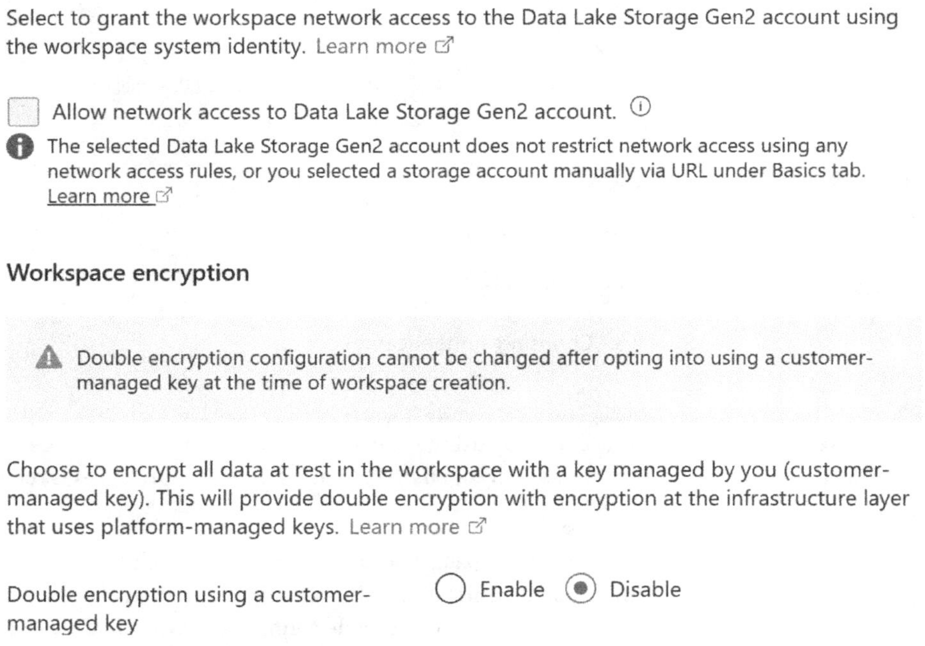

Select to grant the workspace network access to the Data Lake Storage Gen2 account using the workspace system identity. Learn more
Allow network access to Data Lake Storage Gen2 account.
The selected Data Lake Storage Gen2 account does not restrict network access using any network access rules, or you selected a storage account manually via URL under Basics tab. Learn more
Workspace encryption
Double encryption configuration cannot be changed after opting into using a customer-managed key at the time of workspace creation.
Choose to encrypt all data at rest in the workspace with a key managed by you (customer-managed key). This will provide double encryption with encryption at the infrastructure layer that uses platform-managed keys. Learn more
Double encryption using a customer-managed key
Enable
Disable

Figure 2.7 – Additional security options

Let's look at these parameters in detail:

- **System assigned managed identity permission**: If the storage account that you used to configure your new workspace uses security features such as network access rules, you can use this checkbox to provide your workspace's system identity with access to your ADLS Gen2 account.
- **Workspace encryption**: By default, data in Azure disks and storage accounts is encrypted without the need for user configuration. As mentioned in *Chapter 1*, Azure Synapse allows users to use a customer-managed key for the encryption of data at rest. When you configure encryption with a customer-managed key, you add a second layer of encryption to your data, referred to as double encryption in Azure Synapse.

Enabling double encryption requires you to provide the following additional settings:

- **Encryption key**: Azure Synapse workspaces only support encryption keys stored in Azure Key Vault, Azure's enterprise solution for the management of cryptographic keys and secrets. In this field, you must select a key from your Azure Key Vault or type the key identifier.
- **Managed identity**: Determines whether you will use a user-assigned or a system-assigned identity to access your customer-managed key in the key vault. If you select the user-assigned identity option, you will also need to provide the actual identity.

Make sure you understand in advance whether you will need to use a customer-managed key and have all the implementation details to hand (such as your Azure Key Vault or the key identifier) when you create your workspace. These settings cannot be changed after you create your Azure Synapse workspace.

The next step is to provide the network settings for your workspace. Click the **Next: Networking >** button, or simply select the **Networking** tab at the top to move to the next section.

Networking tab

The last steps of configuring Synapse-specific settings are the network settings. Picking your network options allows you to isolate your workspace from other services and protect it from data exfiltration among different sessions. The **Networking** tab is the place to provide these details, and it is illustrated in *Figure 2.8*:

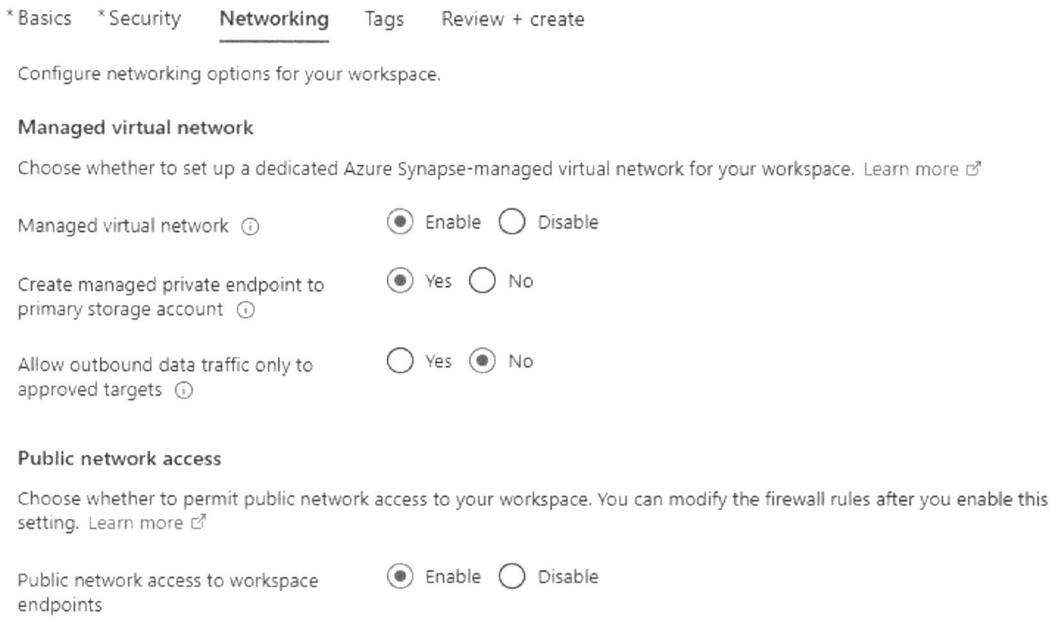

* Basics * Security Networking Tags Review + create

Configure networking options for your workspace.

Managed virtual network

Choose whether to set up a dedicated Azure Synapse-managed virtual network for your workspace. Learn more

Managed virtual network ◉ Enable ○ Disable

Create managed private endpoint to primary storage account ◉ Yes ○ No

Allow outbound data traffic only to approved targets ○ Yes ◉ No

Public network access

Choose whether to permit public network access to your workspace. You can modify the firewall rules after you enable this setting. Learn more

Public network access to workspace endpoints ◉ Enable ○ Disable

Figure 2.8 – Configuration options for a managed workspace virtual network

Let's look at these parameters in detail:

- **Managed virtual network**: An Azure **Virtual Network (VNet)** enables secure communication between your Azure services, with the internet, and with private networks. It provides essential network security options for system administrators, such as network traffic filtering and routing, and is an essential component of cloud deployments in Azure. By enabling the use of a managed VNet in your workspace, Azure Synapse creates and manages the VNet for you, so you don't have to worry about managing it.

  Using a managed workspace VNet ensures your workspace is isolated from other workspaces and creates a security boundary. When combined with managed private endpoints, this feature protects your workspace against data exfiltration.

  If you choose to enable a managed VNet for your workspace, you will be required to provide the following additional parameters:

  - **Create managed private endpoint to primary storage account**: Spark pools need access to data stored in your primary ADLS Gen2 account. This option will send a request to this storage account and requires the account owner to approve the connection request. If you have sufficient privileges in the primary storage account (this is the case in our example, as we are creating the account with the workspace), then Synapse will approve this connection request by itself, and the necessary endpoints will be created automatically.
  - **Allow outbound data traffic only to approved targets**: If you select **Yes**, outbound traffic will go through private endpoints that connect to Azure resources only to Azure AD tenants that you approve.
  - **Public network access to workspace endpoints**: This feature allows you to control whether access to your workspace is allowed from public networks or only using private endpoints.

- **Firewall rules:** Here, you can enable a global workspace setting to allow traffic to your workspace from all IP addresses or only from specific addresses or IP ranges. If you unselect this option, you cannot yet specify the IP addresses or ranges that will have access to your workspace – you will have to do that after your workspace is created. We will cover these settings in *Chapter 12*.

Firewall rules

Azure Synapse Studio and other client tools will only connect to the workspace endpoints if this setting is selected.
Connections from specific IP addresses or all Azure services can be allowed or disallowed after the workspace is provisioned.
Allow connections from all IP addresses to your workspace's endpoints. You can restrict these permissions to just Azure datacenter IP addresses and/or specific IP address ranges after creating the workspace.
Allow connections from all IP addresses

Figure 2.9 – Firewall rules

The final step in the workspace creation process is to provide any necessary tags for your resources. Click the **Next: Tags** > button, or simply select the **Tags** tab at the top.

Tags tab

If you've been using Azure, you are very likely already familiar with tags. They allow you to add custom metadata to your Azure resources to help you and others identify how those resources are being used, who is responsible for them, what applications depend on them, or any other relevant information. They work by allowing you to assign key-value pairs to your resources, where you get to write whatever you want in those pairs. As you would expect, the **Tags** tab is the place to provide your custom tags, as shown in *Figure 2.10*:

* Basics * Security Networking **Tags** Review + create

Tags are name/value pairs that enable you to categorize resources and view consolidated billing by applying the same tag to multiple resources and resource groups. learn more

Note that if you create tags and then change resource settings on other tabs, your tags will be automatically updated.

Name		Value	Resource
Point of contact	:	procha	Synapse workspace
Environment	:	Development	Synapse workspace
Workload	:	Drone telemetry	Synapse workspace
	:		Synapse workspace

Figure 2.10 – Tagging your resources

Common uses for tags in Azure resources include the following:

- **Cost management**: You can set budgets and alerts to monitor Azure usage and spending so that you get notified after your usage reaches a certain threshold.
- **Resource management**: Where I work, we set tags that help everyone understand who is responsible for any Azure resources that are found so that they can easily contact the service owners if needed. Another example is to tag a resource to identify whether it is used in a development, test, or production environment.
- **Security**: Some organizations use tags to apply data classification and alert about the security impact of the data.

When you browse the Azure subscriptions of companies using Azure, you will typically see hundreds or thousands of Azure resources that were created for different projects. Tagging Azure resources is not something that should be overlooked and makes it easier for everyone to identify how resources are used and who is responsible for them.

Now you are finally ready to review your settings and create your Azure Synapse workspace. Click the **Next: Review + create** > button or simply select the **Review + create** tab at the top to review your settings.

Review + create tab

On this page, you have one last chance to review your settings before you submit the Azure Synapse workspace creation request to **Azure Resource Manager** (**ARM**). The important part here is to make sure your settings passed validation and, if they didn't, mitigate any issues. If validation fails, you will see a red banner at the top of the page that says **Validation failed**, and a red icon over the section that needs attention. In *Figure 2.11*, the Azure portal is bringing to your attention that the **Security** tab has settings that are invalid and need to be resolved before you can create your workspace:

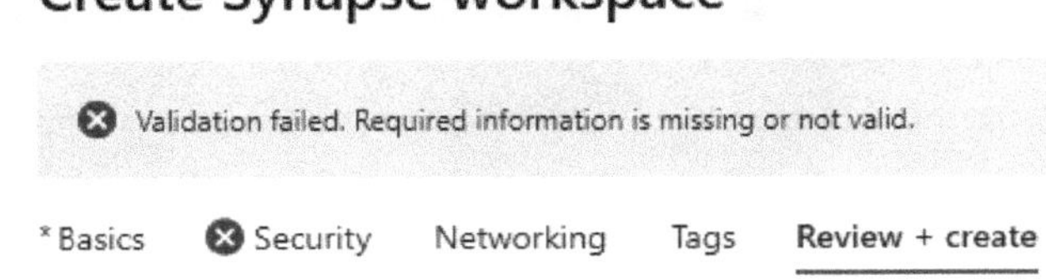

Figure 2.11 – Validation failed in the Security tab

Once you resolve the problems, you can navigate directly to the **Review + create** tab again and review your settings one last time. When you are ready, click the **Create** button. You will be forwarded to a deployment page that shows you the status of the provisioning process, as shown in *Figure 2.12*. You also have the option to download the ARM template that defines your new workspace for automated deployment in the future by selecting the **Template** tab on the deployment page.

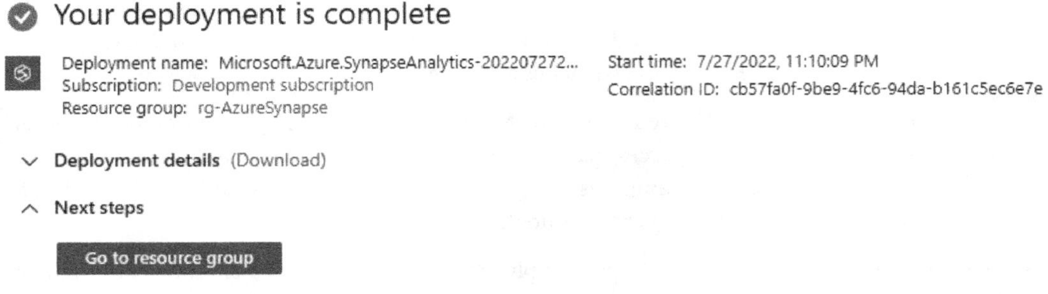

Figure 2.12 – Deployment completion confirmation

> **Note**
>
> To learn more about ARM templates, visit `https://learn.microsoft.com/azure/azure-resource-manager/templates/overview`.

The deployment of your workspace should be complete within 5 minutes, depending on your settings. Once it is complete, you will see a message notifying you of the completion of your deployment and a button inviting you to navigate to the resource group where your services were deployed. This experience is standard across all Azure services. Let's find out how to find and navigate to your new Azure Synapse workspace.

Finding your new workspace

Now that your workspace is ready, let's navigate to the resource page in the Azure portal and review the settings we used by taking the following steps:

1. Click the **Go to resource group** button on the deployment completion confirmation page, as shown in *Figure 2.12*.
2. You will see a page that shows your resource group details, and the bottom of the page shows you all the resources in your resource group. Find the Azure resource of the **Synapse workspace** type that has the name you provided while creating your workspace (`drone-analytics` in my example) and click on it. You will be forwarded to the Azure portal page of your Azure Synapse workspace.

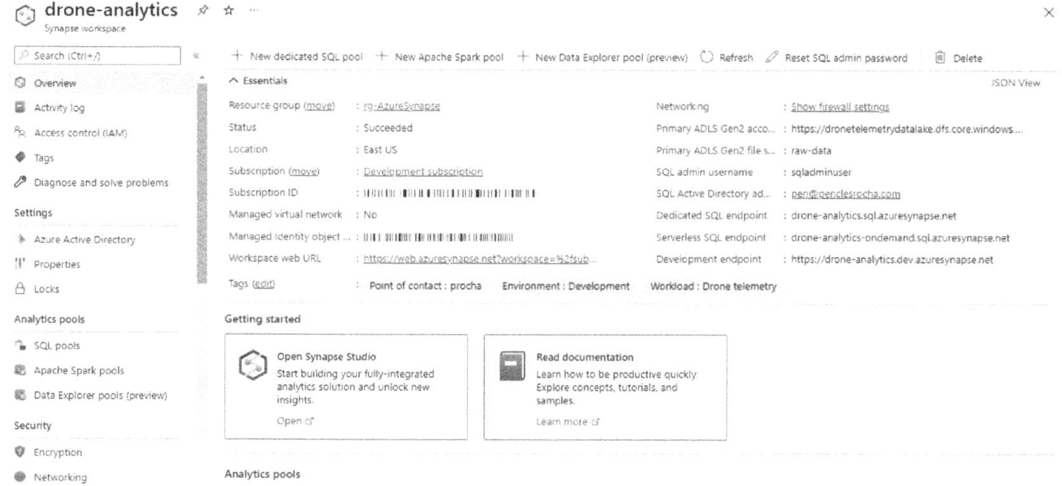

Figure 2.13 – Your new Azure Synapse workspace in the Azure portal

Your Azure Synapse workspace is now ready for use! By default, new Synapse workspaces will offer you access to the storage account you selected in the *Creating an Azure Synapse workspace* section, and you can use the serverless SQL pool to query data in that storage account immediately. However, if you need to utilize dedicated SQL pools, Apache Spark pools, or Data Explorer pools, you will need to create those resources after your workspace is ready.

Keep in mind that you can have several Synapse workspaces, depending on your subscription type. Subscriptions of the pay-as-you-go, free trial, Azure Pass, and Azure for Students type all have a limit of two Synapse workspaces that you can have at any given time. Other subscription types, such as Microsoft Azure Enterprise, allow a default limit of 20 workspaces per subscription, with a maximum limit of 250 subscriptions if you contact Azure Support.

> **Note**
>
> Most Azure services have service limits, also called quotas or constraints. Those limits are often adjustable if you contact Azure support. To see the service limits for Azure services, including Azure Synapse, refer to `https://docs.microsoft.com/azure/azure-resource-manager/management/azure-subscription-service-limits`.

The next step is to create Data Explorer pools. We will do that first using Azure Synapse Studio, next using the Azure portal, and last using the Azure CLI. If your job is to set up the cloud environment for other teams, learning how to provision pools through the Azure portal or the Azure CLI should be a handy skill. On the other hand, if you are more involved with analytics or data science projects and will spend most of your time in Azure Synapse Studio, managing resources from there makes more sense.

Creating a Data Explorer pool using Azure Synapse Studio

Azure Synapse Studio is a unified environment for all your Azure Synapse work, from data integration, through data exploration, to serving data to users. After you create your Azure Synapse workspace, by navigating to this workspace using the Azure portal, you will find your workspace's web URL in the **Essentials** section of the page, under the **Open Synapse Studio** box shown in *Figure 2.14*:

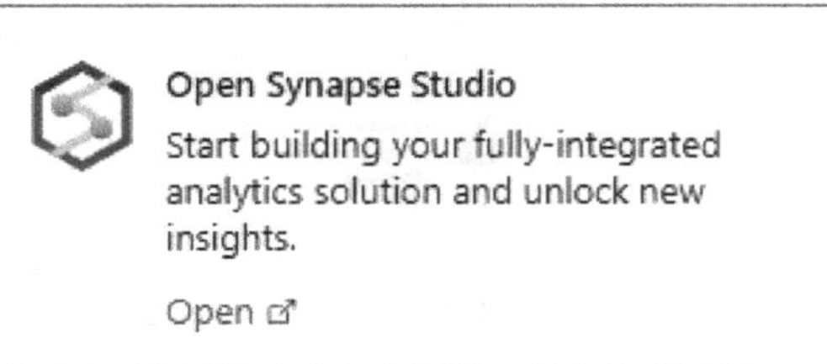

Figure 2.14 – Navigating to Azure Synapse Studio

If you don't want to open the Azure portal to find your workspace URL, you can always navigate directly to `https://web.azuresynapse.net`, provide the directory and Azure subscription where your workspace was created, and then pick the desired workspace from the **Workspace name** dropdown, as shown in *Figure 2.15*:

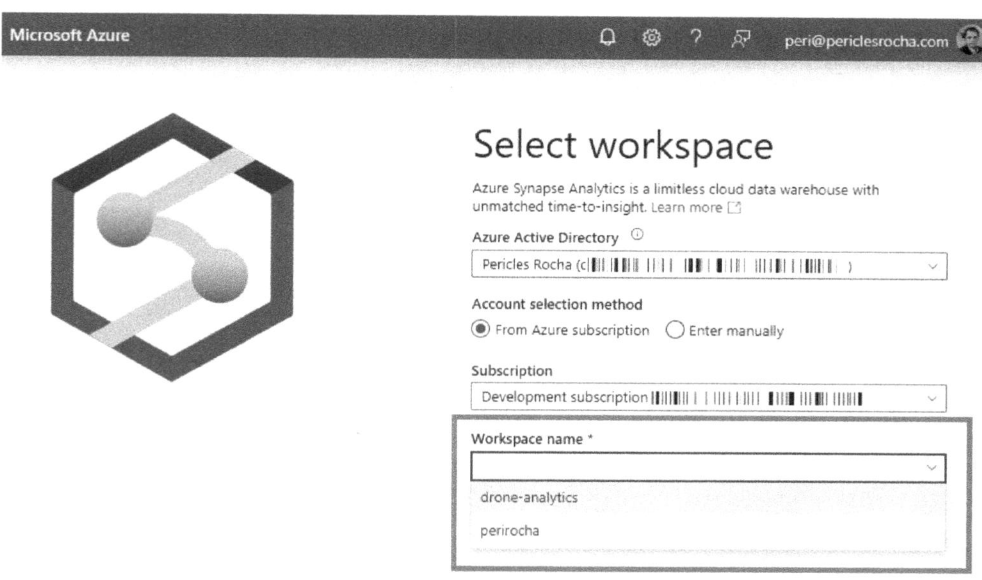

Figure 2.15 – Choosing your workspace in Azure Synapse Studio

Welcome to Azure Synapse Studio! As seen in *Chapter 1*, we can use the Manage hub to create, scale, or delete analytical pools, as well as other management features of Azure Synapse – which we will explore in *Chapter 3* and *Chapters 10* through *13*. Right now, our focus is on creating a Data Explorer pool. So, let's perform the following steps:

1. On the Azure Synapse Studio home page, click the **Manage** option in the left-hand side area – called the Hub.

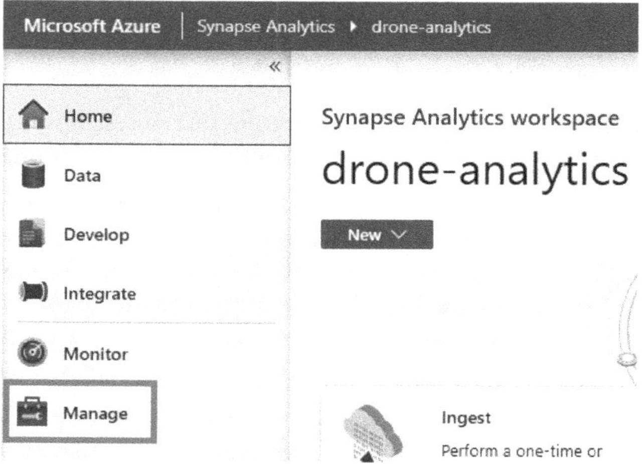

Figure 2.16 – Selecting the Manage hub option

2. Select **Data Explorer pools** under **Analytics pools**.
3. In the **Data Explorer pools** window, click the **+ New** button. You will be presented with the **Create Data Explorer pool** pane.

Let's review the options here in detail.

Basics tab

Let's explore the parameters of the **Basics** tab in detail:

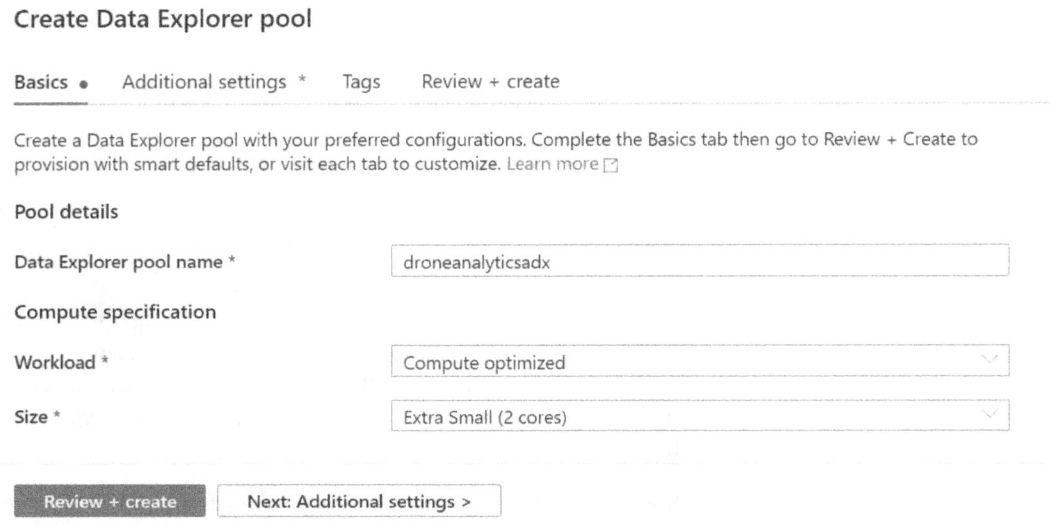

Create Data Explorer pool

Basics • Additional settings * Tags Review + create

Create a Data Explorer pool with your preferred configurations. Complete the Basics tab then go to Review + Create to provision with smart defaults, or visit each tab to customize. Learn more

Pool details

Data Explorer pool name * droneanalyticsadx

Compute specification

Workload * Compute optimized

Size * Extra Small (2 cores)

Review + create Next: Additional settings >

Figure 2.17 – The Basics tab

As seen in the preceding screenshot, the fields are as follows:

- **Data Explorer pool name**: An alphanumeric-only name that must start with a letter and contain only lowercase characters. In this example, I will call our pool `droneanalyticsadx`.
- **Workload**: You should pick whether your pool will be compute optimized or storage optimized. The **Compute optimized** option provides a higher core-to-cache ratio, with a local SSD to reduce latency. It offers the lowest cost per core. The **Storage optimized** option offers better cost options for larger volumes of data. If you have very large volumes of data, you should look to optimize costs using one of the storage-optimized workload size options. For the examples in this book, I'll choose the compute-optimized workload, but you can choose whichever you prefer.
- **Size**: This is the node size of each node in your cluster. To keep costs low, I will pick the **Extra Small (2 cores)** node size, which has plenty of compute for our examples.

At this stage, you can click the **Review + create** button to accept the defaults provided by Synapse Studio, review your settings, and create your Data Explorer pool. But if you are like me, you never accept the defaults – at least not before exploring them.

Additional settings tab

In *Figure 2.18*, you can see the settings that I used in the **Additional settings** tab for our example Data Explorer pool:

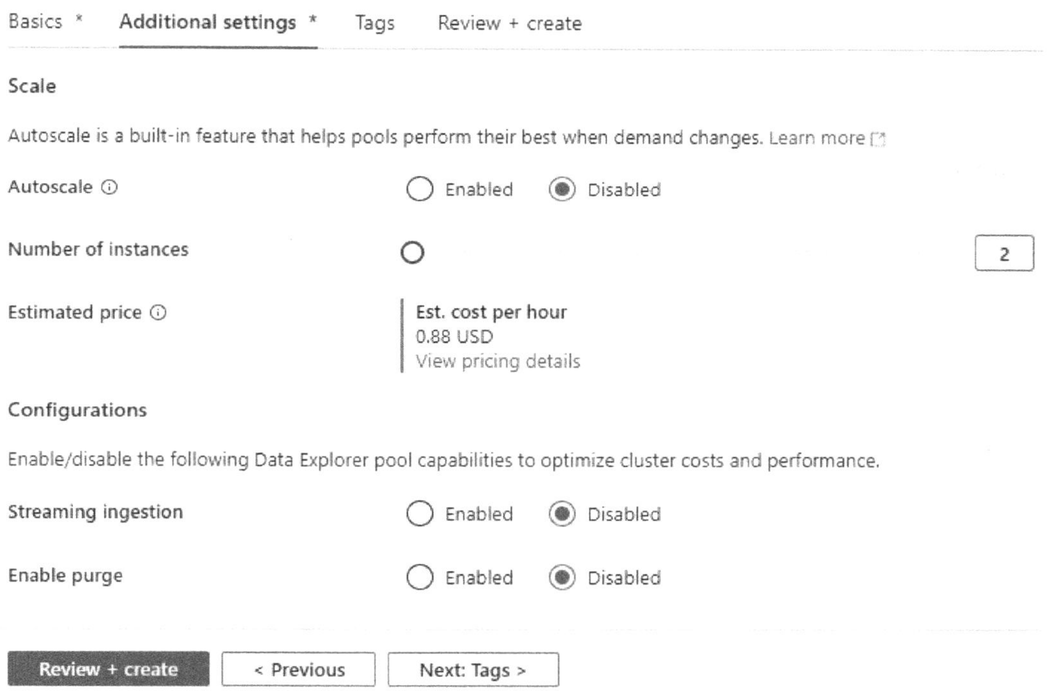

Figure 2.18 – The Additional settings tab

This is the place where you have to make some important decisions about the size of your Data Explorer pool, how you want to scale it, and more. Let's review these settings in detail:

- **Autoscale**: Data Explorer pools can scale out and provision new cluster nodes if demand for your server grows and can scale back in when usage is low again.

  If you set this option to **Disabled**, you will have to monitor the usage of your cluster and scale the service manually when needed. If you set it to **Enabled**, you will have to set the minimum and maximum number of instances that your cluster can scale out and in to.

  Because here I expect to have a predictable and low usage for this Data Explorer pool, I will set this to **Disabled**. In real-life scenarios, make sure you consider how you want to use **Autoscale**.

- **Streaming ingestion**: Enable this option if you require very low latencies – less than a second – between data ingestion and querying. We will discuss streaming ingestion in *Chapter 5*, *Ingesting Data into Data Explorer Pools*. For now, I will not enable it.
- **Enable purge**: Data Explorer supports data purges to protect personal data and satisfy companies' obligations to data protection laws such as the European Union's **General Data Protection Regulation** (**GDPR**). We will discuss data purges in *Chapter 13*. I will leave this option disabled for this Data Explorer pool.

Tags tab

We described tags in detail in the *Creating an Azure Synapse workspace* section of this chapter, and they serve the same purpose for Data Explorer pools. Any tags that we create here will not change the experience described in this book. Feel free to set any tags you need for your Data Explorer pool to adhere to your company's or personal policies.

Review + create tab

This section is also similar to the one we explored when we were creating our Azure Synapse workspace, so we won't get into the details here again. You only need to make sure that your validation passes the tests and that you agree with the estimated cost per hour and the service usage terms. When you are ready, review your settings and click on the **Create** button.

As illustrated by *Figure 2.19*, you will see a notification pop up on Azure Synapse Studio notifying you that your Data Explorer pool is deploying:

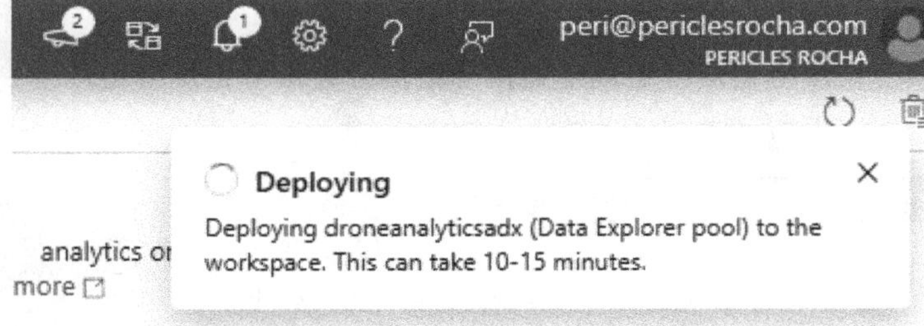

Figure 2.19 – Deploying your Data Explorer pool

That's it! We've provided all the details needed to create a new Data Explorer pool using Azure Synapse Studio. Note that once you click the **Create** button, it can take 10 to 15 minutes to deploy your pool. You can monitor your deployment by clicking the notification area on the top right-hand side of Azure Synapse Studio (the bell icon) or by navigating to **Analytics pools** | **Data Explorer pools** in the Manage hub of Azure Synapse Studio. Here, you should see your new Data Explorer pool listed with a status that says **Creating**, as shown in *Figure 2.20*:

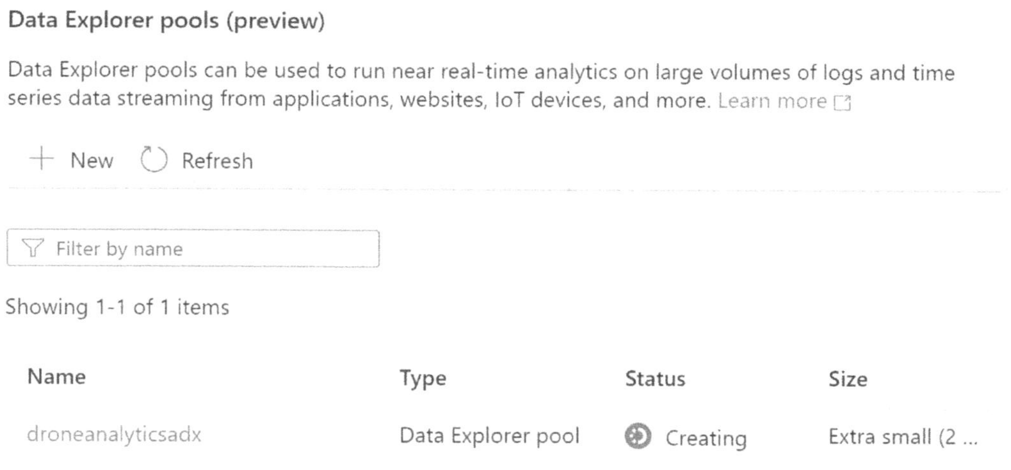

Data Explorer pools (preview)

Data Explorer pools can be used to run near real-time analytics on large volumes of logs and time series data streaming from applications, websites, IoT devices, and more. Learn more

+ New Refresh

Filter by name

Showing 1-1 of 1 items

Name	Type	Status	Size
droneanalyticsadx	Data Explorer pool	Creating	Extra small (2 ...

Figure 2.20 – Monitoring the status of your Data Explorer pool

Once the deployment completes, you will receive a notification that the operation was successful, and you should see the status of your cluster change to **Online**. Your cluster is now ready to start creating databases, loading data, and taking queries.

> **Important note**
>
> The moment your Data Explorer pools are running, Azure starts charging you for the compute hours. Make sure to stop your pool when you don't need it to avoid incurring unnecessary costs.

Next, we will explore how to create the same Data Explorer pool but from the Azure portal.

Creating a Data Explorer pool using the Azure portal

There's no functional difference in creating a Data Explorer pool from the Azure portal when compared to Synapse Studio; the options are the same ones, with minor differences in the user interface. Since we have already explained what the different Data Explorer parameters are, we will quickly navigate through them and focus on the user interface differences.

To create a Data Explorer pool, take the following steps:

1. Navigate to the Azure portal at `https://portal.azure.com`.
2. To quickly find your Azure Synapse workspace, simply type its name in the global search box, as shown in *Figure 2.21*. Click on the name of the service with **Synapse workspace** next to it to be sent to the details page of your workspace.

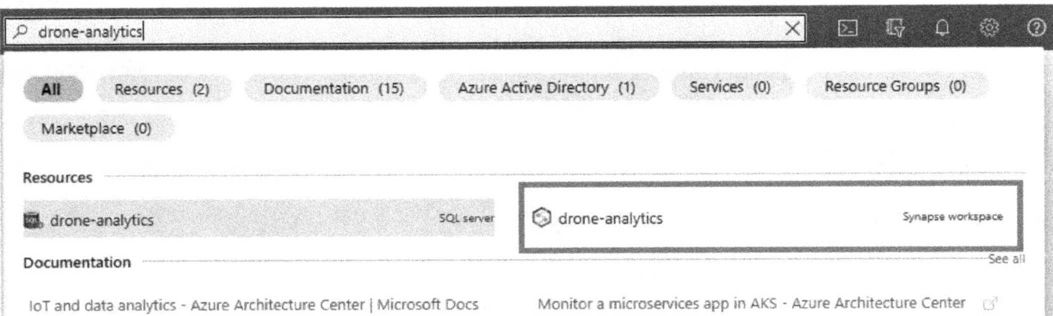

Figure 2.21 – Finding your Synapse workspace

3. On the workspace resource page, right at the top of the resource details, click the **+ New Data Explorer pool (preview)** button.
4. You will now see a page that looks exactly like the first page of the new Data Explorer pool page on Synapse Studio (seen in *Figure 2.17*). Provide the name of your Data Explorer pool, your workload choice, and its size. Click **Next: Additional settings** > when you are ready.
5. The **Additional settings** page is the only one that looks slightly different from the one in Synapse Studio, but functionally, they produce the same results.

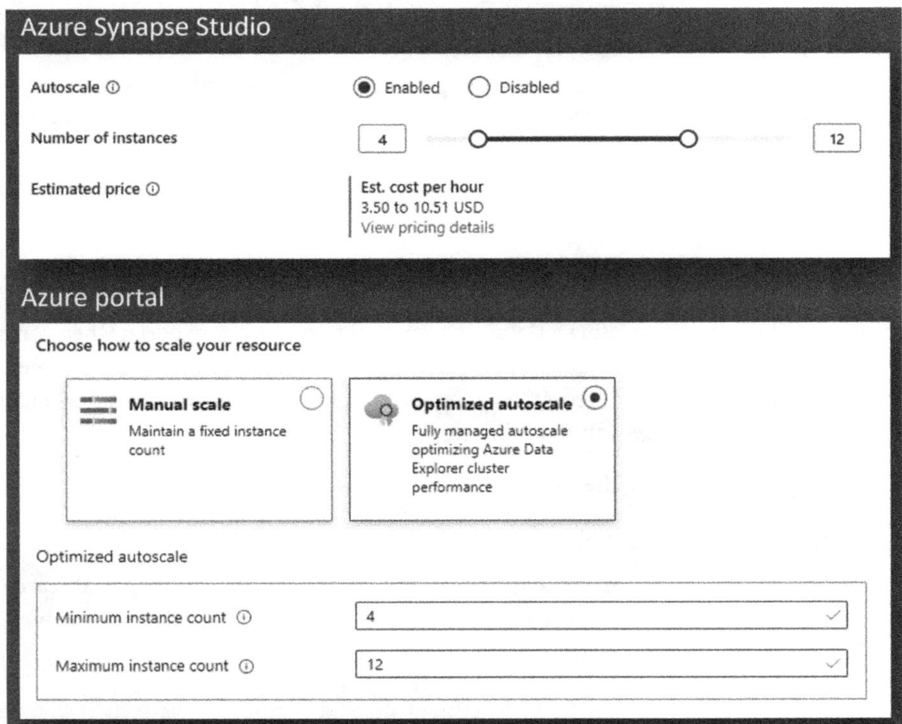

Figure 2.22 – Autoscale settings in Azure Synapse Studio (top) and the Azure portal (bottom)

First, you see a descriptive box to pick **Manual scale** or **Optimized autoscale**. Next, instead of the sliders to choose your minimum and maximum instance count, you are presented with two text boxes to manually type the instance counts for each. Also, note that the Azure portal does not give you a cost estimate based on your cluster size, so be sure to check your estimated costs in the Azure pricing calculator before you make your choice.

6. The **Tags** page is exactly the same as the one in Synapse Studio, so we'll skip the explanation here as well.
7. The **Review + create** page presents the same information as it does in Synapse Studio, except for the price estimation. When you are ready, click the **Create** button.

It takes the same time to provision a Data Explorer pool in the Azure portal as it takes in Synapse Studio – the background operation is the same. When your resource is ready, you will see a confirmation page similar to the one in *Figure 2.12*.

Creating a Data Explorer pool using the Azure CLI

If you are serious about using Microsoft Azure, you might as well get familiar with the Azure CLI. The Azure CLI works cross-platform, regardless of your choice of operating system, and offers a set of commands you can use to create and manage Azure resources. It is extremely helpful for the automation of tasks and running processes repetitively.

To install the Azure CLI on your computer, visit `https://docs.microsoft.com/cli/azure/install-azure-cli` and choose the appropriate link for install instructions based on your operating system. If you are not ready to install the Azure CLI on your computer but would like to experiment with it, you can use Azure Cloud Shell.

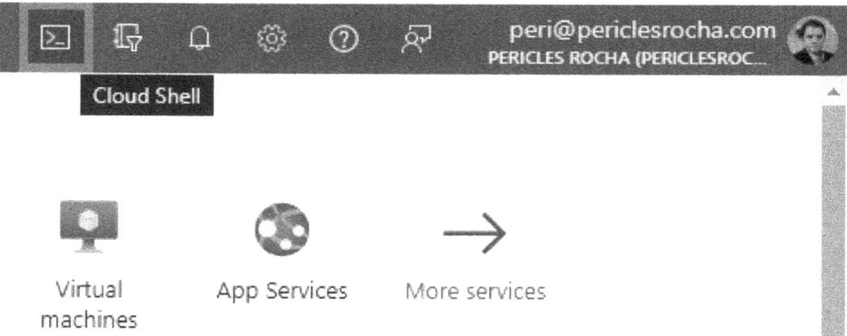

Figure 2.23 – Launching Azure Cloud Shell

Simply navigate to the Azure portal and click the Cloud Shell icon on the top navigation page. For quick access, navigate to `https://shell.azure.com`.

Once you have installed the Azure CLI on your system (or set up Cloud Shell), the first step is to log in to your Azure environment from the CLI. Type `az login` in your terminal to allow the CLI to log in interactively through your browser.

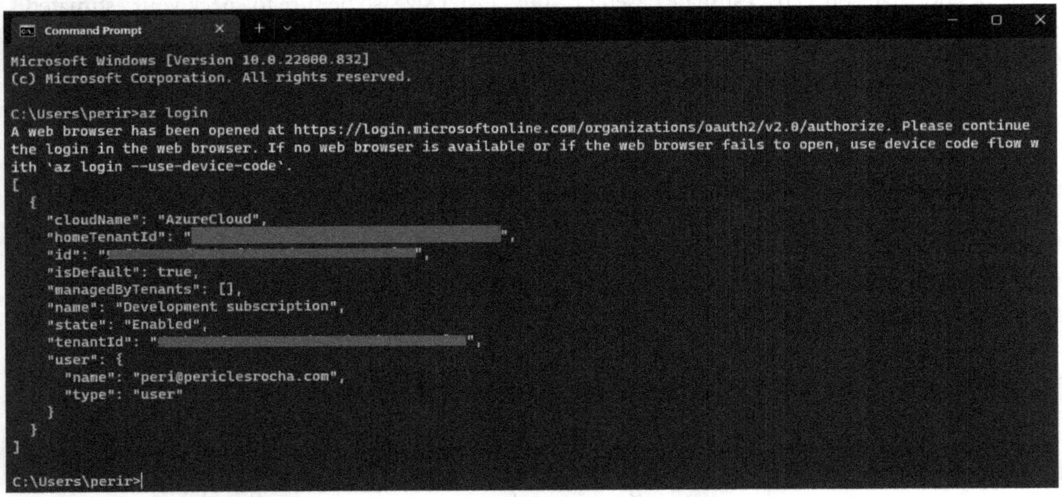

```
Command Prompt
Microsoft Windows [Version 10.0.22000.832]
(c) Microsoft Corporation. All rights reserved.

C:\Users\perir>az login
A web browser has been opened at https://login.microsoftonline.com/organizations/oauth2/v2.0/authorize. Please continue
the login in the web browser. If no web browser is available or if the web browser fails to open, use device code flow w
ith `az login --use-device-code`.
[
  {
    "cloudName": "AzureCloud",
    "homeTenantId": "",
    "id": "",
    "isDefault": true,
    "managedByTenants": [],
    "name": "Development subscription",
    "state": "Enabled",
    "tenantId": "",
    "user": {
      "name": "peri@periclesrocha.com",
      "type": "user"
    }
  }
]

C:\Users\perir>
```

Figure 2.24 – Authenticating to the Azure cloud using the CLI

Your browser will open an Azure dialog to authenticate you. Pick the desired account in the list (or provide your credentials), and your CLI session will show a JSON message with details about your current directory (or tenant) and subscription, as seen in *Figure 2.24*.

You are now authenticated to Azure and can start issuing commands. To create a new Azure Synapse Data Explorer pool, we will use the `az synapse kusto pool create` command. It requires the following parameters:

- `--kusto-pool-name --name -n`: The name of the Data Explorer pool.
- `--resource-group -g`: The name of the resource group where your Azure Synapse workspace was created.
- `--sku`: The compute workload that will be used for this Data Explorer pool. It accepts **Storage optimized** or **Compute optimized**.
- `--workspace-name`: The name of the Azure Synapse workspace where this Data Explorer pool will be created.

Using only these parameters is the same as providing the basic details of the Data Explorer pool through the Azure portal or Azure Synapse Studio and then accepting the defaults. All the other parameters, such as autoscale, purge, and streaming ingestion, are optional and will use default settings if not provided.

> **Note**
>
> For a detailed view of all the commands you can issue via the CLI to manage your Data Explorer pools, refer to `https://docs.microsoft.com/cli/azure/synapse/kusto/pool?view=azure-cli-latest`.

To create a Data Explorer pool named `droneanalyticsadx` with the default settings, you can use the following command:

```
az synapse kusto pool create --name "droneanalyticsadx"
--resource-group "rg-AzureSynapse" --sku name="Compute
optimized" size="Small" --workspace-name "drone-analytics"
```

> **Note**
>
> You can find this command in the `Chapter 02\createPool-CLI.txt` file of the book's code repository.

The output of the CLI is illustrated in *Figure 2.25*:

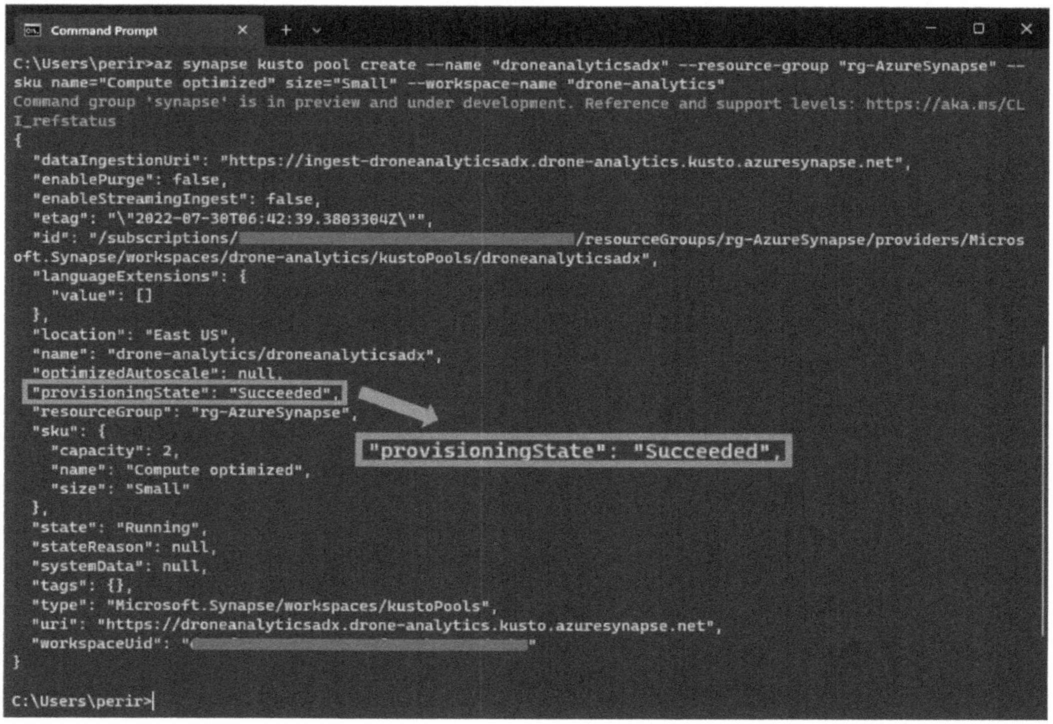

Figure 2.25 – Successful provision of your Data Explorer pool

Note that the JSON message returned to the CLI states `"provisioningState": "Succeeded"`, along with other details about the resource that was created.

Summary

Getting started with Azure Synapse Data Explorer is easy. If you did not have an Azure account yet, this chapter started by explaining how to create one for free – just be aware of the costs incurred by your Azure Synapse experiments, as it is not one of the free services!

We then went through the process of creating an Azure Synapse workspace and explained in detail all the settings that are exposed through the Azure portal.

Next, we covered three different ways to create Data Explorer pools: through Azure Synapse Studio, through the Azure portal, and through the Azure CLI. In each section, we explored which to use and when. We walked through the Data Explorer pool parameters in detail the first time we created it, and then we explored the user experience differences in the Azure portal. Lastly, we created a Data Explorer pool using the Azure CLI, which allows cloud admins to automate the creation of Azure services using shell scripts.

Azure Synapse Studio is the tool we will use most of the time while working with Azure Synapse and throughout this book. Therefore, in the next chapter, we will look at this tool in detail to make sure you understand how to get the most out of it.

3

Exploring Azure Synapse Studio

Now that you have created your first Azure Synapse workspace and Data Explorer pool, you are ready to dive into Azure Synapse Studio – the primary development and management tool for Azure Synapse.

In *Chapter 1*, *Introducing Azure Synapse Data Explorer*, you learned about the basic user experience elements and the purpose of each hub in Azure Synapse Studio. In *Chapter 2*, *Creating Your First Data Explorer Pool*, you learned how to use Azure Synapse Studio to create a new Data Explorer pool and we dug into every parameter in detail. This chapter goes a level deeper and helps you become more familiar with managing and monitoring your environment and exploring data.

First, you will learn about the user experience elements in Azure Synapse Studio, helping you easily find what you need. You will learn where to manage service-level settings, navigate through the hubs, and how to use each of them to manage and query data using Data Explorer pools.

Next, we will put our hands on some data. You will create a database that will be used in some examples throughout this book and load data into it. You will query this data using the query editor, explore how to use Synapse notebooks to work with data stored in Data Explorer pools, and learn how to save your work in your workspace or in your Git repository for group collaboration and source control.

Finally, we will dive into the management aspects of Data Explorer pools. You will learn how to scale your pools, monitor their statuses, and pause and resume them to manage costs.

In a nutshell, here are the topics covered in this chapter:

- Exploring the user interface of Azure Synapse Studio
- Running your first query
- Managing and monitoring Data Explorer pools

Technical requirements

To reproduce the examples and content in this chapter, you will need an Azure subscription and an Azure Synapse workspace. If you choose to load the data we offer as an example in this chapter, you will need to download the drone telemetry dataset from GitHub (approximately 123 MB) on your local machine. This dataset is stored as a **comma-separated values** (**CSV**) file and contains fictional data that was created specifically for this book. It can be found at `https://github.com/PacktPublishing/Learn-Azure-Synapse-Data-Explorer/blob/main/Chapter%2003/drone-telemetry.zip`. Make sure you extract the `drone-telemetry.csv` file from the compressed folder before using it.

The queries demonstrated in this chapter use **Kusto Query Language** (**KQL**). In addition to KQL, we will explore data in Synapse notebooks using Python. While these queries and notebooks are simple and we will describe what they do, teaching KQL or Python is not in the scope of this book. You should have at least a basic understanding of these languages to fully understand the examples.

> **Important note**
>
> As mentioned in *Chapter 2, Creating Your First Data Explorer Pool*, Azure Synapse is not one of the free services available in the Azure free account offering, so running these examples will incur costs in your subscription. Make sure you keep your cluster sizes as small as possible and pause your compute pools when not in use.

Exploring the user interface of Azure Synapse Studio

Azure Synapse Studio offers you a single pane of glass to create, manage, monitor, and develop your Azure Synapse workspaces. If you are going to take Azure Synapse seriously, then you will spend a good amount of time on it. Therefore, let's make sure you get familiar with it.

Figure 3.1 provides a breakdown of the three main user interface elements of Synapse Studio.

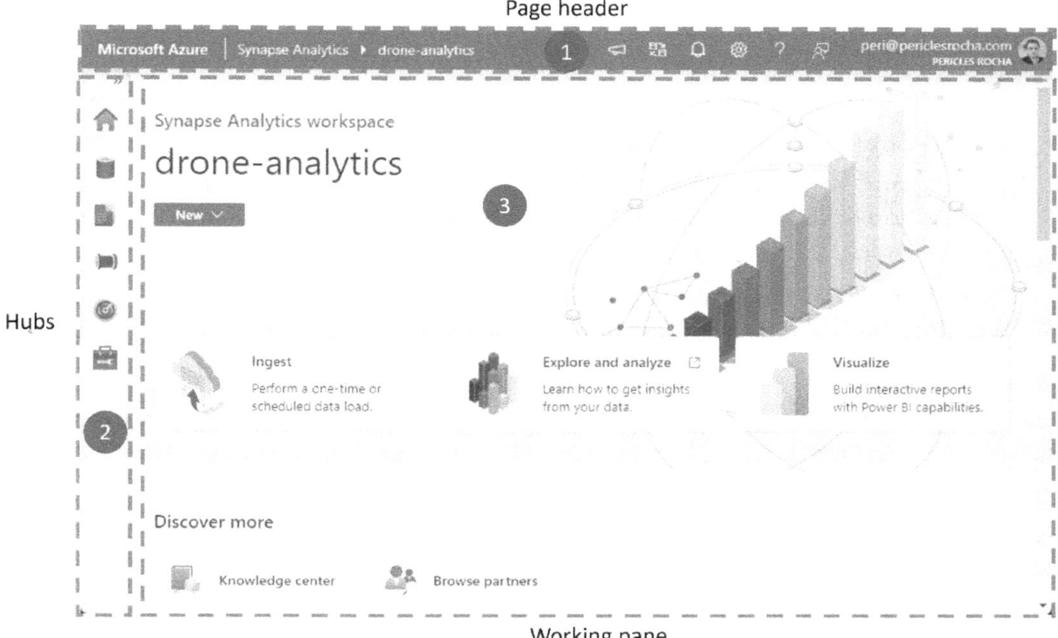

Figure 3.1 – The three main regions of Azure Synapse Studio

We will call each of the areas surrounded by dashed rectangles in *Figure 3.1* a *region*. As you can observe, there are three different regions in Azure Synapse Studio:

- **Page header** (**1**): This region will always be present, regardless of where you navigate within Synapse Studio. It includes the following global controls:
    - Recent release notes and product updates related to Azure Synapse
    - The option to switch to a different workspace
    - A notification area
    - A settings pane that allows you to change the display language of your workspace and its regional format
    - Help resources, including links to a guided tour through Azure Synapse Studio, product documentation, Microsoft Q&A, keyboard shortcuts, and others
    - A feedback form where you can submit not only feedback about your experience but also feature requests

    - Your **Me Card**, which allows you to sign out, view your account details, and manage your cookie preferences
    - A search box that is displayed when you navigate to the **Data**, **Develop**, or **Integrate** hubs

- **Hubs** (**2**): We explained the concept of the hubs area in *Chapter 1, Introducing Azure Synapse Data Explorer*. Note that in *Figure 3.1*, the hubs area is collapsed – you can collapse and expand the hubs area by clicking the arrows over the **Home** icon to make more room in your working pane as needed.
- **Working pane** (**3**): This displays details about the area you selected and provides the working area where you perform queries, create notebooks and pipelines, and perform any actions for assets you build or manage on Azure Synapse.

These regions are the same regardless of what you are doing in Azure Synapse: if you're designing data integration pipelines or writing Data Explorer queries, the experience is the same. Make sure you navigate through each of the regions to familiarize yourself with the user interface.

Next, we will continue exploring Azure Synapse Studio, focusing on the tasks you will typically need to execute using Data Explorer pools.

Running your first query

There is no better way to familiarize yourself with Azure Synapse Studio other than to get your hands dirty – and to get our hands dirty, we need to create a database first.

Creating a database

To create a new database in your Azure Data Explorer pool, proceed with the following steps:

1. In the hubs region, click on the **Data** hub.
2. The working pane region will display the resource explorer area, which shows all your workspaces and linked databases. Click the + sign, and then **Data Explorer database (preview)** as illustrated in *Figure 3.2*.

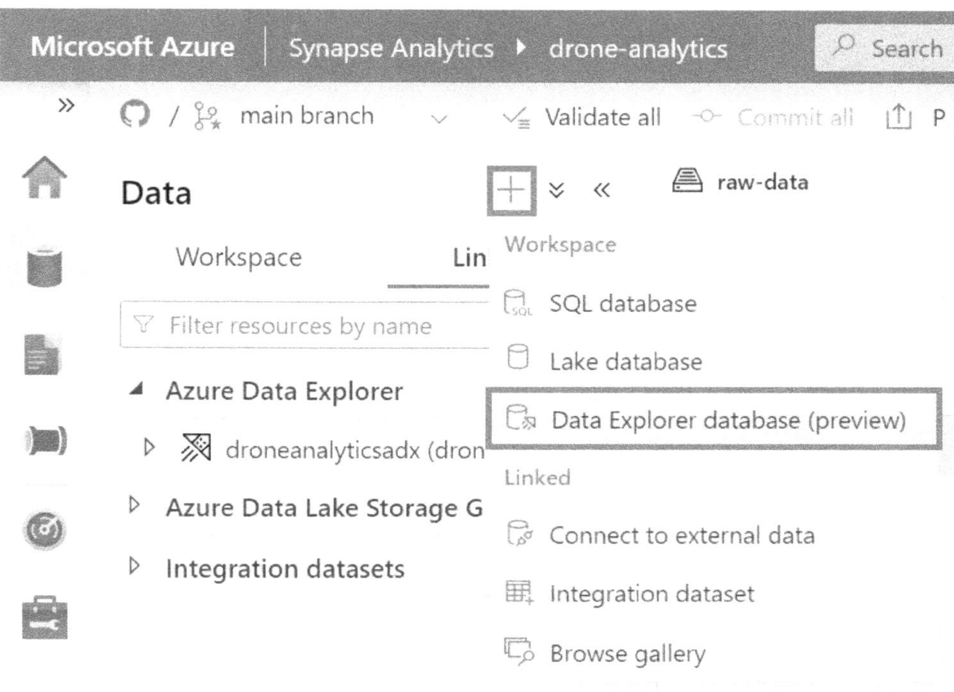

Figure 3.2 – Creating a new Data Explorer database

3. On the right-hand side of the screen, you will be presented with the **Data Explorer database (preview)** pane, which allows you to create a new Data Explorer database. This pane is shown in *Figure 3.3*.

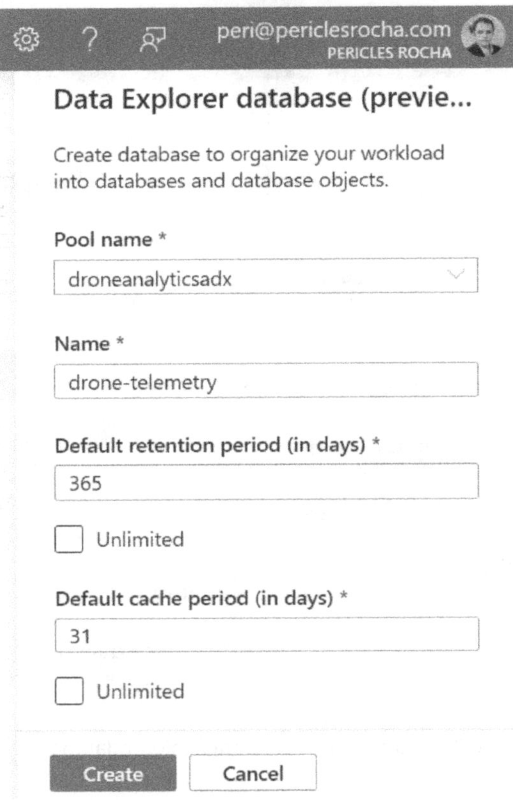

Figure 3.3 – Providing the required parameters

New Data Explorer databases require the following parameters:

- **Pool name**: This is the name of the Data Explorer pool that will host this database. Here, I provided the name `droneanalyticsadx`, the pool we created in *Chapter 2, Creating Your First Data Explorer Pool.*
- **Name**: This is the name of your new database. I will use the name `drone-telemetry`.
- **Default retention period (in days)**: This parameter controls how long data is guaranteed to be available for querying from the time it was ingested. Managing retention policies on databases that continuously ingest data is important to ensure you control costs. For this example, I will leave the default value of 365 days.
- **Default cache period (in days)**: This defines how many days you want to keep data that is frequently queried available in the cache (SSD storage or RAM). Once again, I will accept the default for this value, which is 31 days.

4. Click **Create**. This process should take a few seconds to complete.

Congratulations! You have just created your first Azure Synapse Data Explorer database. To verify that it was created successfully, you can run the `.show databases` control command. To do that, simply head over to the **Develop** hub, click the + sign, and select **KQL script** in the list to start a new script in the query editor. Make sure you select your cluster in the **Connect to** dropdown, as illustrated in *Figure 3.4*.

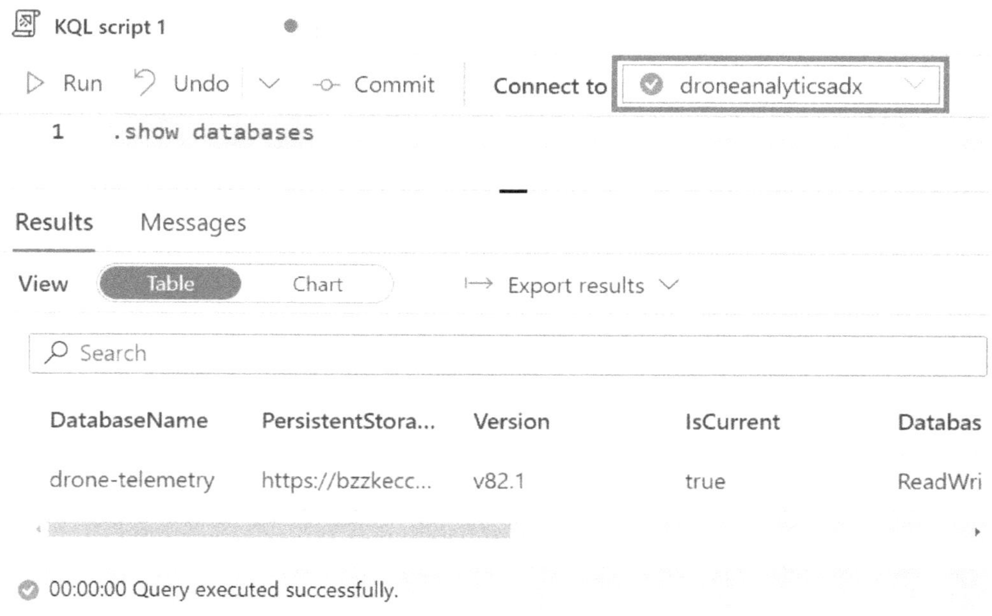

Figure 3.4 – Verifying the creation of your new database

Now that you have created a new database, you will load data into it to be able to finally run your first query.

Loading the data

Your new database is empty – it does not contain any tables or data. To be able to start querying data, you will load the drone telemetry data into the new database using the **Ingest data** wizard, also referred to as **one-click ingestion**.

> **Note**
>
> To run the examples in this book, we've provided the drone telemetry sample dataset, which contains telemetry data from a fictional drone fleet. This data needs to be copied to your local computer in advance. Check the *Technical requirements* section at the beginning of this chapter for instructions on how to download this dataset.

To load the drone telemetry data into your new database, follow these steps:

1. In Azure Synapse Studio, select the **Data** hub (if you are not there yet).
2. Click the ellipsis (**...**) next to the name of your new database and select **Open in Azure Data Explorer**. This step is illustrated in *Figure 3.5*.

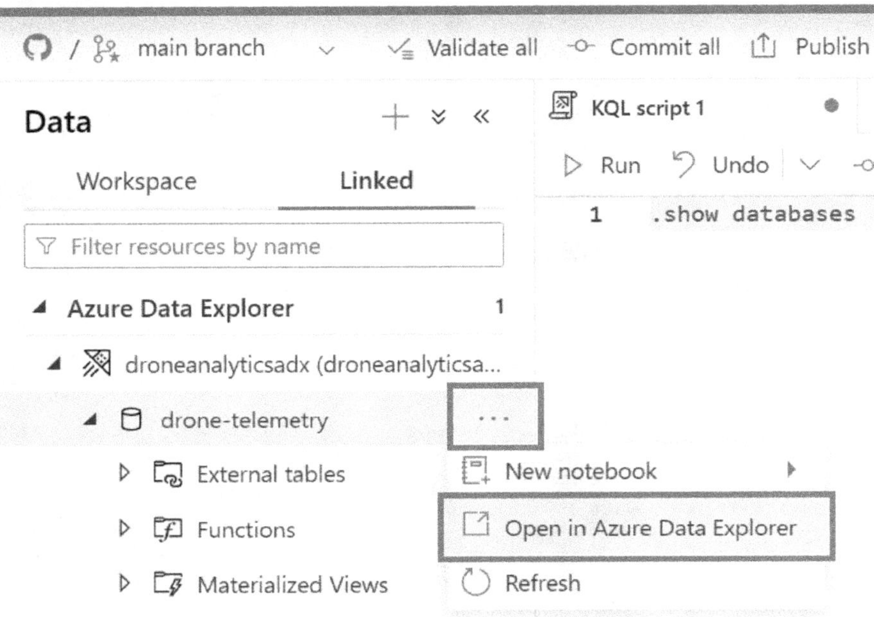

Figure 3.5 – Opening the database in Azure Data Explorer

This action opens a new tab in your browser directing you to the Azure Data Explorer user interface.

> **Note**
> What you see here is the same web experience used for the standalone service Azure Data Explorer. It connects to Data Explorer pools in Azure Synapse seamlessly and allows you to issue queries or control commands at will. At the time of writing this book, the **Ingest data** wizard was not available directly from Azure Synapse Studio.

3. In the Azure Data Explorer web user interface, right-click your database name and select **Ingest data**. Optionally, you could navigate to the **Data** hub and select the **Ingest data** option.
4. You are now directed to the **Destination** tab of the **Ingest data** wizard, as illustrated in *Figure 3.6*.

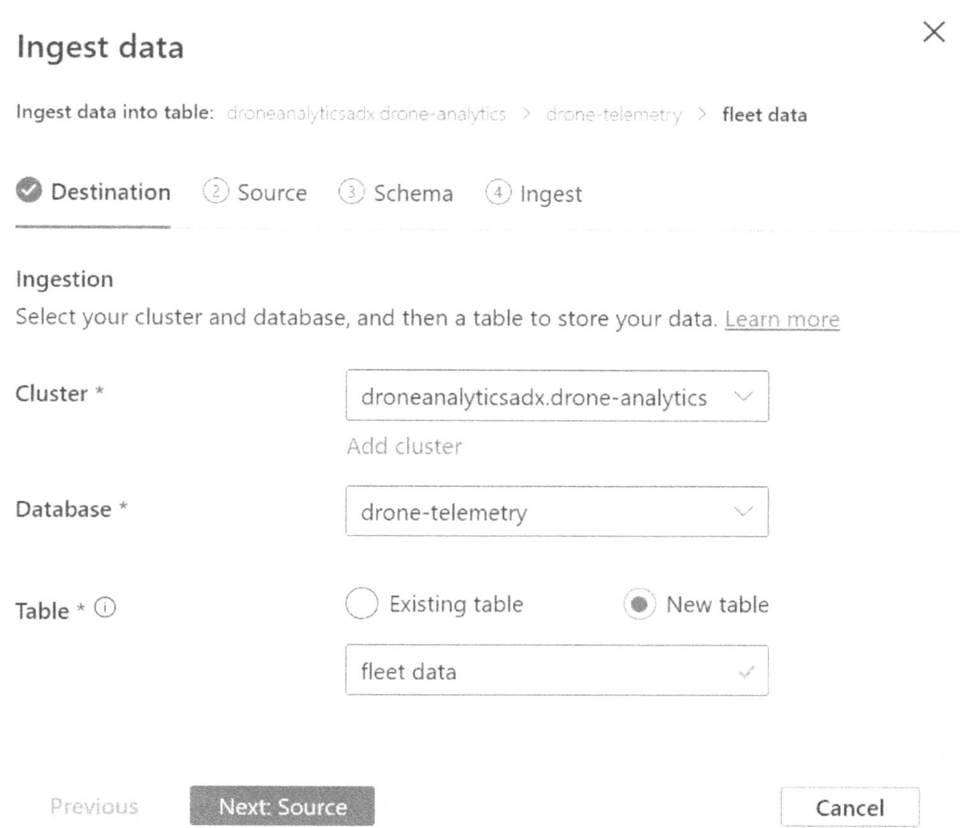

Figure 3.6 – The Destination tab of the Ingest data wizard

Here is a description of these settings:

- **Cluster**: This is the name of your cluster, as seen by the Azure Data Explorer service. This field should be set by default and its value will be the name of your Data Explorer pool and the name of your Synapse workspace, joined by a dot. In our example, the cluster's name is `droneanalyticsadx.drone-analytics`.
- **Database**: This is the target database into which the data will be loaded. We will use the name of the database created in the *Creating a database* section.
- **Table**: This is the name of the target table. Because our new database does not have any tables, you can set this to **New table** and provide a table name in the **New table** name textbox. I will call our table `fleet data`.

5. Click on **Next: Source**. For **Source type**, leave the default option, **From file**, and click the **Select up to 10 files** section. This action will invoke an open file dialog from your browser. Select the drone telemetry dataset file that you downloaded onto your computer to upload it. This action may take a few minutes to complete, depending on your internet bandwidth.
6. When the file upload completes, click **Next: Schema**. The **Ingest data** wizard parses your file and gets its properties. Most importantly, it reads the data and determines the schema definition of your destination table, showing you a preview of what the data will look like in the destination table, as seen in *Figure 3.7*.

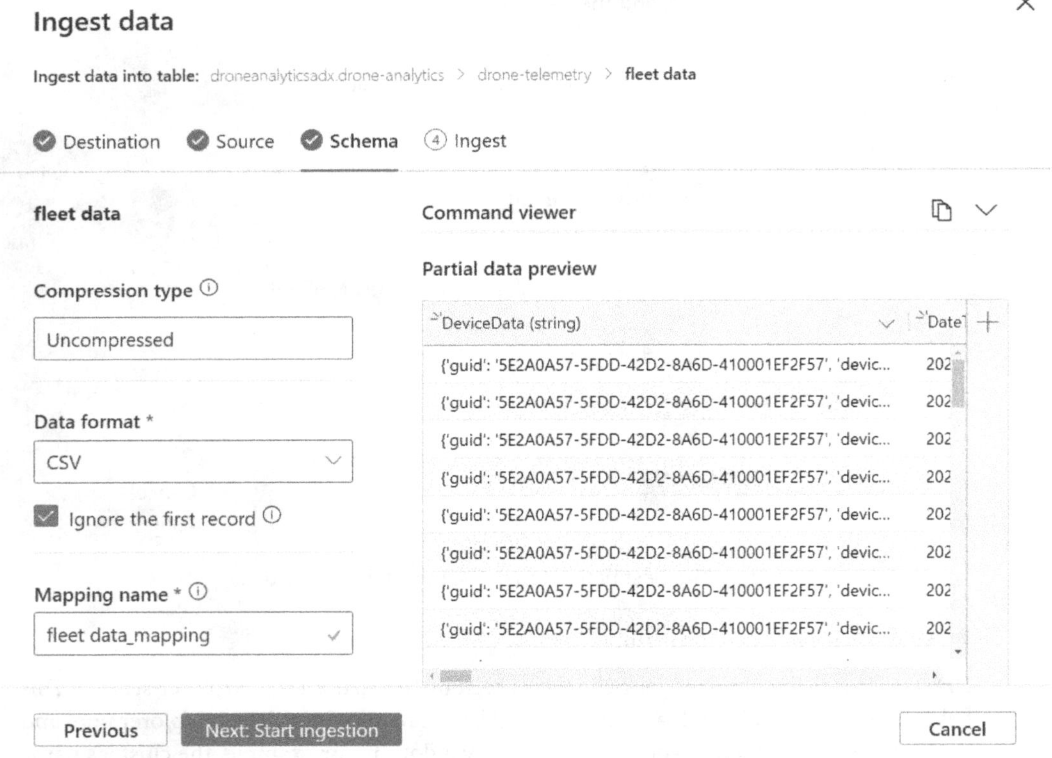

Figure 3.7 – Source file properties and data preview

Let's look at the parameters on this page:

- **Compression type**: This field is immutable. It informs the compression type based on the file you uploaded.
- **Data format**: The **Ingest data** wizard parses your file and tries to understand what format the data is stored in (JSON, CSV, PARQUET, and so on). For our example, the wizard detected that it is a CSV file.

- **Ignore the first record**: This option should be checked if the first row in your file has column names. As that is the case with our file, we will make sure this is checked. The wizard will also use the information in the first row to produce the column names in our destination table.
- **Mapping name**: When you use the **Ingest data** wizard, it automatically creates a data mapping for you. These data mappings are used by the Data Explorer service to map your source data with the columns in your destination table. You can accept the default value for this field or use a different one if you desire.
- **Command viewer**: This section shows you the automatically generated control commands that will be used by the Data Explorer service to create your new table, create the mapping, and perform the actual data ingestion. Feel free to review these commands.
- **Partial data preview**: This allows you to see what your new destination table will look like, change the data type of any columns, add new columns, and delete any columns you don't want to import.

7. When you are done reviewing the parameters, click **Next: Start Ingestion**. This action will trigger the data ingestion task, which should take only a few seconds to complete for our dataset. When you see a **Data ingestion completed** message, click the **Close** button.

At this stage, our new table has been created and data has been loaded into it. To verify that your table was created, you can run a control command to list the tables in your database. To do that, simply go back to Azure Synapse Studio, select the **Develop** hub, delete the `.show databases` command you had before in the query window, type `.show tables` and click **Run**. Your query editor should look like the one shown in *Figure 3.8*.

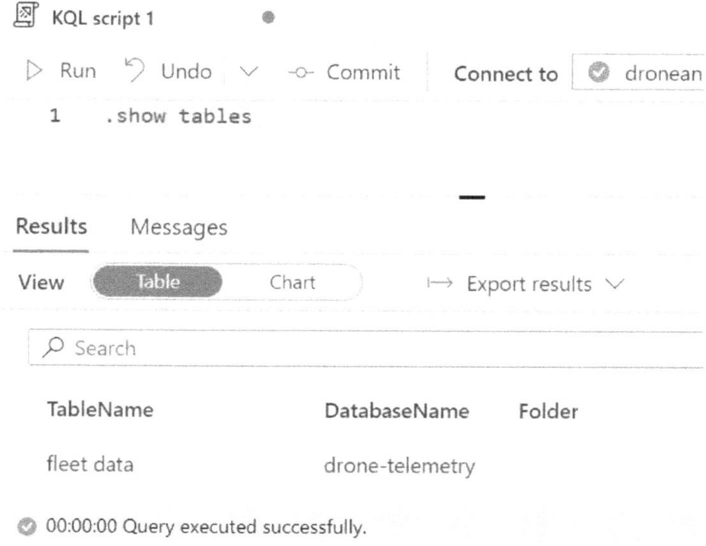

Figure 3.8 – Your newly created table

This proves that your table was created in the destination database as part of the **Ingest data** wizard.

> **Note**
>
> The **Ingest data** wizard is the easiest way to load data into Data Explorer pools. However, it does not satisfy most real-life user requirements to load data at scale. In *Chapter 5, Ingesting Data into Data Explorer Pools*, we will cover additional tools and techniques to load data into Data Explorer databases.

Now that your table is created and you have loaded data into it, it is finally time to run your first query.

Verifying whether your data has loaded successfully

For your first Azure Synapse Data Explorer query, you will first verify that the data was inserted successfully by getting a count of the rows in your table. Then, you will run a simple query to see if data is retrieved successfully.

To run these queries, we need to start a new KQL query:

1. In Azure Synapse Studio, select the **Develop** hub.
2. Click the + sign next to **Develop** and select **KQL script**.
3. In the **Connect to** dropdown, select your Data Explorer pool. Note that the **Use database** option will automatically be set to your database. If you have more than one Data Explorer database at this point, make sure you select the one where we loaded the drone telemetry data.
4. In the query editor, type the following KQL script:

   ```
   ['fleet data']
   | summarize RowCount = count()
   ```

5. Click **Run** or hit the *Shift* + *Enter* keys on your keyboard. The source dataset has exactly 337,500 rows, so your query editor (and results) should look like the one in *Figure 3.9*.

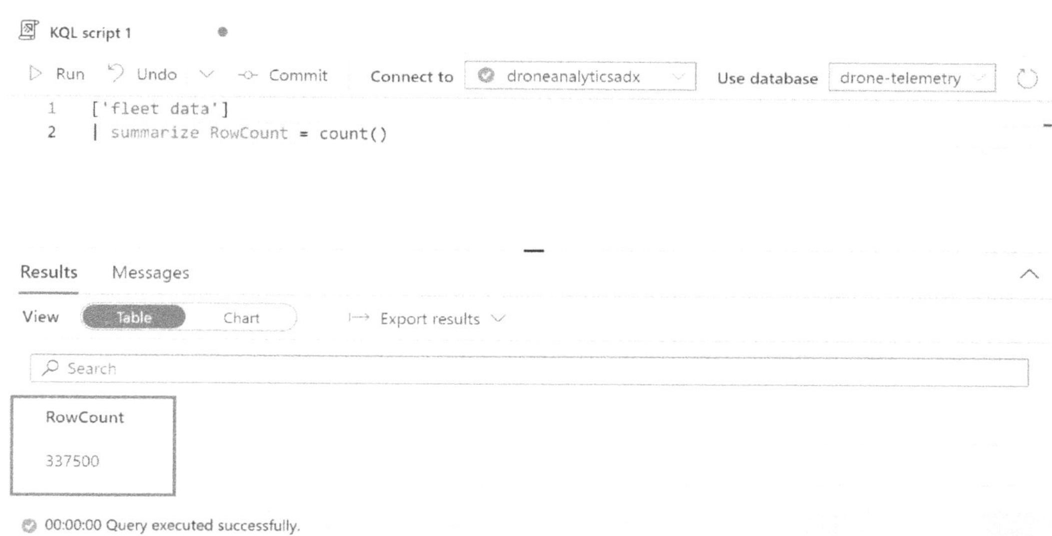

Figure 3.9 – Getting a row count from your fleet data table

Azure Synapse Studio can do more than simply retrieve data and show you results on a table. It can also produce simple charts that allow you to get a visual representation of your data from the query editor. For example, you can run the following query to get the average core temperature recorded by drones, grouped by the state of the device when it was recorded:

```
['fleet data']
| summarize ['Average Temperature'] = avg(CoreTemp) by
DeviceState
| order by ['Average Temperature'] desc
| render columnchart
```

Your chart should look like the one seen in *Figure 3.10*.

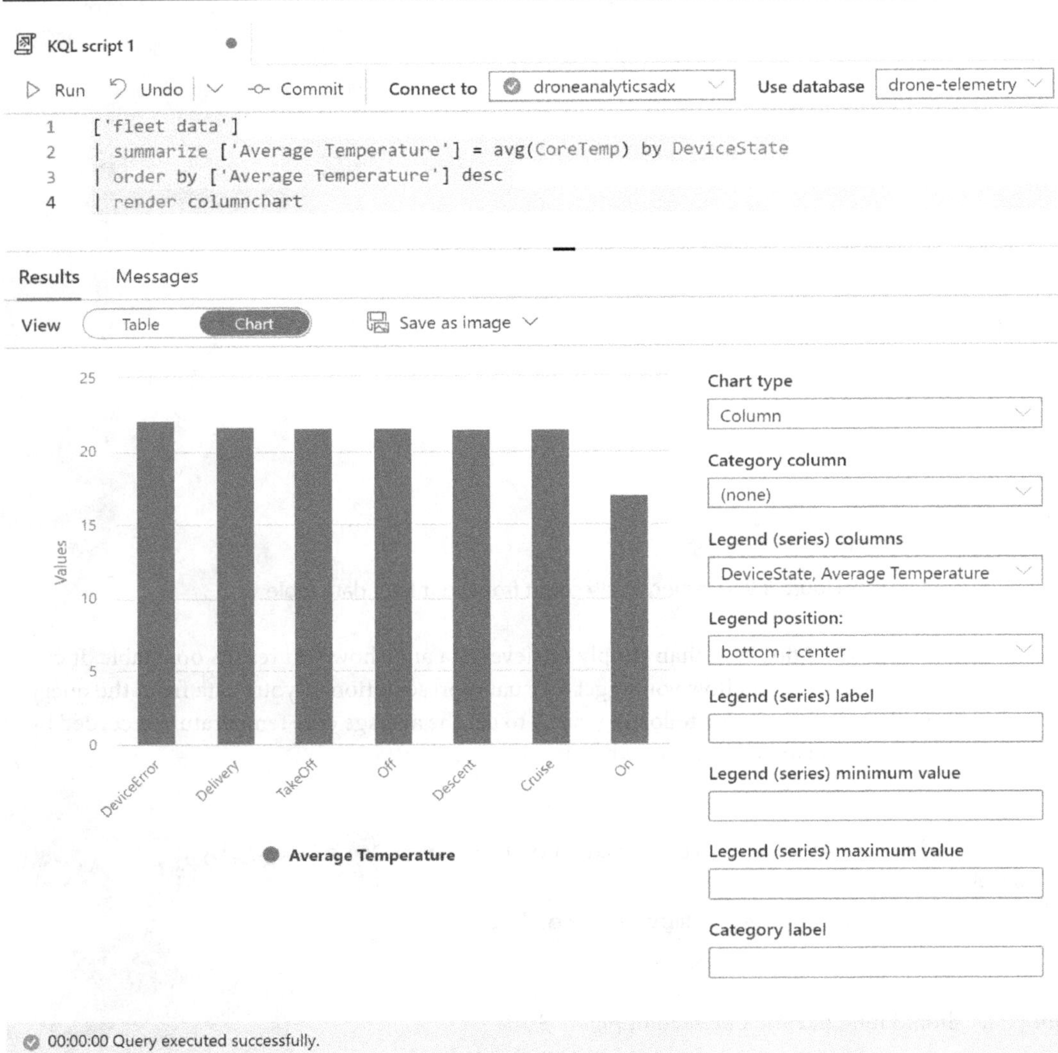

Figure 3.10 – results displayed as a chart in the query editor

Running KQL scripts is one way to explore data stored in Data Explorer pools. Another interesting way to work with data is through notebooks.

Working with data in Azure Synapse notebooks

Notebooks allow you to create narratives around your data by using a mix of code, data processing results, graphics, and text. They have become popular in the industry to help document and present data processing tasks visually, to enable data analysis through rich storytelling, and for learning data science. Azure Synapse notebooks behave like other computational notebook solutions available on the market, mixing text cells, which you can format using the Markdown language, with code cells, where you can run code using a choice between five programming languages (PySpark, Scala, R, C#, or Spark SQL). Azure Synapse notebooks use the Apache Spark analytical engine in Azure Synapse to process notebook jobs.

Azure Synapse notebooks integrate seamlessly with Data Explorer pools. You can conduct rich experiments on notebooks that read data from several different sources (for example, from SQL pools, from data on Azure Data Lake Storage, or from a linked service) and then combine your analysis with data stored on a Data Explorer pool. In *Figure 3.11*, an Azure Synapse notebook is reading data from the `fleet data` table and loading it into a Spark DataFrame.

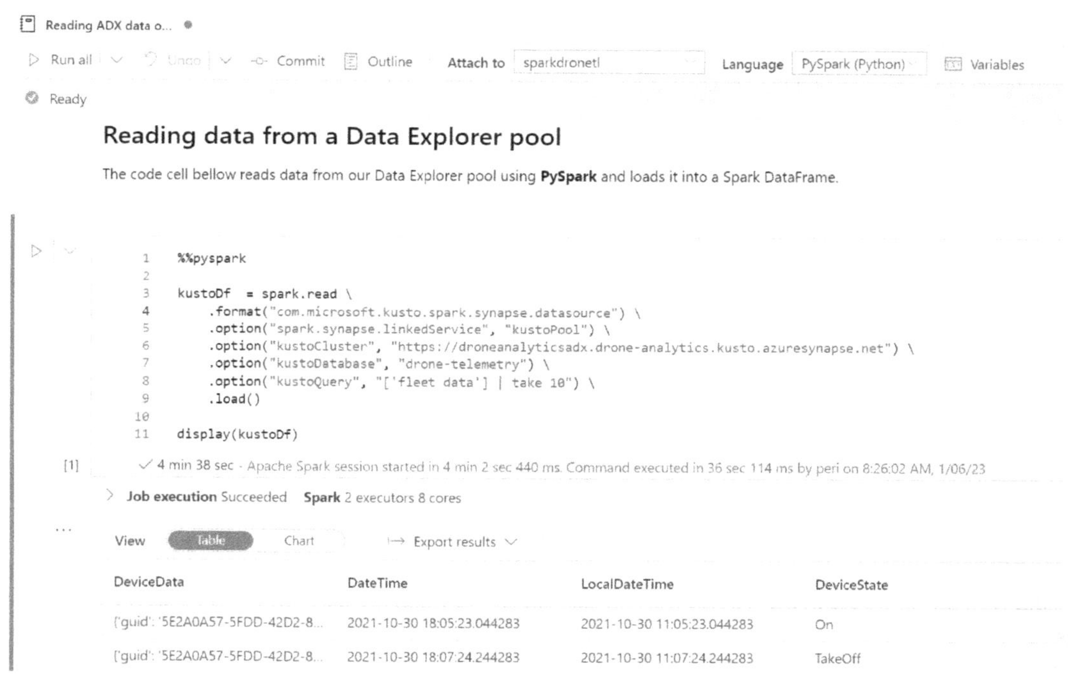

Figure 3.11 – Loading data from Data Explorer pools into Apache Spark DataFrames

From here onward, you can work with the data loaded in your DataFrame just like any other DataFrame. This is a powerful capability, which we will explore more in *Chapter 6, Data Analysis and Exploration with KQL and Python*.

As you start to work in Azure Synapse Studio, you will quickly feel the need to save your work and collaborate with others, so let us now explore how to save your new script for future use.

Saving your work and configuring source control

By now, you may have noticed a **Publish** button in the Azure Synapse query editor. Since you haven't yet configured source control, by clicking this button, your KQL scripts, notebooks, or any assets you produce in Synapse Studio will be saved in the Synapse service, under the context of your Synapse workspace.

You can also create folders to organize your work and move scripts between folders, by clicking the ellipsis (**...**) next to your script name, selecting **Move to**, and picking the desired destination folder. In the example shown in *Figure 3.12*, I created two new folders, `Experiments` and `Production queries`, and moved the first query inside the `Experiments` folder.

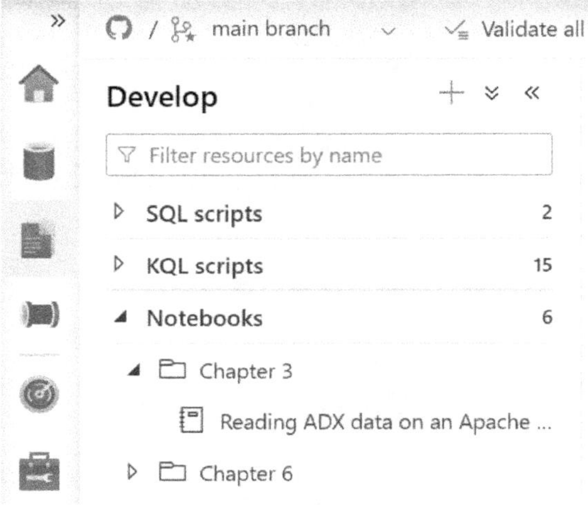

Figure 3.12 – Folder organization under the Develop hub

Saving your work is useful, but how about source control? Azure Synapse Studio comes to the rescue. As discussed in *Chapter 1*, *Introducing Azure Synapse Data Explorer*, Azure Synapse Studio supports collaboration and source control through integration with a Git repository hosted on GitHub or Azure DevOps.

Configuring Git integration in your Synapse workspace is easy:

1. In the top-left corner of Azure Synapse Studio, you should see a drop-down box that says **Synapse live**. Click this drop-down control and select **Set up code repository**.

> **Note**
>
> You can optionally do the same by using the **Git configuration** option on the **Manage** hub and clicking the **Configure** button.

2. The **Configure a repository** dialog will be displayed on the right-hand side of the screen. Select your source control infrastructure on the **Repository type** dropdown. You will have two options:

    - If you select **GitHub** for your source control infrastructure, then type the name of the GitHub organization or GitHub account that owns the repository that you'll use for source control.
    - If you select **Azure DevOps**, then pick the directory where your Azure DevOps organization resides.

    For our example, I will use **GitHub** since that is the repository hosting service that I prefer. Click **Continue**.

3. The next (and final) step is to provide the details about the repository that you will use to collaborate, as seen in *Figure 3.13*.

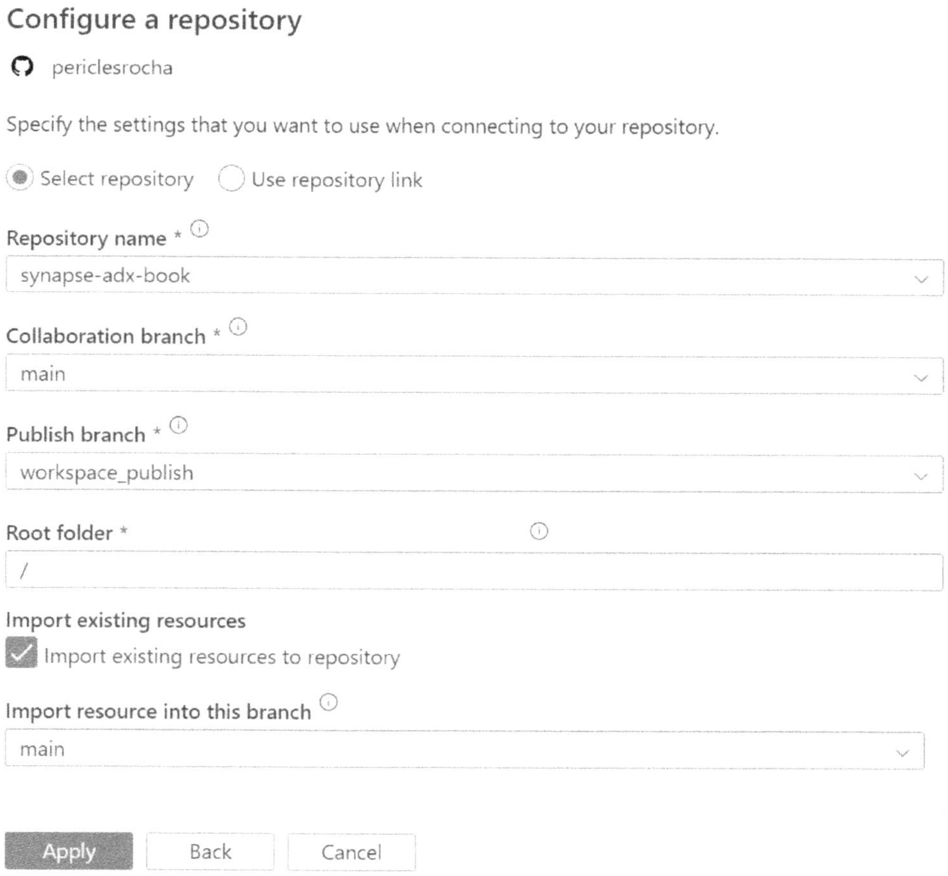

Figure 3.13 – Configuring the GitHub integration

Let's explore the settings on this page:

- **Repository name**: This is the name of the Git repository that you will use to collaborate. You can pick an option from the dropdown or provide a direct link to the repository by selecting the **Use repository link** option.
- **Collaboration branch**: This is the branch that is used for publishing. A branch is a copy of the state of your code that you can use to make your own changes. By default, this is set to **main**.
- **Publish branch**: This is the branch in your repository where publishing-related ARM templates are stored and updated.
- **Root folder**: This is the root folder that you will use for collaboration.
- **Import existing resources**: Choose this option to import any documents that have been saved on your Azure Synapse workspace to your Git repository. This is extremely useful if you saved some scripts when your workspace was configured to save work to Synapse live.
- **Import resource into this branch**: This is the branch the resources will be imported into.

When you are ready, click the **Apply** button. You will be asked to provide a working branch. The working branch is the copy of your repository where you make changes before merging with the main project. Select the desired branch and you are done! From now onwards, you will see that the drop-down box in the top-left corner of Synapse Studio that previously said **Synapse live** now has an icon that represents your source control infrastructure (GitHub or Azure DevOps) and the name of the working branch. This is illustrated in *Figure 3.14*.

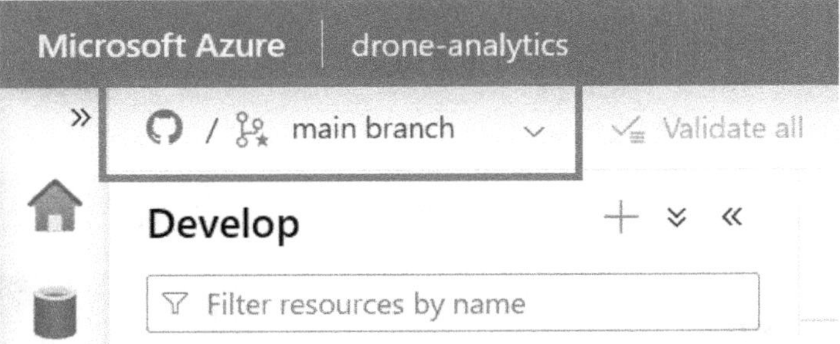

Figure 3.14 – Your workspace is now configured with source control

Configuring source control also implies you now have version control over your workspace assets, allowing you to collaborate on code while also tracking changes. From within Synapse Studio, you can create new feature branches and, when ready to merge your changes, create a new pull request – all done through the same drop-down box highlighted in *Figure 3.13*. Source control *saves* not only your KQL scripts to Git, but also any changes to assets in your workspace, including SQL scripts, notebooks, pipelines, data flows, datasets, linked services, and more.

Now that you know how to use Azure Synapse Studio to launch the **Ingest data** wizard, run queries, and save your work, let us explore how to use it to manage and monitor your Data Explorer pools.

Managing and monitoring Data Explorer pools

To manage Azure Synapse workspace resources, we use the **Manage** hub in Azure Synapse Studio. It allows you to manage key workspace resources and overall workspace security. Let's take a closer look at all the sections of this hub:

- **Analytics pools**: This section allows you to create, delete, pause, resume, and scale your SQL, Apache Spark, and Data Explorer pools. For the workspace's serverless SQL pool (also known as the Built-In pool), you can set budget limits to control your costs. This section also shows the status of your analytics pool, so it's easy to spot if your analytics pools are paused or online.
- **External connections**: This allows you to create and view connections that you have to any external services, such as Azure Data Lake Storage, Power BI, or even links to third-party data sources such as Amazon Redshift, Oracle, and others. It also has a dedicated pane to create and manage your workspace connection with Microsoft Purview.
- **Integration**: This is used to run pipelines. Azure Synapse uses **Integration Runtimes (IRs)** – compute infrastructures that execute the actual integration tasks in data flows, including data movement and transformations. In this pane, you can create a new IR and view the ones you have set up in your workspace. This section also has a **Triggers** pane, where you can create and manage triggers to execute Azure Synapse pipelines.
- **Security**: This allows you to configure your workspace security and create **Active Directory (AD)** credentials that may be used by linked services that support it. We will cover security in more depth in *Chapter 12, Securing Your Environment*.
- **Configurations + libraries**: When you work with Apache Spark, or data flows, it is common to use custom libraries that are not included in Azure Synapse by default. Here, you can upload workspace packages, custom functions for data flows, and custom configurations for Apache Spark.
- **Source control**: This is where you configure and view the Git repository configured with your Synapse workspace, as covered in the *Saving your work and configuring source control* section of this chapter. If you have already set up Git configuration, here, you can view your settings, change the publishing branch, or disconnect your workspace from your Git repository.

These settings are relevant to all assets in your workspace and are not specific to Data Explorer pools. Next, we will understand how to use the Manage hub to perform day-to-day tasks on Data Explorer pools.

Scaling Data Explorer pools

To find your Data Explorer pools, select **Data Explorer pools (preview)** under **Analytics pools** in the Manage hub. A common task when managing Data Explorer pools is to scale them in and out. To scale a Data Explorer pool, hover your cursor over the desired pool to reveal the action buttons and click the **Scale** button. This brings up the **Scale** page, shown in *Figure 3.15*.

Scale

Autoscale is a built-in feature that helps pools perform their best when demand changes. Learn more

Autoscale

Enabled Disabled

Number of instances

6

Estimated price

Est. cost per hour
2.63 USD
View pricing details

Apply Cancel

Figure 3.15 – Scaling Data Explorer pools

The scale options exposed on the **Scale** page are the same ones that we already covered in the *Creating a Data Explorer pool using Azure Synapse Studio* section of *Chapter 2, Creating Your First Data Explorer Pool*, so you may refer to that section to refresh your knowledge. Note that your Data Explorer pool must be online to perform scale operations, which we'll discuss in the following section.

Pausing and resuming pools

Another frequent task is to pause and resume your Data Explorer pools. Because Azure Synapse charges you for the time your Data Explorer pools are online regardless of whether you're using them or not, it is a good idea to pause your analytics pools when they are not being used.

You can also see if your pools are online or paused from the **Manage** hub. To pause or resume your Data Explorer pools, hover your cursor over the desired pool, as seen in *Figure 3.16*, and select the resume or pause button (depending on its current state).

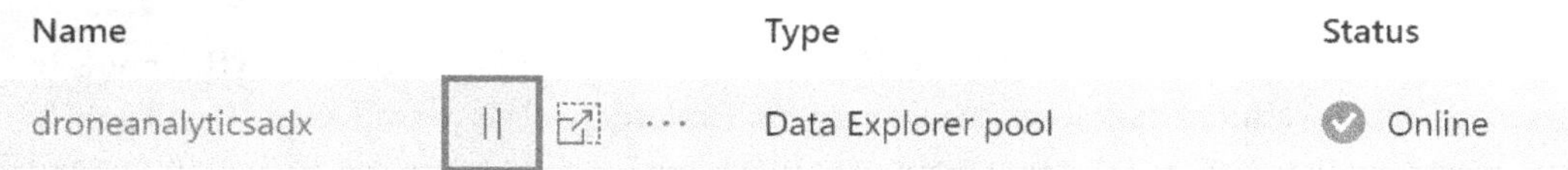

Name
Type
Status
droneanalyticsadx
Data Explorer pool
Online

Figure 3.16 – Pausing and resuming Data Explorer pools

Pausing and resuming Data Explorer pools can take several minutes. Make sure you check the status to verify that your request is completed before you consider using your pools. You can refresh the pool's status by clicking the **Refresh** button on this page.

Monitoring Data Explorer pools

The **Monitor** hub offers resources to help monitor the status of Data Explorer pools, the execution of query requests, pipeline runs, and more. As we did with the **Manage** hub, let's look at the sections of the **Monitor** hub in detail:

- **Analytics pools**: This section lists your SQL pools, Apache Spark pools, and Data Explorer pools. Here, you can easily glance at details of your Data Explorer pools, such as their current status, size, instance count, the count of virtual CPUs they utilize, and the allocated cache and RAM sizes. By selecting a specific Data Explorer pool here, you can see up to a 30-day view of query results over time, as illustrated in *Figure 3.17*.

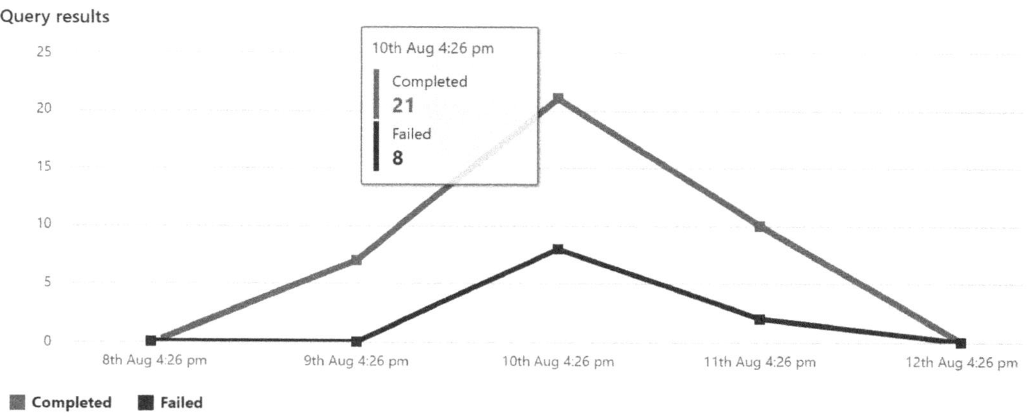

Figure 3.17 – A historical view of queries processed by Data Explorer

  Besides recent queries, on this page, you can also see instance count over time and ingestion results over time, in the same fashion as recent queries. Analyzing resource consumption over time can help you determine whether you need a larger or smaller Data Explorer pool.

- **Activities**: Here, you can see all KQL scripts processed by your Data Explorer pools by any user, or by pipeline runs. This includes any KQL requests processed by the cluster. Similarly, you can do the same for your SQL pools and Apache Spark pools.
- **Integration**: This section allows you to monitor Azure Synapse pipeline runs, including ones executed by automated triggers. You can also monitor any link connections and integration runtimes you created on the **Manage** hub (or ones that are automatically created for you) here.

The monitoring resources found in Azure Synapse Studio offer you a high-level view of your environment, where you can see Data Explorer pool resource allocation, requests submitted, and more. *In Chapter 10, System Monitoring and Diagnostics*, we will take a more in-depth view of monitoring, and use control commands to monitor the environment.

In your day-to-day work with Azure Synapse Data Explorer, you will often use the **Data**, **Develop**, **Monitor**, and **Manage** hubs, which we have explored in this chapter. You may be asking, how about the **Home** and **Integrate** hubs? These hubs do not expose activities that are relevant to the daily tasks you will perform while managing Data Explorer pools, but you may use them while working in other areas of Azure Synapse. In a nutshell, you can perform the following activities in these hubs:

- **Home**: In the **Home** hub, which is also the landing page of Azure Synapse Studio, you will have easy access to recently utilized workspace assets such as KQL scripts or notebooks, links to learning resources, and easy access to Azure Synapse samples.
- **Integrate**: The authoring of Azure Synapse pipelines is done in this hub. While we will briefly work here in *Chapter 5, Ingesting Data into Data Explorer Pools*, building data integration pipelines in Azure Synapse is a broad topic, which is not in the scope of this book. If you would like to learn more about Azure Synapse pipelines, visit `https://learn.microsoft.com/en-us/azure/data-factory/concepts-pipelines-activities?tabs=synapse-analytics`.

Azure Synapse is an end-to-end analytics solution that goes beyond Data Explorer. Even though these hubs are less relevant to Data Explorer, make sure you navigate through them and get familiar with them as you may need to use them in your daily work on Azure Synapse.

Summary

In this chapter, we explored Azure Synapse Studio, the primary development and management tool for Azure Synapse. If you are planning to use Azure Synapse Data Explorer, learning about the core elements and how to navigate Azure Synapse Studio is important.

We began by exploring the key user interface elements in Azure Synapse Studio to help you navigate and find what you need. Learning about the three main regions of Azure Synapse Studio helped you to understand where to manage service-level settings, navigate through the hubs, and realize what you can do in each of them.

Next, we took a hands-on approach to learn how to use the key areas of Azure Synapse Studio by creating a database, ingesting data into it, and running our first queries. We also explored how to configure Git integration to enable team collaboration and source control.

Finally, we explored the **Manage** and **Monitor** hubs to learn where to find, manage, and monitor your Data Explorer pools. We learned how to perform common tasks such as pausing and resuming pools, scaling pools, and monitoring recent activities.

In previous chapters, we introduced Azure Synapse Data Explorer and learned how to get started with the service. Now that we understand how the service works and how to take our first steps, in the next chapter, we will learn about real-world usage scenarios for Azure Synapse Data Explorer.

4
Real-World Usage Scenarios

By now, you should have a good understanding of Azure Synapse Data Explorer and how it fits into the overall architecture of Azure Synapse. In *Chapter 1*, *Introducing Azure Synapse Data Explorer*, we learned when to use the service by approaching common analytics scenarios.

In most cases, Azure Synapse Data Explorer is not the sole solution in your analytics environment. You need to consider which types of sources you have, if you will need to process information in real-time or in batches, where the data will be stored, how you will process and transform the data, and how the data will be served to end users. Azure Synapse may be able to cover most of the technological challenges in these architectures, but chances are that, in real life, you will find a mix of solutions being used by companies to address their business requirements.

This chapter offers a more in-depth view of real-world scenarios and solution blueprints that combine additional Azure services to deliver an end-to-end solution to business problems. We will examine industry-specific examples and architecture patterns that you can leverage in your projects.

There is not one solution to every problem. Therefore, in each scenario, you may find alternatives and other services that could have been utilized in favor of the ones proposed in the examples we will provide. We will focus on applying the services available in Azure Synapse to address the business problems in the examples as much as we can, with additional Azure services complementing the architecture – especially for streaming data ingestion. We will discuss the following scenarios:

- Building a multi-purpose end-to-end analytics environment
- Managing IoT data
- Processing and analyzing geospatial data
- Enabling real-time analytics with big data
- Performing time series analytics

Technical requirements

Some of the architectures listed in this chapter contain Azure services that are outside the scope of this book, such as Azure Event Hubs, Azure Machine Learning, and others. It would be impossible to explain in detail how each Azure service that is mentioned here works, as well as the technical details of the service's implementation. You should pick the example architectures that are of interest to you and research each component in depth, including the cost implications, before you build a solution architecture. This book provides only a quick overview of the role of such services in the architectures we propose.

Some of the solutions presented here were adapted from existing ones in the Azure documentation found at `https://docs.microsoft.com/azure/architecture`. Rest assured that they have not been copied from this source: instead, this chapter includes core scenarios that are commonly seen in the industry and adapts these solution architectures to lean toward an Azure Synapse Data Explorer-oriented approach. They also reflect the real-life experiences of the author.

Building a multi-purpose end-to-end analytics environment

When you do not know where to start, you should start with the basics. As mentioned previously, Azure Synapse Analytics is a broad platform that combines different services to deliver an end-to-end solution for most analytical demands. This scenario proposes a blueprint that addresses common analytical requirements for most organizations. Some use cases where this blueprint can be used include the following:

- **Categorizing users based on their product usage behavior**: You can use clustering algorithms on a machine learning model to classify your users based on historical log data that records how they interact with your product. Identifying user cohorts helps companies build more customized experiences for their users.
- **Detecting credit card fraud**: By performing real-time analysis of data streamed from financial transactions, you can use machine learning models to classify a transaction as fraudulent or legit in just a few seconds from the moment the transaction was attempted.
- **Creating real-time dashboards of service metrics**: You can combine data from sources such as product telemetry, sales transactions, and customer sentiment to get a 360-degree view of your business.
- **Interactive experimentation**: This should be an environment where analysts and data scientists can have access to all the relevant data and perform experiments using the language of their choice. Creating such an environment for analysts opens doors for new findings that may not have been anticipated and provide insights that can realize the real value of your data. Companies that want to adopt a data culture for decision-making to unlock all the value from their data need to consider offering an analytical environment for interactive experimentation that offers tools that encourage you to work with data from any source, using the tools of your choice.

Every architecture described in this chapter is derived from the blueprint illustrated in *Figure 4.1*, so we will describe it in more detail than the others:

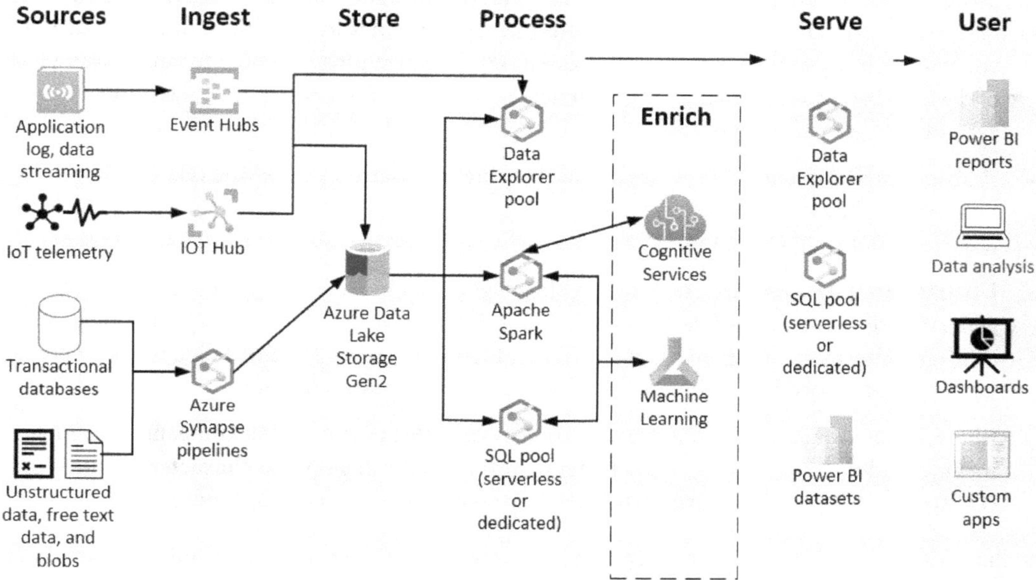

Figure 4.1 – Multi-purpose end-to-end analytics environment

This diagram may make things look more complicated than they are. It follows a simple flow from data originating from their sources (left) until it gets consumed by users at the other end of the chart (right) and everything that needs to happen in between. Let's describe each of the stages on this diagram in detail.

Sources

This is where data originates. The source systems in the diagram are as follows:

- **Application log**: This includes data generated by events on an application. It may contain errors, informational events, warnings, user actions, and more. Performing analytics on application logs can help companies build better products and increase software reliability by analyzing trends in usage behaviors, error events that correlate, and more.
- **Data streaming**: This involves data that is continuously generated and ingested in real time. In data streaming scenarios, we typically see low latencies between the time data is generated until it gets ingested into the analytical store. Examples of systems that stream data include financial applications that calculate risk in real time based on stock market transactions, or click-stream data from e-commerce websites that use this information to adjust ads and product placement as customers navigate a retail website.

- **IoT telemetry**: IoT devices constantly generate all sorts of telemetry events, such as information about where they are, information from their sensors, actions they performed, their status, and more – all of which are typically associated with a timestamp of that event. Companies use real-time dashboards to monitor the status of their machines in real time and use predictive analytics techniques to anticipate failure events and perform proactive maintenance. Examples of devices that generate telemetry events include machines in manufacturing, elevators, racing cars, drones, and much more.
- **Transactional databases**: These are classical **online transaction processing** (**OLTP**) databases that store application data used by businesses. Examples include a Microsoft SQL Server database that stores data of a CRM application, an Oracle Database that stores sales data, and others.
- **Unstructured data and free text data**: This category can include user-generated or machine-generated documents such as Microsoft Office documents or web pages. By employing text mining techniques, companies can use these documents to build knowledge bases, analyze user sentiment based on feedback forms or social media posts, and more.
- **Blobs**: Images, videos, and audio files can contain valuable information for companies. Examples include reading image content and extracting text from it using **optical character recognition** (**OCR**) algorithms, reading bar codes, identifying objects in images, and more.

Note that this is not an exhaustive list of possible source types. Other sources include existing data warehouses, **online analytical processing** (**OLAP**) databases, applications that expose data via an API, and others. The sources listed here are sources that are typically used with Azure Synapse Data Explorer. Next, we will understand how data from these sources is ingested into the data stores in Azure Synapse.

Ingest

The next step is to build the infrastructure that will move data from our sources to our data stores. Three main components are proposed in this solution blueprint:

- **Event Hubs**: Azure Event Hubs is Microsoft's platform for real-time data streaming. It can process millions of events by streaming data every second and writing it into an **Azure Data Lake Storage** (**ADLS**) **Gen2** container, or directly into a Data Explorer pool for real-time analytics.
- **IoT Hub**: IoT Hub processes millions of events per second, just like Azure Event Hubs, but also offers infrastructure and **application programming interfaces** (**APIs**) for bi-directional communication with IoT devices.
- **Azure Synapse pipelines**: For all data that does not need to be streamed in real time into the analytics store, Azure Synapse pipelines come to the rescue. Azure Synapse supports more than 90 data sources and offers a robust platform for data flow orchestration, error handling, and performance.

When you are ingesting data into Azure Synapse, it is very common to ingest data into ADLS Gen2 first, and later copy this data to Data Explorer pools or SQL pools. Even though data can be written directly from a source into a Data Explorer pool, enabling data exploration on the data lake is a key scenario in Azure Synapse and all three analytical engines (Data Explorer, Apache Spark, and SQL) can benefit from that. Therefore, the typical ingestion pattern for Azure Synapse is to write data to the data lake first and then copy it to tables on Data Explorer pools or dedicated SQL pools. For scenarios that require real-time data analysis, you can use a *hot path*, which copies data directly to Data Explorer pools and significantly reduces latency, while writing a copy of the data to ADLS in parallel to enable the benefits of working with the data lake.

Store

This is the data lake where ingested data is stored. This is a critical aspect of this blueprint as it enables data exploration from the analytical engine of your choice. Any of the analytical engines in Azure Synapse can read data stored on the data lake.

So, why the data lake? Because it allows you to store data in its original form. Regardless of the format of the data you acquired from the source systems, from structured tables stored as **comma-separated values** (**CSV**) files, through semi-structured JSON documents, or even images and videos, ADLS Gen2 allows you to combine data from all sources and store it in the same format it was captured in. This also makes data ingestion fast since normally, minimal transformation is done with data at this stage (although it can happen). Once you've ingested the source data into the data lake, you can decide how you want to transform, further process, and use the data before you serve it to end users.

Process

In this stage, we analyze the data, curate it, and transform it to make it more useful to the end user. For the multi-purpose end-to-end analytics blueprint, this can be achieved in the following ways using the analytical pools in Azure Synapse:

- **Data Explorer pool**: In real-time analytical scenarios, data will land in Data Explorer pools directly from Azure Event Hubs or Azure IoT Hub. Data is then stored in Data Explorer pools as it arrives, enabling users to process and analyze this data in near real time. In non-real-time scenarios, Data Explorer pools can import or directly process data that was copied into ADLS Gen2. In cases where you ingest data from ADLS Gen2 into Data Explorer pools, a data transformation may occur to make the data more useful to the end user.
- **Apache Spark**: The key scenarios for Apache Spark are around data preparation, experimentation, and machine learning. As discussed in *Chapter 1*, *Introducing Azure Synapse Data Explorer*, Apache Spark in Azure Synapse allows users to use one of five languages to work with data stored in the data lake.

- **SQL pool (serverless or dedicated)**: Microsoft offers a convenient `COPY` **transact-SQL** (**T-SQL**) statement that reads data from ADLS Gen2 and copies it into SQL tables. Serverless SQL pools can be used to query data directly in ADLS using T-SQL queries.

So, when should you use each analytical engine? You should use Data Explorer pools when working with a mix of structured, semi-structured, and unstructured data at large volumes. You should use Apache Spark for data transformation, training machine learning models using data stored on the data lake, creating new calculated columns that require complex processing and logic, or any work that requires robust data processing at scale. Finally, dedicated SQL pools are the best fit for structured data that can be stored efficiently in SQL tables. They allow blazing-fast queries to be performed on structured data.

Data processing is one of the stages where you will spend the most time with data. I spend a good amount of time here understanding the data that I have at hand, curating and cleaning it, and figuring out how I will use future stages in the process (such as **Enrich** and **Serve**) given the user requirements. This stage also leverages the components in the **Enrich** stage to apply artificial intelligence techniques that make your data even more valuable. We will look at this next.

Enrich

Azure Synapse not only offers native services for data ingestion, storage, processing, and serving data, but it also integrates with external services such as Azure Machine Learning and Azure Cognitive Services. This integration allows you to consume models available in an Azure Machine Learning registry, perform cognitive analysis tasks, and much more. For example, you can train a model that predicts the fuel consumption of cars based on historical data and then use that model to score values with new data at runtime using the `PREDICT` T-SQL statement. Alternatively, you could determine a customer's sentiment based on free-text product feedback in your dataset.

You can think of the **Enrich** stage as part of the **Process** stage, in the sense that you are working with your data to make it more valuable and useful for the end user. Once you've finished enriching and processing the data, the next step is to think about how you will make this data available to end users.

Serve

This is where the data is made available to be consumed by dashboards, reports, user queries, applications, or any other resource connecting to Azure Synapse to retrieve data. Data can be served to users by using one of the analytical pool types in Azure Synapse, or even through a Power BI dataset.

To serve data to clients, Azure Synapse exposes public endpoints for Data Explorer pools (and to SQL pools). Users consuming data that is being served by an analytical pool in Azure Synapse do not necessarily need access to the Azure Synapse Workspace; they only need valid credentials to connect to the analytical pool directly.

In some cases, the actual data that is served in this layer can be the same as the data we are working on in the **Process** stage, and no data duplication is needed. In most cases, however, as data is processed in the **Process** stage, it gets copied to a table that contains the final version of the data, which is what is served to users.

User

At this stage, the data is ready and all we need to do is connect to the datasets exposed by the **Serve** stage to consume it through the means desired: a dashboard, a report, a custom application, or any client.

The public endpoints that are exposed by Data Explorer pools use a **uniform resource identifier** (**URI**) that follows the `https://<POOL_NAME>.<WORKSPACE_NAME>.kusto.azuresynapse.net` format, where `<POOL_NAME>` is the name of your Data Explorer pool and `<WORKSPACE_NAME>` is the name of your Azure Synapse Workspace. When connecting from a client, you will need to provide the URI of your Data Explorer pool, the database name, and the authentication details.

Summary

This blueprint offers a comprehensive reference architecture that supports most use cases for analytical workloads in Azure Synapse. Next, we will explore some specific solutions while using subsets of this blueprint. Since this section described all the layers of the blueprint in depth, in the next four scenarios, the focus will be on the differences from this reference blueprint, and on describing scenarios where they can be useful.

Managing IoT data

As discussed in the *When to use Azure Synapse Data Explorer* section of *Chapter 1, Introducing Azure Synapse Data Explorer*, we learned that Data Explorer pools are optimized for unstructured data, which typical IoT sensors or devices generate. Companies that make use of sensors to monitor manufacturing processes, fleet management, and many other scenarios that work with sensor-generated data commonly deal with this sort of unstructured data. This section demonstrates a blueprint that describes how to manage big data in IoT scenarios.

This blueprint can be useful in scenarios such as the following ones:

- **Predictive maintenance**: Using data from sensors, you can estimate when you should perform maintenance on factory equipment. This is a key strategy that's adopted by companies to save on operating costs as they can perform maintenance at the optimal time in the life cycle of machines.
- **Fleet management**: Vehicles of any type (cars, drones, boats, or others) constantly generate large amounts of data reporting their status, health, location, failures, events, and more. This blueprint can be used in a solution that captures this data to help you analyze conditions regarding how vehicle errors occur, allowing you to correlate events for diagnostics and vehicle improvement, determine factors of fuel consumption, and much more.

- **Smart buildings**: By capturing data from sensors, power systems, water consumption, **heating, ventilation, and air conditioning** (HVAC), and other data sources, building managers can use machine learning algorithms to *learn* user habits and make building systems respond to the needs of their users. For example, by learning typical times of high traffic in buildings throughout the week, HVAC systems can blow fresh air into common areas ahead of such high-traffic moments.
- **Video surveillance**: Using machine learning and cognitive services, a video surveillance solution can use images to detect packages, individuals, vehicles, or other objects to send personalized alerts to security personnel.

> **Note**
>
> This solution was adapted from `https://docs.microsoft.com/azure/architecture/solution-ideas/articles/iot-azure-data-explorer`.

As you can see in *Figure 4.2*, the blueprint for this solution is very similar to our multi-purpose blueprint described at the beginning of this chapter:

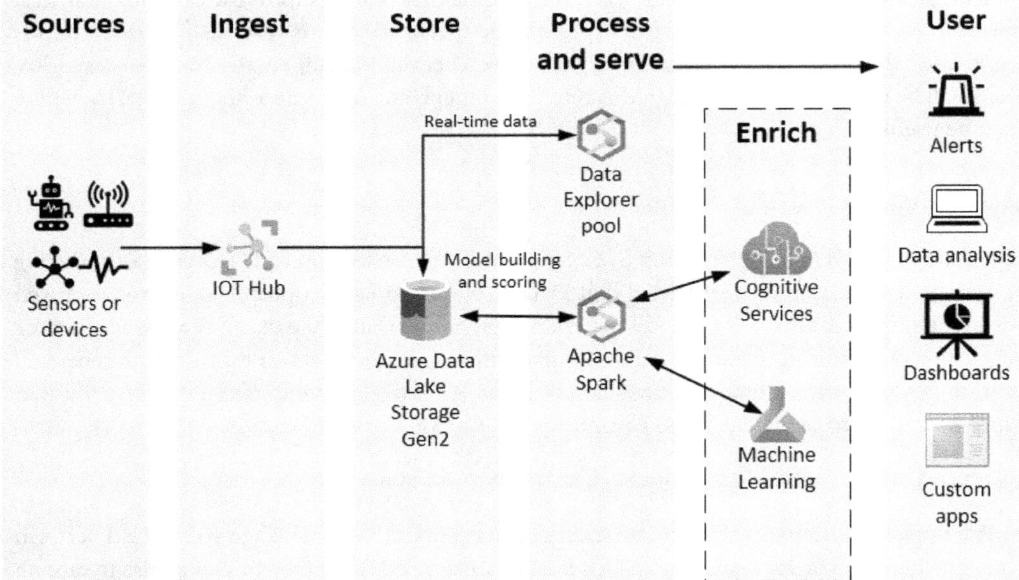

Figure 4.2 – IoT analytics

This blueprint has been simplified so that it works with IoT sensor and device data only. Therefore, most of the data sources listed in the multi-purpose blueprint were removed, and the **Ingest** layer was adjusted to use **Azure IoT Hub** only. Additionally, the following changes were made:

- **Store**: Here, we can see a *real-time data path*, which transfers data acquired from **IoT Hub** directly to the Data Explorer pool for immediate analysis, and a *model building and scoring path*, which uses Apache Spark to connect to Cognitive Services and Azure Machine Learning to build models and score new values. As an example, in the video surveillance use case described in the introduction of this use case, any video files would be stored in ADLS Gen2 and processed using Apache Spark and Cognitive Services. All the work that results from categorizing or predicting new values using Azure Machine Learning (or Cognitive Services) is stored back in ADLS Gen2 to be consumed by the Data Explorer pool.
- **Process and serve**: In this example, the **Process** and **Serve** stages are combined into one to demonstrate that you can serve data from the same database instance you are using to process data. All upcoming scenarios in this chapter use this same pattern. In addition to this, in *Figure 4.2*, the SQL pool has been removed from the blueprint in favor of using only Data Explorer pools, which are optimized for storing and querying IoT data. SQL pools are not needed.
- **User**: The only change here was the addition of an alert, which can be triggered if certain events occur, such as an imminent device failure.

This blueprint offers a robust and highly scalable solution for analytics on IoT data. It can process large amounts of unstructured data at a low latency by following the *real-time data path*, while still allowing data to be enriched through the use of Azure Machine Learning and Cognitive Services.

Processing and analyzing geospatial data

Being able to support geospatial data and functions is a key requirement in most IoT scenarios. It allows you to determine proximity from one device to another, store and process vector and raster **geographic information system** (**GIS**) data to produce rich visualizations on top of geospatial data, determine device routes, and much more. This blueprint builds on the IoT analytics scenario to expand support for geospatial data.

> **Note**
>
> A useful article to learn more about vector data and raster data in the context of geographic information systems can be found at `https://101gis.com/vector-data-vs-raster-data`.

Here are some of the usage scenarios for this blueprint:

- **Building rich map visualizations**: Raster and vector data allows you to digitally store representations of the real world in the form of geometries and continuous blocks of pixels that represent vast areas, such as bodies of water, fields, and forests. You can combine raster and vector data with device information that provides humidity, temperature, and other data to build rich 3D visualizations plotted over a map for site surveys or virtual tours.
- **Optimizing routes**: By looking at device locations and available routes for different vehicle types, you can create optimized routes, plan for stops, determine the best time to travel to a destination, and more.

> **Note**
> This solution was adapted from `https://docs.microsoft.com/azure/architecture/example-scenario/data/geospatial-data-processing-analytics-azure`.

This scenario allows you to join IoT device data with vector and raster data, which represents real-world features such as lakes, trees, bridges, and more, to build rich end user map visualizations based on your device data:

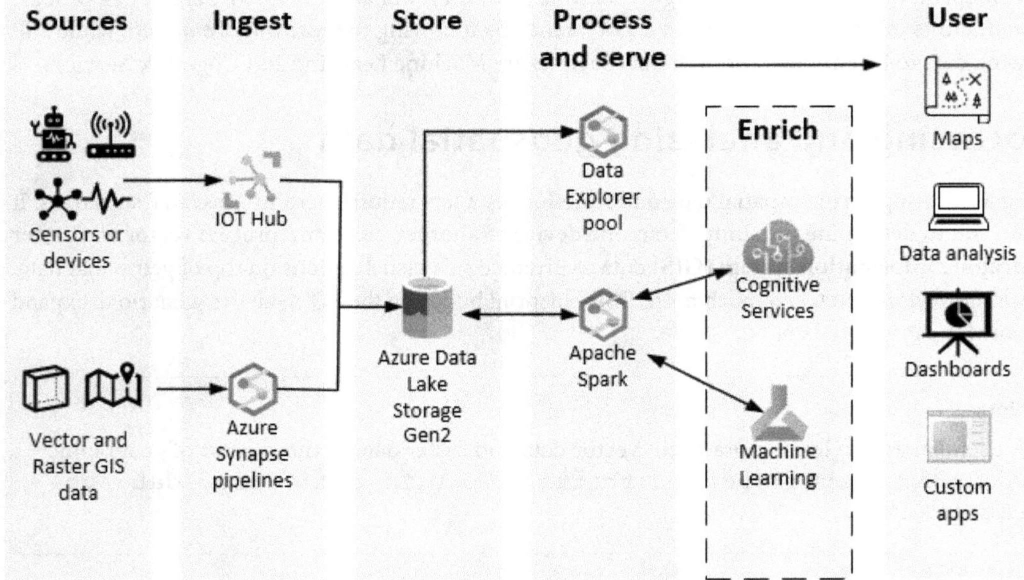

Figure 4.3 – Geospatial data processing and analytics

Note that because this blueprint joins device data with vector and raster GIS data, we do not have the **real-time** and **model and scoring** paths. In this case, this blueprint offers greater latency to the IoT scenario as data needs to be processed before it gets served to the user.

Enabling real-time analytics with big data

Performing analytics by streaming data from applications at large volumes is a common scenario for companies that need to make quick decisions based on events that are happening in real time. The idea behind real-time analytics is that companies can make decisions soon after the data event has been captured. Here are a few examples:

- **Detecting patterns on financial transactions**: The first blueprint in this chapter discussed using Data Explorer pools and Azure Machine Learning to detect credit card fraud. In financial services, however, there are several other opportunities to employ real-time analytics and machine learning for other scenarios, such as providing financial advice and risk assessments based on current market situations.
- **Optimizing retail websites based on user flow**: This scenario can be used to read log data from web applications that describe how an e-commerce website is being used in response to promotions and, with user analysis, perform corrective actions such as adjusting the user experience, promotions, and more.
- **Monitoring application health**: Similar to the retail website scenario, application log monitoring can help you identify issues in real time or identify if, based on a series of current log events, a website is suffering some sort of attack or degradation of service.
- **Performing live sentiment analysis**: It is common for companies to monitor live user feedback to live events, announcements, responses to television ads, and more. This solution blueprint can take unstructured, free-text data from customer feedback and perform sentiment analysis tasks to monitor customer responses to these live events.

This solution blueprint, illustrated in *Figure 4.4*, proposes an architecture that works with Data Explorer pools to process and serve user queries at low latency but also works with Apache Spark to enrich data:

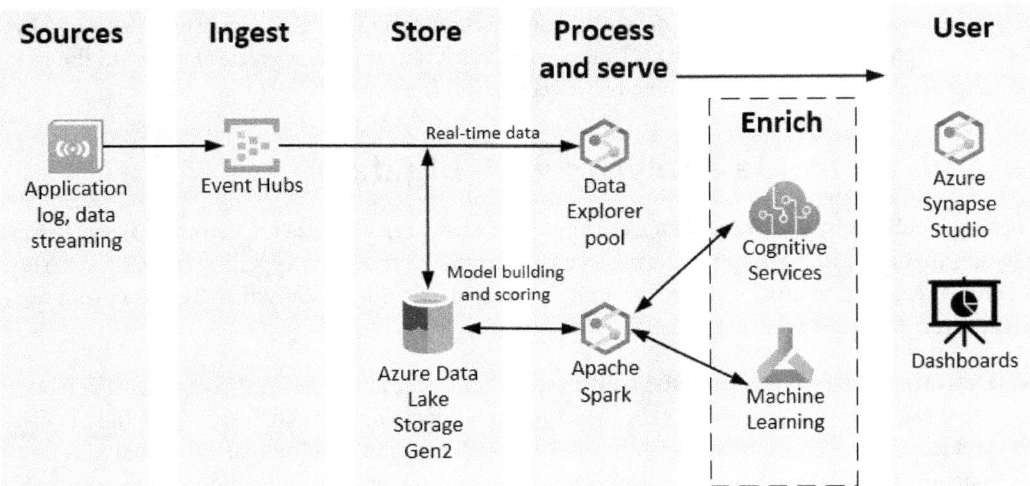

Figure 4.4 – Real-time analytics on big data

The solution is similar to the IoT analytics one, though that shouldn't be much of a surprise since both scenarios generate high data volumes at low latency. The only difference here is the use of Event Hubs (instead of IoT Hub) to stream data to a Data Explorer pool, and at the same time store data on ADLS Gen2 for data enrichment.

Performing time series analytics

As you start to capture and store IoT or application log data, an opportunity arises to perform time series analysis, a technique in statistics that helps you perform trend analysis, forecasting, detect anomalies, and more. Time series analysis can be a helpful asset whenever you have time series data – that is, data that is produced over a certain period. Here are a few real-world scenarios where it can be useful:

- **Stock price analysis**: Every stock trade operation records a date and time when the event occurred. By analyzing stock market transactions, you can determine trends for a given stock or industry, or detect outliers in the middle of millions of transactions.
- **Energy consumption forecasting**: Energy consumption for buildings varies based on several factors, including the season, day of the week, building traffic, and others. Time series analysis can help you understand energy consumption patterns over time to plan expenditure, equipment maintenance, and personnel.
- **Sales forecasting**: You can use time series techniques to forecast product sales based on historical sales. This helps you understand how much investment you can provision for marketing, sales workforces, product inventory, and more.

- **Cloud application resource needs per seasonality**: Managing costs in cloud environments is a key concern of organizations. For applications that are web-based and run on public cloud offerings (such as Microsoft Azure), having the right balance between how much compute resources you allocate to your services and how much you are willing to spend on cloud consumption costs may directly determine your customer's experience. By using time series analysis, you can process historical log data from your application to determine which part of the day, the month, and the year your application is most used, and provision additional cloud resources for your application during those times, while scaling down when not needed.

The solution blueprint for this scenario, seen in *Figure 4.5*, includes data captured from IoT devices, as well as any application log data, or even data that is streamed into the system. So long as these sources produce time series data, other data sources may apply to this scenario as well:

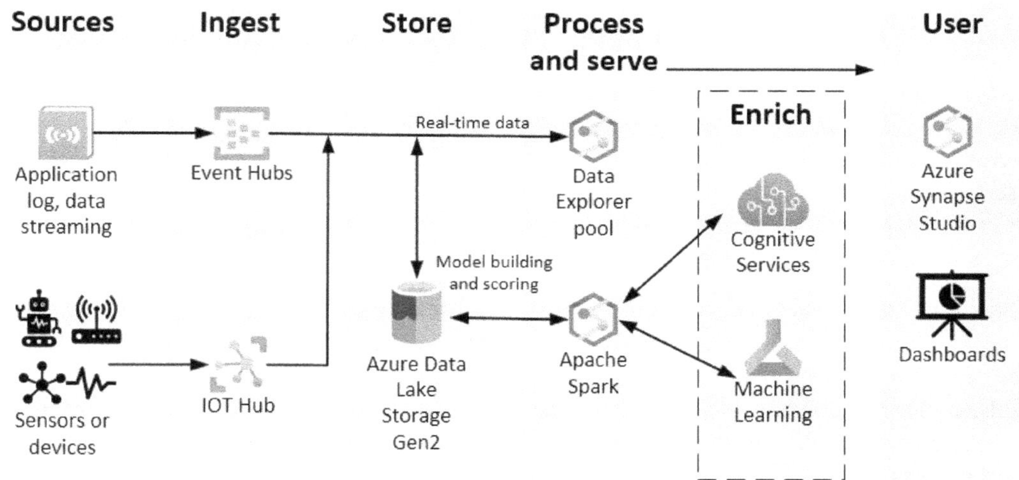

Figure 4.5 – Time series analytics

Time series analysis is a powerful tool in analytics. It is used even more now that we have robust platforms, such as Azure Synapse, to process and store large volumes of historical data.

Summary

This chapter concludes *Part 1* of this book, which provided an introduction to Azure Synapse Data Explorer. By now, you should be able to describe Azure Synapse Data Explorer, navigate through the Azure portal to create an Azure Synapse workspace, find what you need in Azure Synapse Studio, create Data Explorer pools, load data, and even perform simple queries.

This chapter offered five blueprints that you can use today in your implementation of Azure Synapse Data Explorer. These examples were based on the author's real-life experience and leverage formal solution architectures published by Microsoft in the Azure Architecture Center. They were adjusted to be built around Azure Synapse Data Explorer, with diminished dependency on external Azure services.

We started by breaking down the blueprint for a multi-purpose, end-to-end analytics environment. This is a generic blueprint that addresses most of the needs for analytical solutions that work with big data. Besides listing potential use cases and scenarios for this blueprint, this section explored each layer of the architecture, its role, and how the components in each layer are used.

The multi-purpose architecture is not, however, appropriate for every single analytical scenario that works with big data. It lists several services that may not be needed for more targeted scenarios. Therefore, after breaking down the details of the multi-purpose architecture, we listed additional blueprints that target more specific tasks, such as performing analytics on IoT data, working with geospatial data, enabling real-time analytics with big data, and performing time series analysis.

When you think about the solution architecture for your next project, make sure you understand what the data flow will be and how users will interact with your data. Clarify all the business requirements in terms of latency, data volume, sources, and any other relevant details. To save on costs, use only the services you need. For example, if you need a solution for IoT analytics, maybe you won't need a dedicated SQL pool. These decisions are important if you wish to save on your Azure consumption costs, as well as the ongoing costs to maintain these services.

Finally, keep in mind that there is not a single solution to every problem. As you analyzed the solution architectures proposed in this chapter, odds are you had some ideas of how you would have built them differently. That is normal, and every project will have tradeoffs to help you make the right decision to address your requirements. Also, note that these blueprints don't include satellite services that may be part of your final solution, such as Azure Active Directory, GitHub integration, Azure Monitor, and others.

The next chapter kicks-off *Part 2* of this book, where we will focus solely on working with data. You will learn more about data ingestion strategies and take a deep dive into data analysis with Power BI, KQL, and even Python through Azure Synapse notebooks. Finally, you will learn about advanced analytics scenarios that leverage Azure Synapse services such as Apache Spark for machine learning workloads, and powerful scenarios that use Azure Cognitive Services. To finish *Part 2*, you will learn how to export data from Data Explorer pools.

Part 2
Working with Data

The second part of the book is all about getting your hands busy working with data. You will use the provided sample dataset to ingest data into an Azure Synapse Data Explorer database and use different tools to explore, analyze, transform, and present data, from KQL, through Python and Microsoft Power BI. You will also get a primer on machine learning, and leverage different services within Azure Synapse to build machine learning experiments that help get deeper insights into your log and telemetry data. Last but not least, you will learn all about exporting data from Azure Synapse Data Explorer to other destinations, including why you should have a strategy to export data in the first place.

The second part contains the following chapters:

- *Chapter 5, Ingesting Data into Data Explorer Pools*
- *Chapter 6, Data Exploration and Analysis with KQL and Python*
- *Chapter 7, Data Visualization with Power BI*
- *Chapter 8, Building Machine Learning Experiments*
- *Chapter 9, Exporting Data from Data Explorer Pools*

5

Ingesting Data into Data Explorer Pools

Welcome to *Part 2, Working with Data*. From here until *Chapter 9, Exporting Data from Data Explorer Pools*, you will learn how to bring data to Data Explorer pools, analyze it through different means, leverage other Azure Synapse services for advanced analytics scenarios (such as machine learning), and export data.

There are different ways to ingest data into Data Explorer pools. Your strategy may vary based on the format of the source data, frequency of updates, real-time needs, and cost constraints. The *Running your first query* section of *Chapter 3, Exploring Azure Synapse Studio*, walked through the steps to use one-click ingestion to quickly import data from a **comma-separated values** (**CSV**) file into your Data Explorer pool. In this chapter, you will learn different strategies to copy data from databases and files, and how to continuously collect data from streaming sources, **internet of things** (**IoT**) devices, and more.

This chapter starts with an overview of the data loading process, comparing batch versus streaming ingestion. Next, you will learn about different tools you can use for ingestion, including Azure Synapse pipelines, and the ability to build your own programs using the Data Explorer **software development kit** (**SDK**). Finally, as discussed in *Chapter 4, Real-World Usage Scenarios*, acquiring data from streaming sources is a common requirement when you are working on real-time analytics. Therefore, you will learn how streaming ingestion works and how to implement continuous data ingestion.

At a glance, here are the topics covered in this chapter:

- Understanding the data loading process
- Defining a retention policy
- Choosing a data load strategy
- Performing data ingestion

By the end of this chapter, you should be able to describe the different data ingestion mechanisms in Azure Synapse Data Explorer, how to continuously ingest data from a variety of data sources including databases, files, application logs, telemetry systems, and others, and how to implement continuous ingestion.

Technical requirements

In case you haven't yet, make sure you download the book materials from our repository at `https://github.com/PacktPublishing/Learn-Azure-Synapse-Data-Explorer`. You can download the full repository by selecting **Code**, and then **Download ZIP**, or by cloning the repository using your Git client of choice.

For the examples in this chapter, we will use a simplified version of the drone telemetry dataset. This simplified version contains the same number of rows as the original dataset but with only 10 columns (compared to 30 in the full dataset). This version of the dataset will make it simpler to create all the tables and data mappings we will need in the examples in this chapter.

You can find this dataset in the `Chapter 05\drone-telemetry-simplified.zip` compressed file of the book repository. Make sure you extract the content of this file, `drone-telemetry-simplified.csv`, to a local folder on your computer.

In addition to the simplified dataset, this chapter also provides the same dataset split across four files: `Chapter 05\drone-telemetry-simplified - slice 1.csv` to `Chapter 05\drone-telemetry-simplified - slice 4.csv`. These files, when combined, add up to the full simplified dataset. They will be used in one of the data ingestion examples, where you ingest data continuously as new files are added to a container in **Azure Data Lake Storage** (**ADLS**) **Gen2**. The split files can be found in the `Chapter 05\drone-telemetry-simplified - split.zip` file of the book repository. You need to extract the contents of this file to a local folder on your computer.

After you have the files locally, upload the `drone-telemetry-simplified.csv` file (not the split flies – we will do that later!) to your Azure Synapse workspace's primary storage account using the following instructions:

1. Open the Azure portal at `https://portal.azure.com`. Find your Azure Synapse workspace's primary storage account by typing the name of the storage account in the global search box.

> **Note**
> If you don't know the name of your workspace primary account, you can find it on the **Overview** page of your Azure Synapse workspace in the Azure portal.

2. You'll land on your storage account's management page in the Azure portal. On the left-hand-side menu, select **Containers**. Then, select the **+ Container** button to create a new container. Use a familiar name, such as `simplified-data`.
3. Once your container is created, select it from the list to navigate to it.
4. Select the **Upload** button. The **Upload blob** page will show up on the right side of the Azure portal. Select the button with a folder icon to browse your local computer files, and select the `chapter 05\drone-telemetry-simplified.csv` file.
5. Click **Upload**.

You will see the `drone-telemetry-simplified.csv` file listed on the container.

Understanding the data loading process

In *Chapter 4, Real-World Usage Scenarios*, we discussed sample architectures that you can use as blueprints in your own projects. For every single one of them, on the left-hand side of the diagrams, you could see data sources, followed by the **Ingest** stage. Your data load strategy, and the services you will use to perform data ingestion, will depend on your data source types and your latency requirements.

Simply put, the data loading process can be summarized into four steps:

1. Defining a retention policy.
2. Choosing a data load strategy.
3. Creating destination tables and data mappings.
4. Performing data ingestion.

Note that before you perform the data ingestion task, you should create your destination tables and data mappings. Since we will explore different ways to perform data ingestion, the steps to create destination tables and data mappings in this chapter (when needed) will be presented in topics where we will perform the data ingestion tasks.

Defining a retention policy

Working with large volumes of data can become an expensive operation over time. As your data volume grows, so do your storage costs. Additionally, in some cases, working with aging data may provide undesired results. When we are working with machine-generated data, especially from application logs and IoT devices, data volumes grow quite quickly. This begs the question, how long do you need to keep your data?

In machine learning, the more data you have in your hands to train new models, the better. This is based on the law of large numbers, which states that *the results obtained from a large number of trials should be close to the expected value and tends to become closer to the expected value as more trials are performed*. Translating that to the data verbatim, the more samples you have of a certain measurement, the closer you are to predicting what that measurement should be in the future. Some researchers, however, such as Dr. Andrew Ng, one of the pioneers in the area of deep learning, believe in a smart sizing approach that suggests artificial intelligence should use the least amount possible of data.

> **Note**
> To learn more about Dr. Andre Ng's data-centric artificial intelligence vision, visit `https://fortune.com/2022/06/21/andrew-ng-data-centric-ai/`.

Managing how much data to keep and for how long is something you should consider as you build your analytical environment. In my experience, I worked in a product team that kept only 1 year of telemetry data and constantly purged everything past that. I also worked in a different product team where the data volume was so large that they only kept the last 3 months of telemetry data, and even that was enormously expensive to maintain. Anything past those 3 months had to be retrieved from cold storage.

To help manage these challenges, Data Explorer pools support a retention policy configuration that automatically removes all data from your tables that's older than what's specified in your policy. The retention policy can be set at the database level, at the table level, or for individual materialized views. When applied at the database level, all tables and materialized views in this database inherit the database policy. Policies set at the table and materialized view levels override the database policy.

When you create Azure Synapse Data Explorer databases using Azure Synapse Studio, you are asked to provide the default retention policy for that database, which is 365 days (1 year) by default. You can change this to any number you would like, or check the **Unlimited** box, as shown in *Figure 5.1*. However, note that these defaults are provided by the graphical user interface in Azure Synapse Studio only. If you create a database or a table using KQL control commands, no retention policy is applied to your database or table.

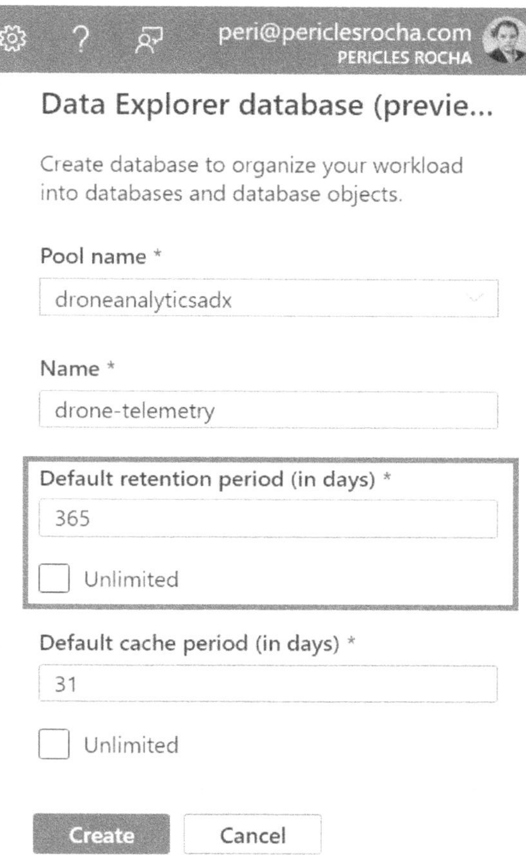

Figure 5.1 – Defining your data retention period

To calculate the age of your data, retention policies consider the date your data was ingested. Keep in mind that retention policies ensure your data will be available for the time specified in the policy, but they do not necessarily guarantee that the data will be deleted the moment the policy date is reached. The actual deletion time is imprecise.

Retention policies also contain a recoverability property that defines whether your data can be recovered after deletion. If this property is set to **Enabled**, you will be able to recover any deleted data in a period of up to 14 days after its deletion.

To see the retention policy of a database, you can use the `.show database policy retention` control command. As an example, running the following command shows the retention policy of our database:

```
.show database ['drone-telemetry'] policy retention
```

Figure 5.2 illustrates the results produced by this command.

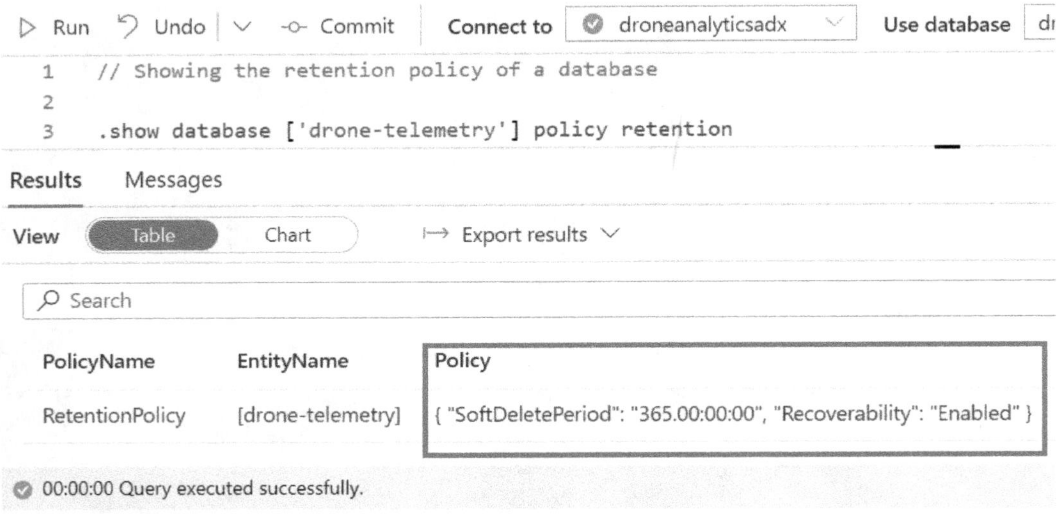

Figure 5.2 – Showing an object's retention policy

Note that this database has a retention policy with a soft-delete setting of 365 days, which is what we set when we created the database using Azure Synapse Studio, as shown previously in *Figure 5.1*. To change the policy for this database, we can use the `.alter database policy retention` control command. The following example sets a new retention policy for our database, with a soft-delete period of 90 days and the recoverability property disabled:

```
.alter database ['drone-telemetry'] policy retention
"{\"SoftDeletePeriod\": \"90.00:00:00\", \"Recoverability\":
\"Disabled\"}"
```

Running this command changes the retention policy in our database and produces the output illustrated in *Figure 5.3*.

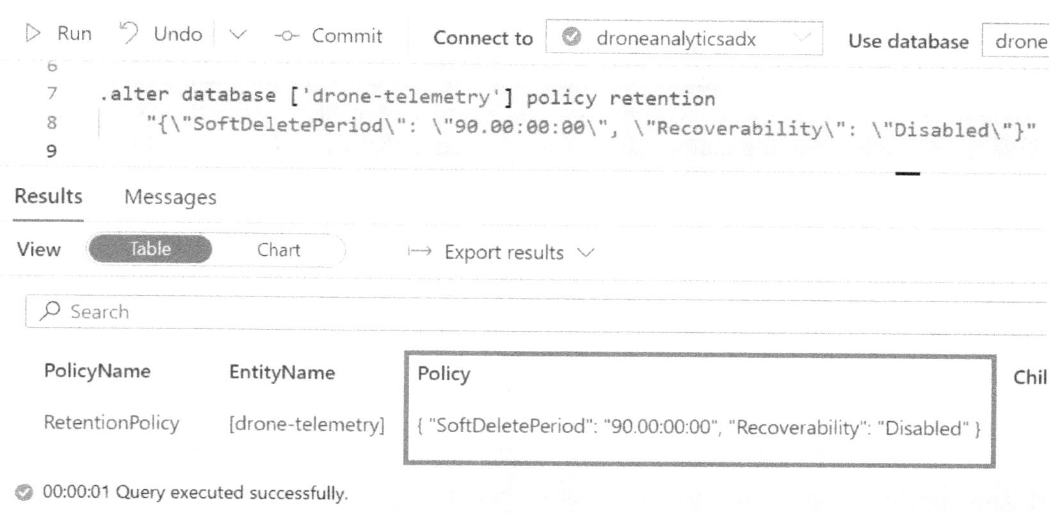

Figure 5.3 – Applying a retention policy to a database

> **Note**
>
> These commands can be found in the `Chapter 05\RetentionPolicies.kql` file of the book repository.

Now that you have defined the retention policy of your database, the next step is to choose your data load strategy.

Choosing a data load strategy

Two of the key parameters you have to consider when you design your data load strategy are your tolerance for latency and the data volume that you will process.

In some scenarios, your business may require seeing the latest data available as soon as possible to enable quick decision-making, while in others you may be able to work with data from a day, or a few days ago, to perform analysis on historical data. Working with low latency in data ingestion implies inserting smaller chunks of data more frequently, while at a higher latency, you will insert larger volumes of data maybe once a day, or a few times per day.

To address these scenarios, Azure Synapse Data Explorer works with two ingestion strategies: **streaming ingestion** and **batching ingestion**. Let's look at these strategies in detail.

Streaming ingestion

When real-time analytics is a hard business requirement, you will most likely implement a streaming ingestion strategy. This strategy allows continuous data ingestion of smaller sets of data into your Data Explorer pools from streaming sources, such as an application that writes files into a cloud storage solution, an event processing service (such as Azure Event Hubs), or an IoT device (through Azure IoT Hub). Streaming ingestion tolerates latencies of less than a second for data ingestion, which enables real-time analytics scenarios. You can also use streaming ingestion when you have larger volumes of data to insert at one time, as long as this data is processed to different tables in parallel and each table processes a few records per second.

In fact, parallelism plays a big role in ingestion overall, and data load performance scales as you grow the size of your Data Explorer pools. Data Explorer limits ingestion requests to a maximum of six requests per core in your compute specification; therefore, as an example, if your compute specification has a 16-core workload, Data Explorer will process 96 ingestion requests in parallel at any given time. Note that for streaming ingestion, each one of these ingestion requests needs to be limited to a maximum data size of 4 MB.

Staying on the topic of performance and resource usage, keep in mind that enabling streaming ingestion in your Data Explorer pool will reduce the amount of storage you have available for the hot cache used for queries. The reason for this is that both streaming ingestion and the hot cache use the local **solid-state drive** (**SSD**) of the Data Explorer pool machines in your cluster. This is the reason why streaming ingestion needs to be enabled in your Data Explorer pool if you wish to use it. If you don't plan on using streaming ingestion, don't enable it.

However, if you do plan on using streaming ingestion in your Data Explorer pools, there are two ways of enabling it:

- **During the creation of your Data Explorer pool**: As part of the **Create Data Explorer pool** wizard, select **Enabled** next to **Streaming ingestion**, as shown in *Figure 5.4*.

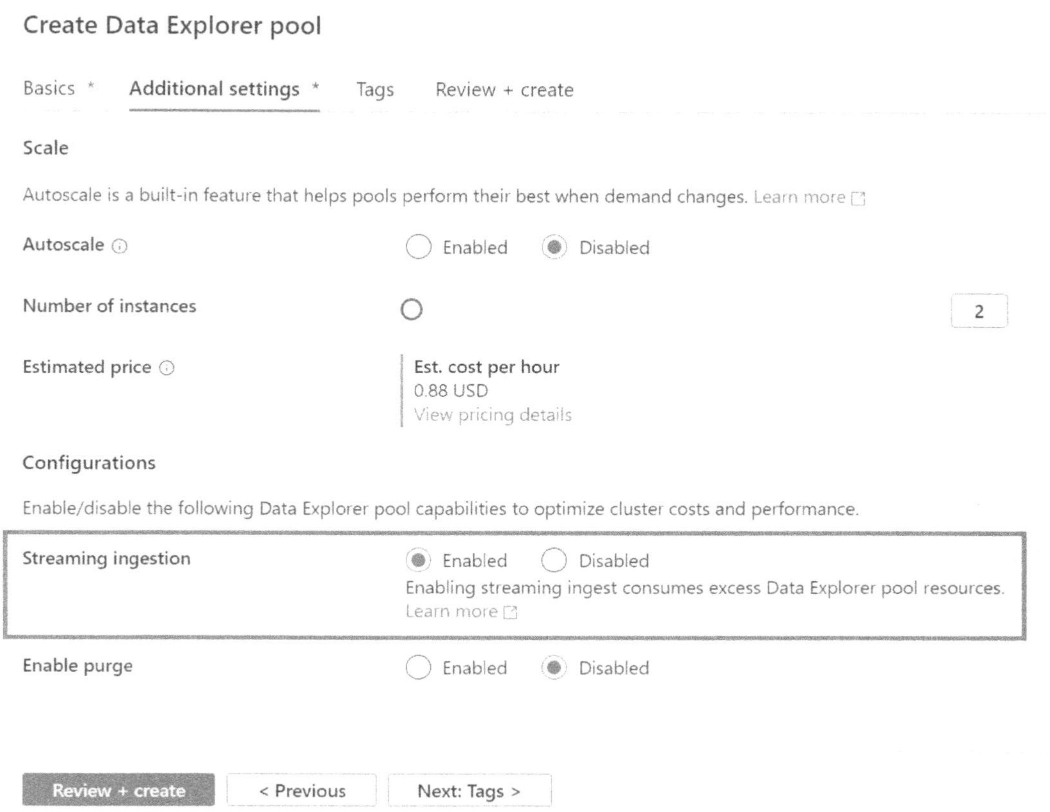

Create Data Explorer pool

Basics * Additional settings * Tags Review + create

Scale

Autoscale is a built-in feature that helps pools perform their best when demand changes. Learn more

Autoscale ⓘ Enabled Disabled

Number of instances 2

Estimated price ⓘ Est. cost per hour 0.88 USD View pricing details

Configurations

Enable/disable the following Data Explorer pool capabilities to optimize cluster costs and performance.

Streaming ingestion Enabled Disabled
Enabling streaming ingest consumes excess Data Explorer pool resources. Learn more

Enable purge Enabled Disabled

Review + create < Previous Next: Tags >

Figure 5.4 – The Create Data Explorer pool wizard

If you need a reminder of how to get to this screen, remember that this was discussed in the *Creating a Data Explorer pool using Azure Synapse Studio* section of *Chapter 2*, *Creating Your First Data Explorer Pool*.

- **After your Data Explorer pool has been created**: If you didn't enable streaming ingestion when you first created your Data Explorer pool, you can still enable it on your existing pools. At the time of writing, it is only possible to enable streaming ingestion on Data Explorer pools through the Azure portal.

> **Note**
>
> Changing this setting will cause your Data Explorer pool to restart, disconnecting any users that are using the service. Make sure you plan this change in advance.

To enable streaming ingestion on an existing Data Explorer pool, follow the following steps:

1. Go to the Azure portal and find your Azure Synapse workspace. Select your workspace to see its details.
2. On the left-hand-side menu, under **Analytics pools**, select **Data Explorer pools (preview)**.
3. Select your Data Explorer pool.
4. You will land on the details page of your Data Explorer pool. In the left-hand menu, under **Settings**, select **Configurations**.

> **Note**
> The **Configurations** menu option is only available if your Data Explorer pool is running.

5. Select the **On** toggle option next to **Streaming ingestion**, as shown in *Figure 5.5*, and click the **Save** button.

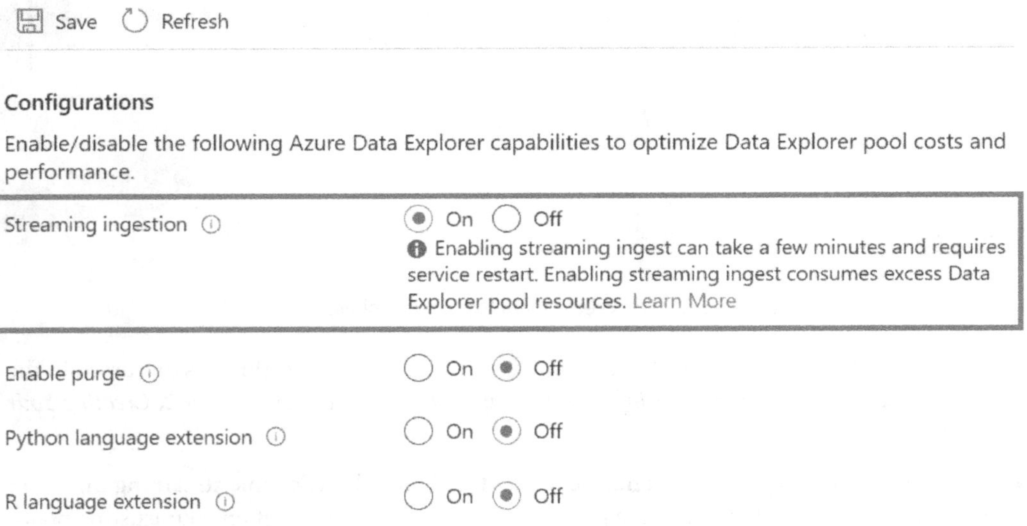

Figure 5.5 – Enabling streaming ingestion on an existing Data Explorer pool

This operation may take several minutes to complete.

After your Data Explorer pool is configured to use streaming ingestion, you need to configure the streaming ingestion policy to the tables that will receive data from streaming ingestion. As an example, to enable streaming ingestion on a table called `fleetdata`, you could use the following command:

```
.alter table fleetdata policy streamingingestion enable
```

You can also set this policy at the database level to apply to all tables in your database. When this is done, if you also have a streaming ingestion policy at the table level for some tables, the policy for the table takes precedence over the policy for the database. This is useful when you want to enable streaming ingestion at the database level but not use it for a few tables.

As an example, to enable the streaming ingestion policy at the database level, you can use the exact same syntax that is used to enable it at the table level but replace `.alter table` with `.alter database`. Your new command will be as follows:

```
.alter database drone-telemetry policy streamingingestion
enable
```

After enabling the policy at the database level, you can optionally disable the policy at the table level for some tables using the following command:

```
.alter table myothertable policy streamingingestion disable
```

These steps prepare your Data Explorer pool, databases, and tables to use streaming ingestion. The missing part here is performing the actual data ingestion. We will cover an example of continuous data ingestion using streaming ingestion later on in this chapter, but first, we will explore the second – and most common – data ingestion strategy, batching ingestion.

Batching ingestion

Batching ingestion is optimized for high ingestion throughput and is the recommended strategy in most cases. If you don't have real-time analytics requirements for your analytics environment, then you should use a batching ingestion strategy. By default, this strategy performs data ingestion in batches (as the name suggests) after one of the following three conditions is met:

- A maximum delay time of 5 minutes since the last batch was processed. This is the `time` parameter of batching ingestion.
- The number of items to process reaches the default limit of 1,000. This is the `count` parameter.
- The batched data reaches a total size of 1 GB. This is the `size` parameter.

The data ingestion process can consume significant compute resources from your cluster, depending on your data volume and frequency of data updates. If you set a lower `time` limit and don't have large volumes of data to process, you may be processing less data at every batch and consuming unnecessarily excessive compute resources. If you set a higher `time` limit and a higher data `size` limit, it may take time to process your batch, and you may see outdated results in your queries because fresh data has not been processed yet. Finding the right balance between how much time you want to wait until you process the next batch and how much data you want to process immediately is the key to success here.

> **Note**
>
> The `time`, `count`, and `size` parameters can be controlled by defining an ingestion batching policy. These policies can be defined individually at the table level, or globally at the database level. New tables that don't have an ingestion batching policy set inherit the database policy. To learn more about this policy, refer to `https://docs.microsoft.com/azure/data-explorer/kusto/management/batchingpolicy`.

Now that we understand streaming ingestion and batching ingestion, the next step is to explore how to perform data ingestion by using either of these strategies.

Performing data ingestion

You have several options to load data into Azure Synapse Data Explorer. You can use one-click ingestion to quickly load some data and automatically create tables and data mappings as needed, with a few clicks of a mouse. For quick data exploration, you can use **Kusto Query Language** (**KQL**) commands to ingest data. If you have complex Azure Synapse pipelines or Azure Data Factory pipelines, you can use them for anything from robust data integration jobs to simple data movement tasks, leveraging rich logging, error handling, and more. Finally, if you need to build a custom application that connects to your Data Explorer pool using Data Explorer's REST API for data ingestion, you can do that too.

The tool that you will use for data ingestion depends on your business needs and scenario. You can refer to the *Running your first query* section of *Chapter 3*, *Exploring Azure Synapse Studio*, where we already walked through the process to load data using the **one-click ingestion** feature. In this chapter, we will focus on using all the other data ingestion tools and methods.

Using KQL control commands

KQL offers convenient control commands that you can use to quickly insert data into your Data Explorer pools for experimentation and exploration. The `.ingest into table` command reads data from a source file stored on a storage location, such as ADLS Gen2, and inserts this data into a table stored on a Data Explorer pool.

Before we explore ingesting data using KQL, we will create a target table to ingest data into.

Creating a destination table

We will create a new table specifically for this example, which you can delete when you are done. To load data, follow the following steps:

1. Navigate to Azure Synapse Studio at `https://web.azuresynapse.net` and log in to your Azure Synapse workspace. Optionally, navigate directly to your workspace URL if you have it (which you can find by reviewing the **Overview** page of your Azure Synapse workspace in the Azure portal).

2. If your Data Explorer pool is paused, make sure you resume it by going to the **Manage** hub, selecting **Data Explorer pools (preview)** on the left-hand-side navigation menu, and then selecting the **Resume** command by hovering your mouse over your Data Explorer pool name.
3. Select the **Develop** hub.
4. Hover your cursor over **KQL scripts** and select **New KQL script**, as shown in *Figure 5.6*.

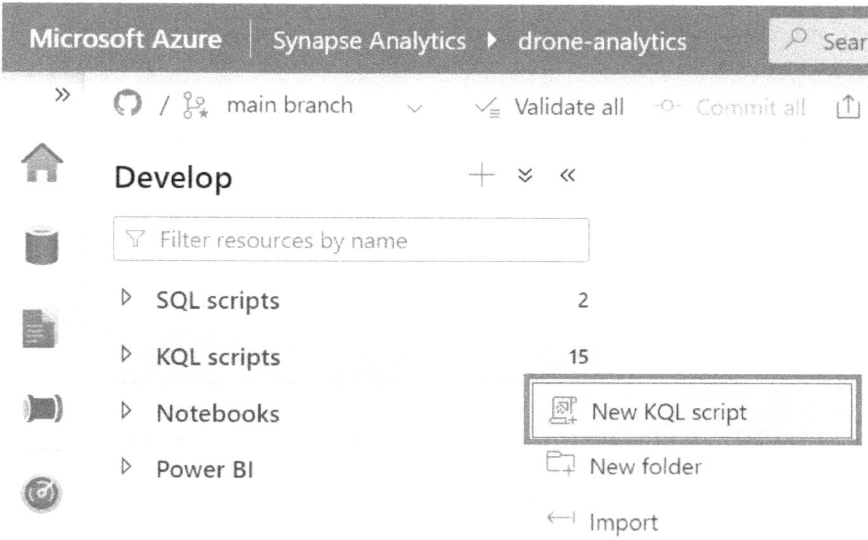

Figure 5.6 – Creating a new KQL script

5. In the query window, type the following command to create a new table:

```
.create table ['fleet data simplified'] (
    DeviceData: string,
    LocalDateTime: string,
    DeviceState: string,
    CoreTemp: long,
    CoreStatus: string,
    Compass: long,
    Altitude: long,
    Speed: long,
    PayloadWeight: long,
    EventData: string
)
```

> **Note**
>
> The code to create this table can be found in the `Chapter 05\Ingestion-KQL.kql` file in the book repository.

6. Select your Data Explorer pool in the **Connect to** dropdown if it is not yet selected. Similarly, make sure you select your database in the **Use database** dropdown.
7. When you are ready, click **Run** to execute the script. If your script worked normally, the bottom of the query editor window should show the message query executed successfully, as shown in *Figure 5.7*.

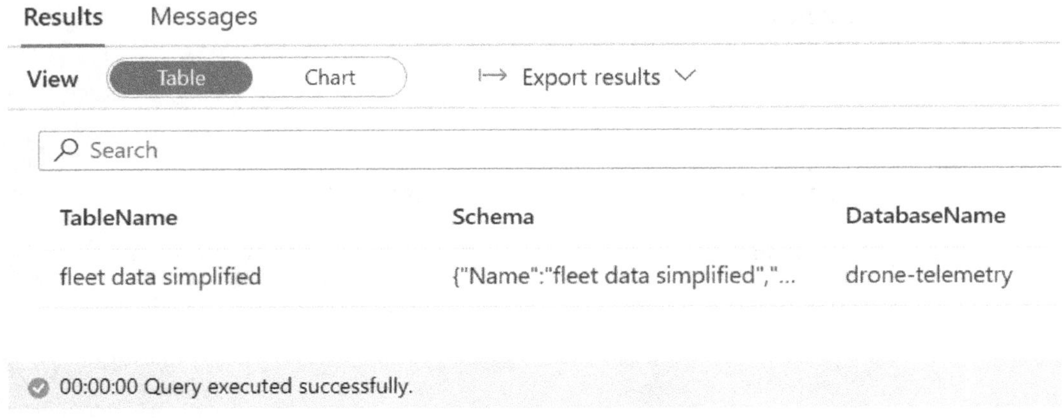

Figure 5.7 – The query editor shows the successful execution of your script

Your new table is now created and ready for data ingestion.

Ingesting data

For this example, we will use the simplified telemetry data file, `drone-telemetry-simplified.csv`. First, we need to generate a **shared access signature** (**SAS**) token that provides secure access to your file in your ADLS Gen2 container. To generate a SAS token, execute the following steps:

1. Find and select your storage account in the Azure portal.
2. Under **Data storage**, select **Containers**.
3. Select the container where the `drone-telemetry-simplified.csv` file is stored, as shown in *Figure 5.8*.

dronetelemetrydatalake | Containers
Storage account
Search
Overview
Activity log
Tags
Diagnose and solve problems
Access Control (IAM)
Data migration
Events
Storage browser
Data storage
Containers
File shares
Container
Change access level
Search containers by prefix
Name
$logs
data-exports
external-tables
raw-data
simplified-data

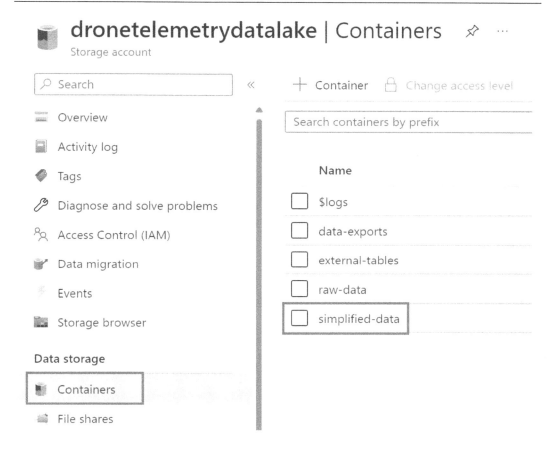

Figure 5.8 – Finding the container where your data file is stored

4. Hover the mouse over the `drone-telemetry-simplified.csv` file and select the ellipsis to reveal the menu options. Select the **Generate SAS** option, as shown in *Figure 5.9*.

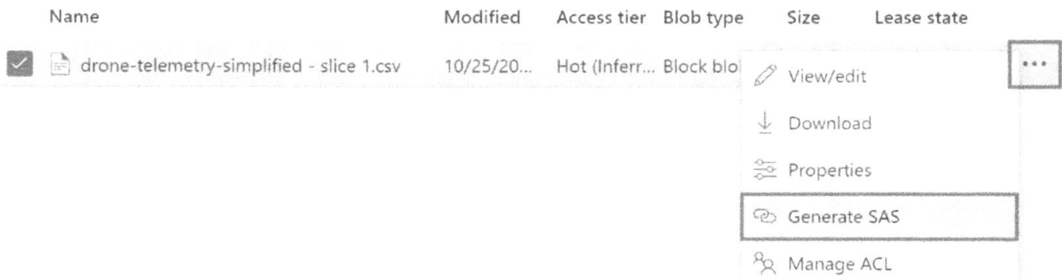

Figure 5.9 – Generating a SAS token

5. Review the options, and when you are ready, select the **Generate SAS token and URL** button.
6. Copy the value of the **Blob SAS URL** field. This is your SAS token.

As mentioned, we will use the `.ingest into table` KQL command to insert data from the file in your ADLS Gen2 container (referenced through the SAS token) into the table we just created. To achieve that, run the following command, but replace the URL after the table name with your own SAS token:

```
.ingest into table ['fleet data simplified']
    'https://dronetelemetrydatalake.blob.core.windows.net/
raw-data/drone-telemetry-simplified.csv?sp=r&st=2022-09-
14T05:25:46Z&se=2022-09-14T13:25:46Z&spr=https&sv=2021-06-08&sr
=b&sig=0CC1u1wSdu9IL4Am%2F6QUF7kcUY90fExt4MIooKZJnVs%3D'
    with (ignoreFirstRecord=true)
```

> **Note**
>
> The code used in this section can be found in the `Chapter 05\Ingestion-KQL.kql` file in the book repository.

Keep in mind that our source data file contains column headers in the first row, so we're adding the `with (ignoreFirstRecord=true)` parameter with our command to ensure that the first row gets skipped.

To verify that the data successfully loaded into your table, run the following simple query:

```
['fleet data simplified']
| project DeviceData, CoreTemp, Altitude
| take 10
```

Your editor should produce a result similar to what's shown in *Figure 5.10*:

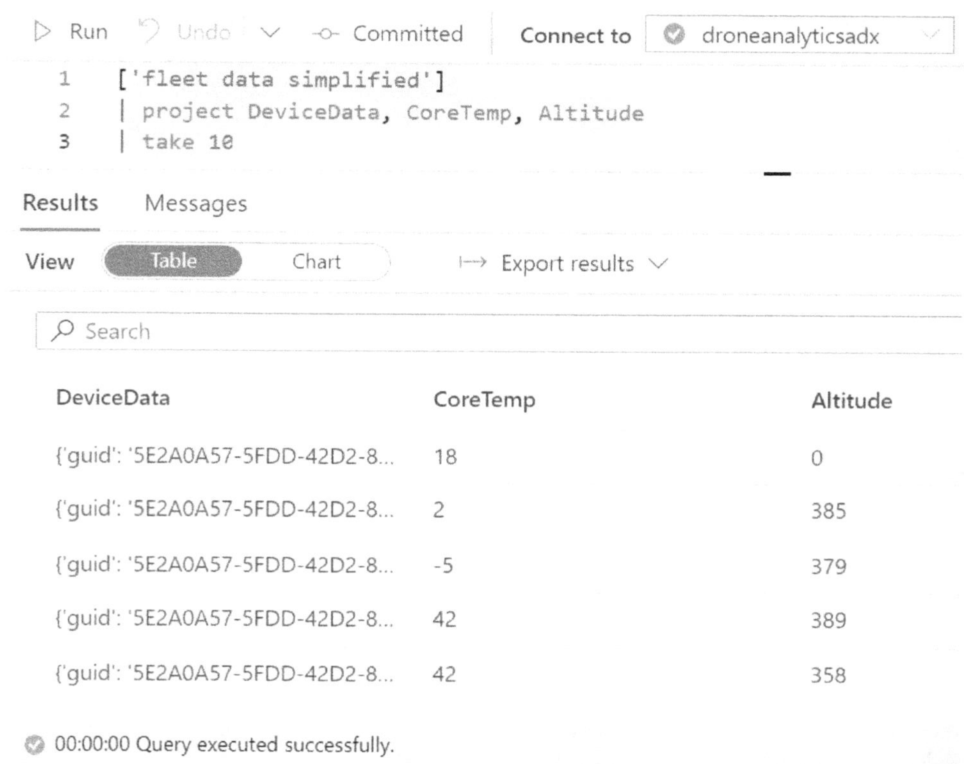

Figure 5.10 – The query results show that the data ingested successfully

Note that in this example, we did not create a data mapping to map the source data to the destination table. In this case, since the column order of the source file and the columns in the destination table are the same, and the data types are also the same, data mapping was not needed. Inserting data without a data mapping is called *identity mapping* – data gets mapped in real time by deriving the table's schema. For some file formats, such as JSON, Parquet, and Avro, if your file contains column names, the columns will be mapped to the destination table as long as it has columns with the same name.

> **Note**
>
> To learn more about identity mapping, visit `https://docs.microsoft.com/azure/data-explorer/kusto/management/mappings?context=%2Fazure%2Fsynapse-analytics%2Fcontext%2Fcontext#identity-mapping`.

In addition to the `.ingest into table` command, KQL supports the `.ingest inline` command, which inserts data that is provided as part of the command itself, and a set of commands that can ingest data from a query. For details on these commands, refer to `https://docs.microsoft.com/azure/synapse-analytics/data-explorer/ingest-data/data-explorer-ingest-data-overview`.

As a data professional, you may be familiar with using **data manipulation language** (**DML**) commands to insert, select, update, and delete data in your database tables. In Data Explorer, KQL control commands are not technically called DML commands. You should not use KQL control commands the way you would use DML commands in a SQL language, such as Transact-SQL. In fact, Microsoft does not recommend loading data at scale using KQL control commands, as they bypass the data management service in Data Explorer. Use these commands in the context of data ingestion only for experimentation and quick exploration of data.

Building an Azure Synapse pipeline

As discussed in *Chapter 1, Introducing Azure Synapse Data Explorer*, Azure Synapse pipelines offer a robust platform for data integration with connectors to more than 90 data source types. Azure Synapse pipelines are not limited to moving data from a source to a destination. In fact, when used with data flows, they can perform several tasks required by data integration jobs, such as data cleaning, data transformation, filtering, data lookups, and much more. Pipelines can be scheduled to run at certain times or follow a certain recurrence by using triggers. Data flows that integrate pipelines allow you to view data interactively as you transform it, helping you visualize the results and debug any problems as needed.

> **Note**
>
> For a detailed view of Azure Synapse pipelines, visit `https://docs.microsoft.com/azure/data-factory/concepts-pipelines-activities`.

Creating a destination table

To configure and test our pipeline, we will use a copy of the fleet data simplified table, which will be used exclusively for this pipeline. Use the following command to create the new table:

```
.create table ['fleet data pipeline'] (
    DeviceData: string,
    LocalDateTime: string,
    DeviceState: string,
    CoreTemp: long,
    CoreStatus: string,
    Compass: long,
```

```
    Altitude: long,
    Speed: long,
    PayloadWeight: long,
    EventData: string
)
```

> **Note**
>
> The code used in this section can be found in the `Chapter 05\Ingestion-Pipelines.kql` file in the book repository.

By using this new table, you will keep the previous ones you created intact if you want to further use them for your learning objectives.

Ingesting data

Before we can build a pipeline that writes data into a Data Explorer pool, we need to create a new linked service. In Azure Synapse, linked services store the connection information to certain external services, such as ADLS Gen2, Azure Data Explorer, and Power BI.

Creating a linked service

Use the following steps to create a linked service:

1. First, we need to find out the query and data ingestion endpoints of your Data Explorer pool. To find this information, navigate to Azure Synapse Studio, select the **Manage** hub, select **Data Explorer pools (preview)**, and select your Data Explorer pool from the list. This will show the properties page for your pool, which shows the **Query endpoint** and **Data ingestion endpoint** URLs, as shown in *Figure 5.11*. Make sure you copy the query endpoint.

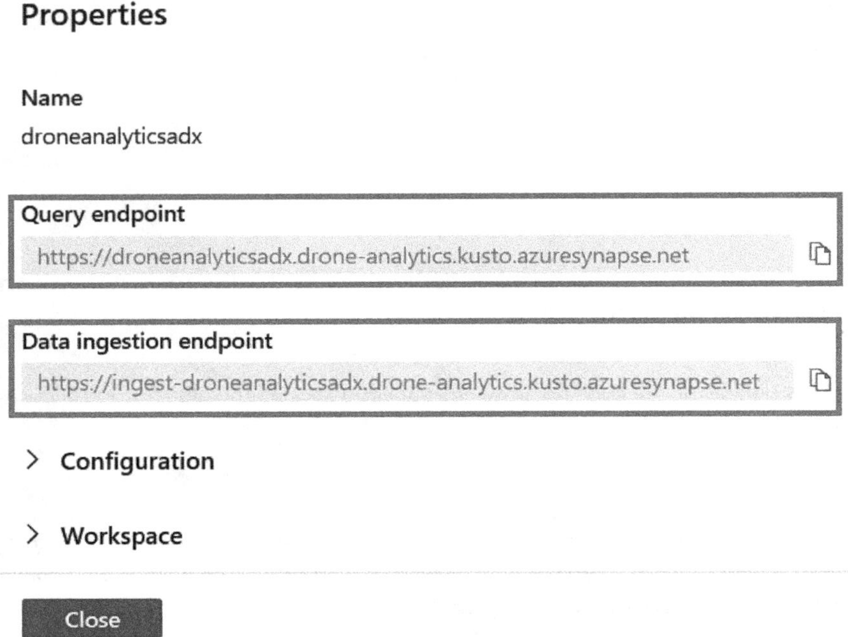

Figure 5.11 – Getting the endpoint addresses

2. Next, we will create the linked service. In the **Manage** hub, on the left-hand-side menu, select **Linked services** under the **External connections** group.
3. Select the + **New** button to create a new linked service. In the **New linked service** page, type `Data Explorer` in the search box to find the **Azure Data Explorer (Kusto)** option. Select the **Azure Data Explorer (Kusto)** box and click **Continue**.
4. The **New linked service** page will now ask for details about your Azure Data Explorer (Kusto) instance. Provide the following values:

   - **Name**: You can provide any name you want, as long as it contains only letters, numbers, underscores, or spaces. I've provided the name `droneanalyticsadx`.
   - **Description**: This is free text that helps you identify this linked service in the future. You can leave this blank if you wish.
   - **Connect via integration runtime**: This is the compute infrastructure that executes the data integration tasks. By default, Azure Synapse uses **AutoResolveIntegrationRuntime**. This integration runtime tries to detect the Azure region where you will copy data and allocates the compute infrastructure to perform the data integration job in that same region. Optionally, you can host an integration runtime on your own premises. For this exercise, we will select the default **AutoResolveIntegrationRuntime** option.

- **Authentication method**: Select **System Assigned Managed Identity**. This will use your workspace identity to authenticate and list the available Data Explorer pools.
- **Account selection method**: Select **Enter manually**. This reveals the **Endpoint** textbox and the **Database** drop-down box.
- Paste the address of your query endpoint (the one you copied from *step 1*) into the **Endpoint** textbox.
- In the **Database** drop-down box, select your destination database.

5. You can verify that your connection information is correct by selecting the **Test connection** button. When you are ready, select the **Commit** button. Your **New linked service** page should look similar to *Figure 5.12*.

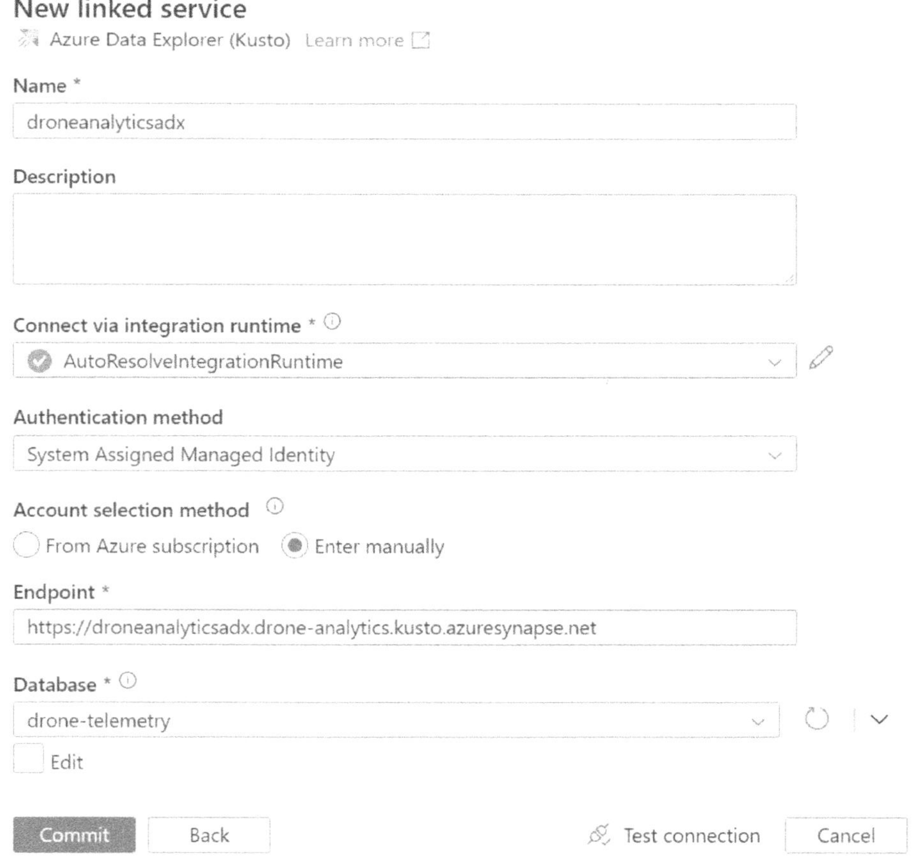

Figure 5.12 – All the parameters of the New linked service page

Now that you have the linked service ready, you are ready to create the pipeline.

Creating a pipeline

Proceed with the following steps:

1. Select the **Integrate** hub from the hub area.
2. Under the **Integrate** area, select the + button and pick **Pipeline** from the menu options, as shown in *Figure 5.13*.

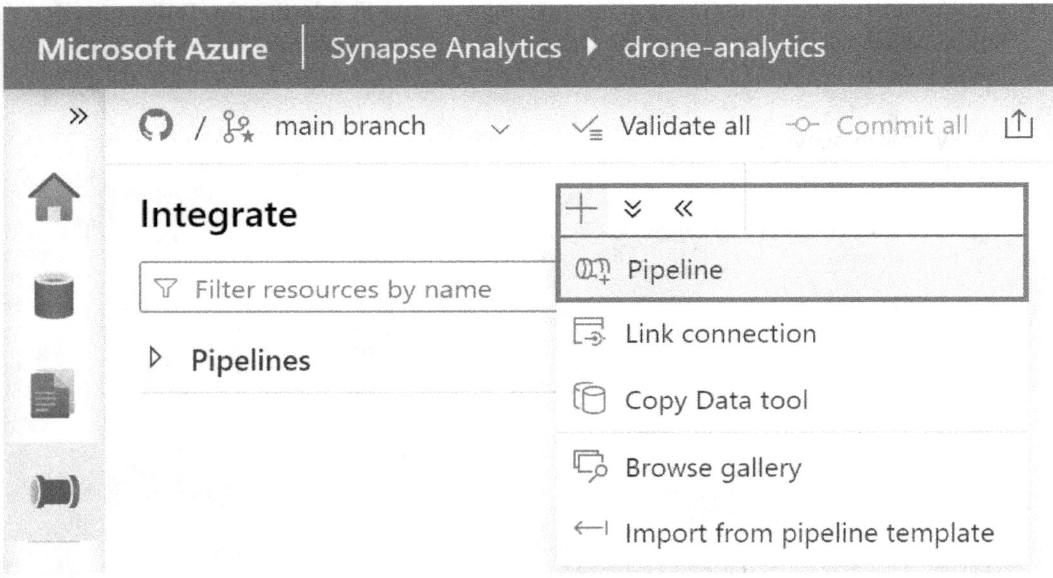

Figure 5.13 – Creating a new pipeline

3. This action will show the pipeline editor. Under **Activities**, expand **Move & transform**. Select the **Copy data** activity and drag it to the canvas. The rectangle and arrow in *Figure 5.14* show you which component you should drag, and where.

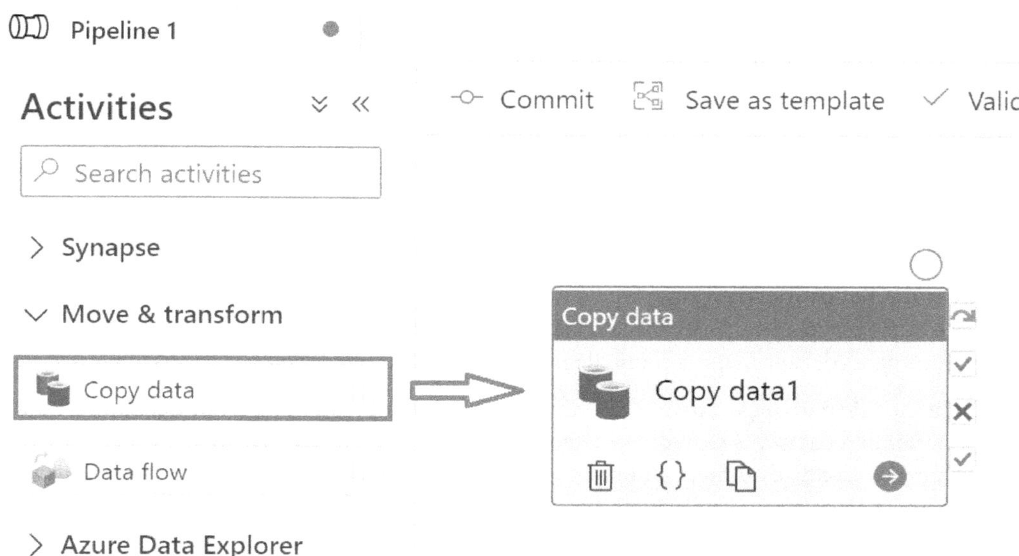

Figure 5.14 – Creating a Copy data activity

4. When you select the **Copy data** activity, you will see the list of properties of it under the pipeline workflow authoring area, as shown in *Figure 5.15*. Select the **General** tab and provide a familiar name for your **Copy data** activity. As an example, I've provided the name `Copy telemetry data` for mine.

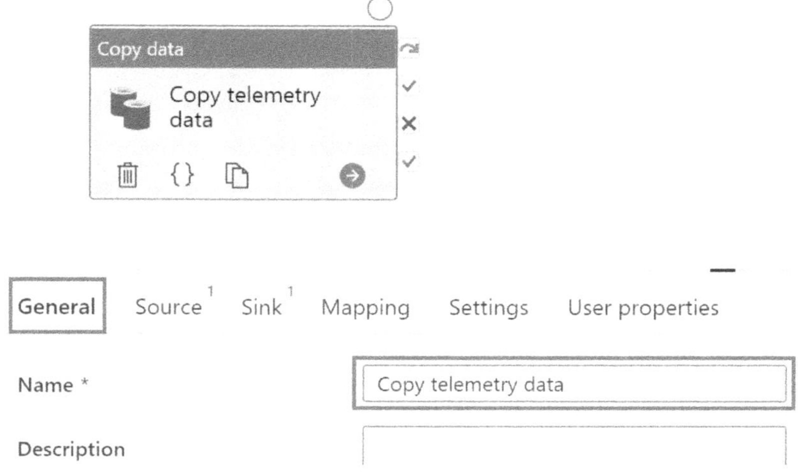

Figure 5.15 – Configuring the Copy data activity

5. Select the **Source** tab. This is where we specify the location of the source data. In our case, it will be the `drone-telemetry-simplified.csv` file located in the ADLS Gen2 container. To select it, click on the + **New** button to reveal the **New integration dataset** page. Provide the following details on this page:

    - Type `Data Lake` in the search box and select **Azure Data Lake Storage Gen2** when it shows up in the list. Click **Continue**.
    - On the **Select format** page, select **CSV Delimited Text** and click **Continue**.
    - On the **Set properties** page, type a familiar name for your integration dataset, such as `telemetrydata`. Under **Linked service**, select the linked service that connects to your workspace's default ADLS account. Remember that this is where you saved the sample dataset.
    - On the **File path** page, provide the **File system**, **Directory**, and **File name** details for where the sample dataset is stored. For this example, we will use the `drone-telemetry-simplified.csv` file. Optionally, you can click the folder icon and browse to your file to let Azure Synapse Studio fill in the details for you.
    - Make sure you select **First row as header**.
    - The **Set properties** page should look similar to what is shown in *Figure 5.16*. When you are ready, select **OK**. Your source is now configured and ready for use.

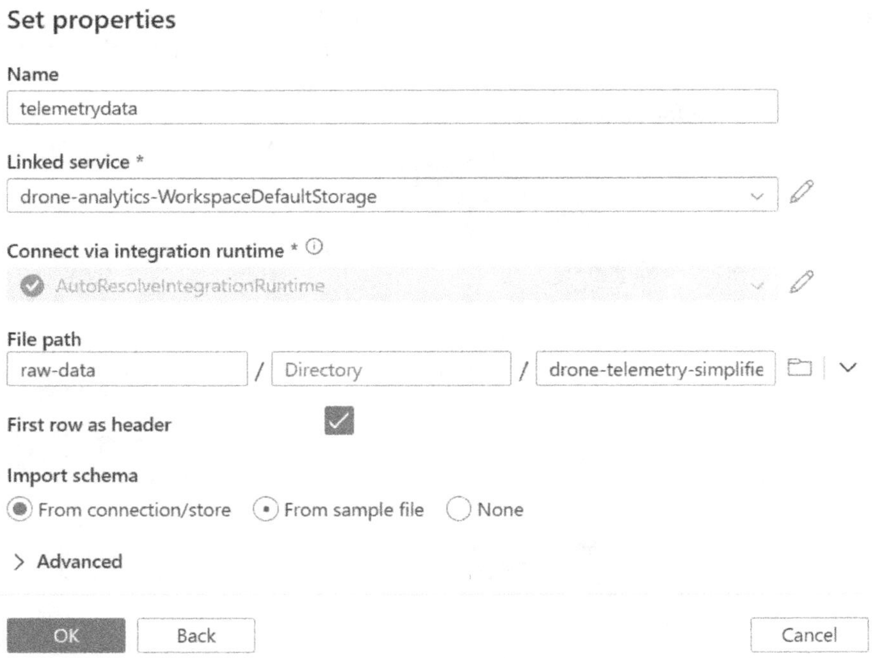

Set properties

Name

telemetrydata

Linked service *

drone-analytics-WorkspaceDefaultStorage

Connect via integration runtime *

AutoResolveIntegrationRuntime

File path

raw-data / Directory / drone-telemetry-simplifie

First row as header

Import schema

From connection/store From sample file None

> Advanced

OK Back Cancel

Figure 5.16 – Setting the properties of your source

In the data domain, a sink is a location that's capable of receiving data. Next, we will configure the `Copy telemetry data` activity to use the Data Explorer table we created previously as its sink.

Configuring the destination details

Select the **Sink** tab to provide the details for the destination store that we will copy data to. Next to the **Sink dataset** dropdown, select the + **New button**. This reveals the **New integration dataset** page. Provide the following details on this page:

1. On the search box, type `Data Explorer` and select **Azure Data Explorer (Kusto)**. Select **Continue**.
2. On the **Set properties** page, provide a name that will help recognize this sink in the future. For this example, I used the name `fleet_data_pipeline_table`. In the **Linked service** dropdown, select the **droneanalyticsadx** linked service that you created. Finally, select the **Edit** checkbox under the **Table** textbox, and type `['fleet data pipeline']`. It's important that you type the table name here and not select the table from the list because, at the time of writing, there is a bug in Azure Synapse Studio that causes issues with composite table names. Your **Set properties** page should look similar to the one shown in *Figure 5.17*. Select **OK** when you are done.

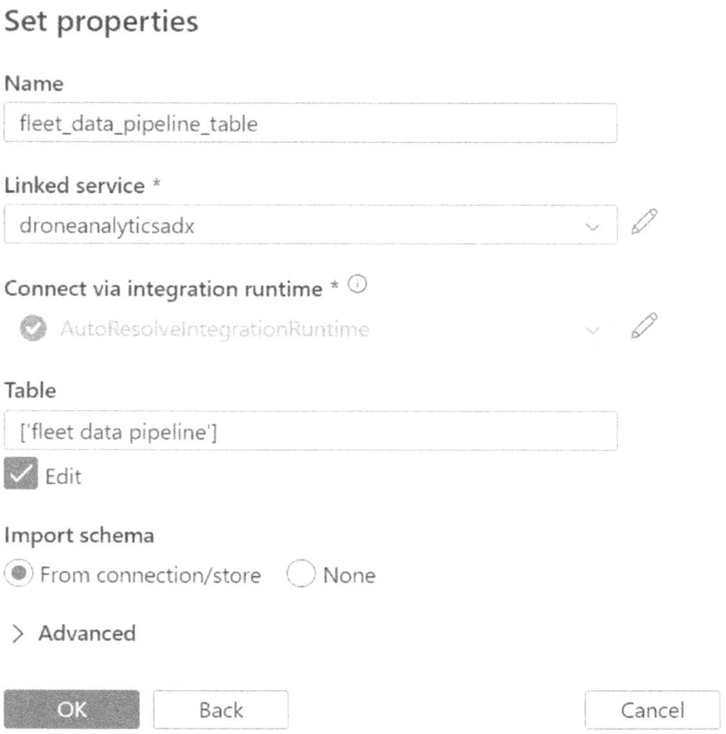

Figure 5.17 – Setting the properties of your sink

Now that both our source and sink are ready, we need to configure the **Copy data** activity to match the columns from the source file with the columns in the destination table. We will use the **Mapping** tab to perform the following steps:

1. Select the **Mapping** tab.
2. Select the **Import schemas** button. Azure Synapse will detect the schema from the source and the sink and attempt to create mappings between the source and destination columns automatically. Review the column mappings to make sure they are accurate.

We're done! Feel free to navigate through the other tabs and options of the `Copy telemetry data` activity. When you are ready, click the **Debug** button to test your pipeline. This task should take less than a minute to complete. When your task finishes, you will see results similar to *Figure 5.18*.

Figure 5.18 – The Copy data activity results

This completes the data ingestion process. Next, let's verify whether the activity worked as expected.

Verifying data ingestion

To verify that data was inserted into your table, go to the **Develop** hub, select the + button, pick **KQL script**, and type the following query into the query editor:

```
['fleet data pipeline']
| take 100
```

You should see the query results showing 100 rows of the data you just ingested.

Pipelines, even simple ones, are immensely helpful to ingest data periodically. Now that your pipeline is created, you can create a trigger that executes it on a certain recurrence. *Figure 5.19* illustrates an example where the pipeline would be triggered to run every hour.

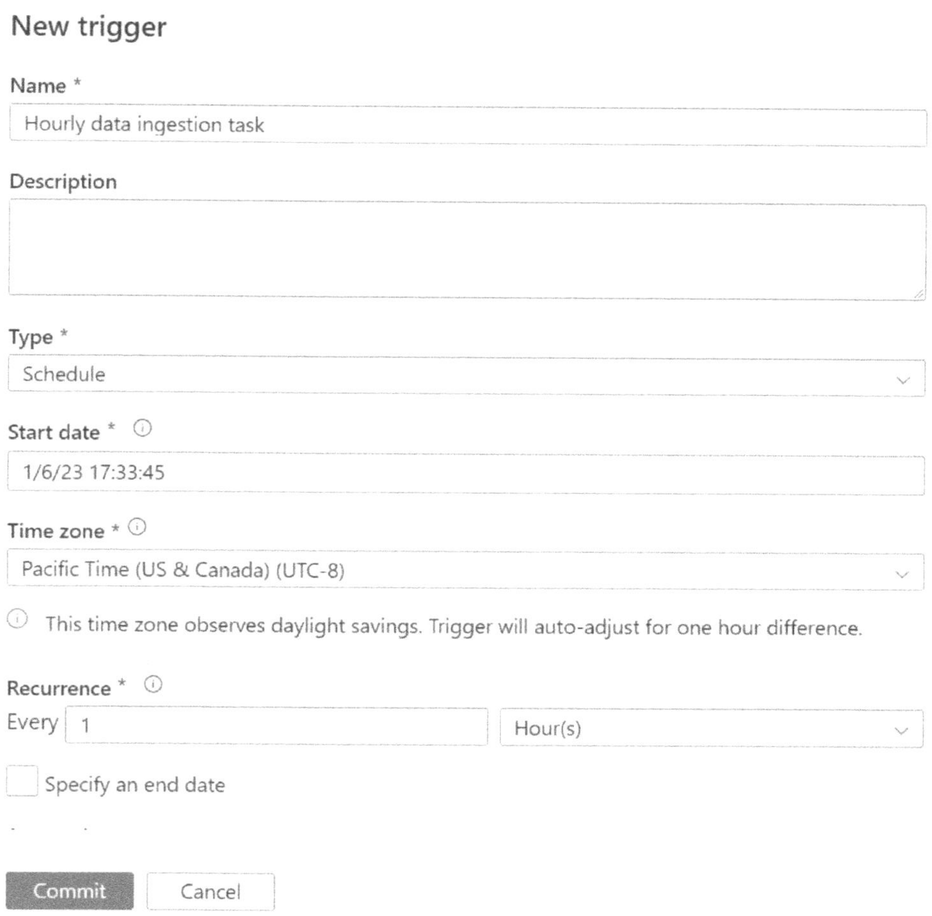

New trigger

Name *

Hourly data ingestion task

Description

Type *

Schedule

Start date *

1/6/23 17:33:45

Time zone *

Pacific Time (US & Canada) (UTC-8)

This time zone observes daylight savings. Trigger will auto-adjust for one hour difference.

Recurrence *

Every 1 Hour(s)

Specify an end date

Commit Cancel

Figure 5.19 – Configuring a new trigger

We just created a simple pipeline with just a **Copy data** activity that copies data from a CSV file into a table, sitting on a Data Explorer pool. It's worth mentioning here that, even though we used a CSV file as a source, it could have been any other data source, such as a transactional database, and the process would be almost the same.

In addition, this pipeline simply reads from a source and copies data to a destination. An alternative to this simple pipeline would be to create a more robust data flow that reads data from one or more sources, transforms data to accommodate the destination requirements, and ingests data into this destination. A whole book could be written about data flows in Azure Synapse. To learn more about data flows in Azure Synapse, visit `https://learn.microsoft.com/azure/synapse-analytics/concepts-data-flow-overview`.

Implementing continuous ingestion

In the previous example, we built a pipeline that reads one or more files stored in a blob storage location and ingests the content in these files into a destination table. This is useful for one-time operations, or to run as a batch operation on a schedule.

The next scenario also ingests data from files in a blob storage location, but instead of running as a batch job, it automatically detects any new files that are written into the container and processes them instantly as they are created in the container. This solution uses Azure Event Grid to subscribe to new file events in ADLS Gen2 and then continuously routes these events to Azure Event Hubs, which finally ingests the new data into your Data Explorer pool.

You do not need to know Azure Event Grid and Azure Event Hubs in detail to implement this ingestion method. In fact, as you will see next, the process to create a new data connection takes care of creating the required resources on Azure Event Grid and Azure Event Hubs. With that said, it's useful to know how Azure Event Grid and Azure Event Hubs work at a high level to fully understand the data flow, from its source until it gets ingested into your Data Explorer database.

> **Note**
> To learn more about Azure Event Grid and Azure Event Hubs, visit `https://learn.microsoft.com/azure/event-grid/overview` and `https://learn.microsoft.com/azure/event-hubs/event-hubs-about` respectively.

As usual, before we proceed with ingesting data, we will create the destination table. This example also requires the creation of a data mapping and, optionally, a new Azure storage account.

Creating the required resources

Before we get started, there are a few things that we need to do. First, we will create a new table to receive the data that will be ingested, as we did for the other examples. We will call this table `fleet data continuous`.

To create the table, execute the following script in a new KQL script window:

```
.create table ['fleet data continuous'] (
    DeviceData: string,
    LocalDateTime: string,
    DeviceState: string,
    CoreTemp: long,
    CoreStatus: string,
    Compass: long,
    Altitude: long,
```

```
    Speed: long,
    PayloadWeight: long,
    EventData: string
)
```

> **Note**
>
> The code used in this section can be found in the `Chapter 05\Ingestion-Continuous.kql` file in the book repository.

Next, we will create a data mapping that, as you may have guessed, maps the columns in the incoming data with the columns in the destination table. Data mappings are an important part of the ingestion process. Even though it is not technically required to create a data mapping in all cases, it is a good practice to create and use them in big data projects when you work with unstructured data, as it can help mitigate problems with schema changes:

```
.create table ['fleet data continuous'] ingestion csv mapping
'fleetDataContinuousMap'
'['
       '{"Column": "DeviceData", "Properties": {"Ordinal":
"0"}},'
        '{"Column": "LocalDateTime", "Properties": {"Ordinal":
"1"}},'
        '{"Column": "DeviceState", "Properties": {"Ordinal":
"2"}},'
        '{"Column": "CoreTemp", "Properties": {"Ordinal":
"3"}},'
        '{"Column": "CoreStatus", "Properties": {"Ordinal":
"4"}},'
        '{"Column": "Compass", "Properties": {"Ordinal":
"5"}},'
        '{"Column": "Altitude", "Properties": {"Ordinal":
"6"}},'
        '{"Column": "Speed", "Properties": {"Ordinal": "7"}},'
        '{"Column": "PayloadWeight", "Properties": {"Ordinal":
"8"}},'
        '{"Column": "EventData", "Properties": {"Ordinal":
"9"}},'
']'
```

In this data mapping, each line defines the mapping for a column that you are mapping between the source data and the target table. The `Column` value specifies the name of the column on the target table, while the `Properties` value specifies properties about the source data. In this case, the `Ordinal` property, which is specific to CSV file formats, specifies the zero-based column index in the source file. Depending on the file format (CSV, Parquet, Avro, or others that are supported), you may see different properties available, such as a constant value for the column, or the path to the column inside the file.

> **Note**
> To learn more about data mappings, visit `https://learn.microsoft.com/azure/data-explorer/kusto/management/mappings`.

Now that we have the new table and data mapping created, you need to specify a storage account and a container where you will drop the files for ingestion. You can use the primary workspace storage account for this purpose, or you can create a new one. For this example, since we will be ingesting any new files present in the storage account, I decided to create a separate storage account. You can create a new storage account by following the tutorial at `https://learn.microsoft.com/en-us/azure/storage/common/storage-account-create` using the default values.

I named my new storage account `droneanalyticssimplified` so that I know that I can only upload copies of the drone telemetry data with the simplified schema here. I also created a new container on this ADLS Gen2 account named `simplified-data`. We will later upload the files of the sliced simplified dataset here.

Performing data ingestion

We are finally ready to create the data connection, which will configure our data ingestion job. To implement continuous data ingestion using Azure Event Grid notifications, run the following steps:

1. Navigate to the Azure portal at `https://portal.azure.com` and find your Azure Synapse workspace. You can find it easily by typing its name in the global search box, or by visiting the Azure Synapse Analytics page. Once you've found your workspace, select it to see your workspace details.
2. In the left-hand menu, select **Data Explorer pools (preview)**. You will see all the Data Explorer pools you have created in your workspace. Select the one that contains the database where the destination table for data ingestion is.
3. On the details page of your Data Explorer pool, select **Databases** and then the database where you will ingest data.
4. You will land on the database's detail page inside the Azure portal. In the left-hand menu, select **Data connections**.

5. Select the **+Add data connection** button and pick the **Event Grid (Blog storage)** option, as shown in *Figure 5.20*.

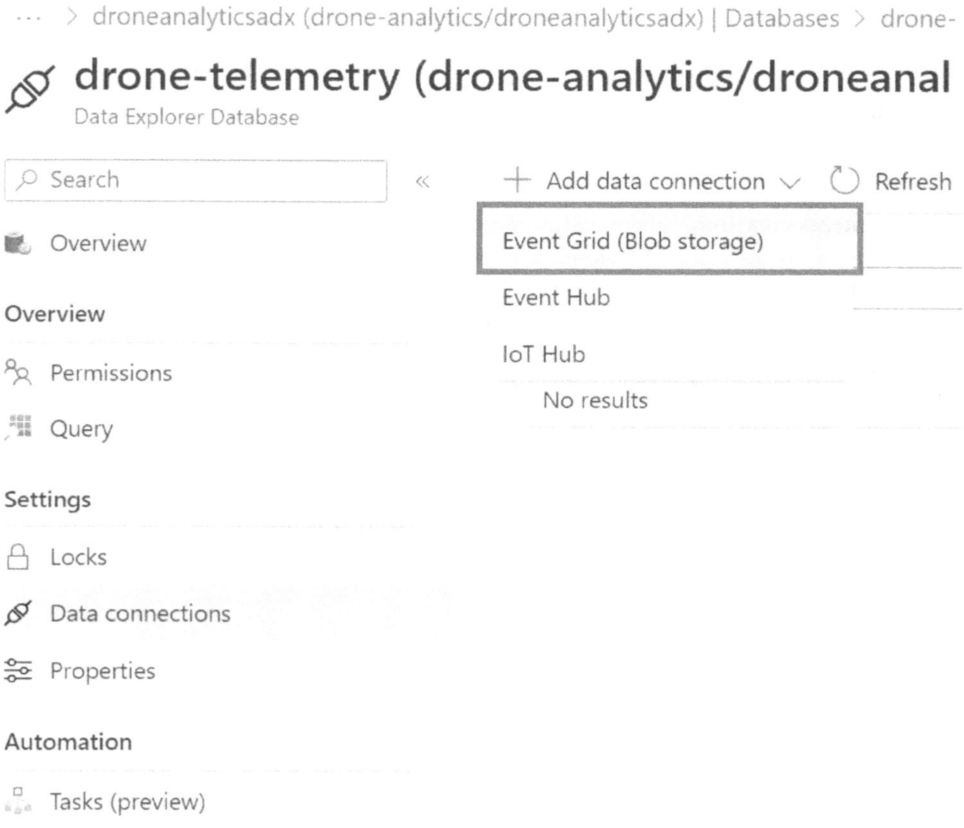

Figure 5.20 – Creating a data connection in the Azure portal

6. On the **Event Grid (Blob storage) Create data connection** page, fill in the required fields with the following details:

   - **Data connection name**: Use a familiar name to help you identify this data connection in the future. For this example, I used `new-files-ingestion`.
   - **Storage account subscription**: Select the Azure subscription where your storage account was created.

- **Storage account**: The storage account where we will find the source files for ingestion. This is the location where new files will be saved and processed for ingestion. As mentioned, I created a new storage account for this purpose and provided a name for this new storage account.
- **Event type**: You can create your event type to be triggered when new blobs are created or renamed. For this example, I selected **Blob created** so that we will process new files only.
- **Assign managed identity**: This is the managed identity that will be used to access your storage account. The recommendation is to use the system-assigned identity, which is the workspace identity, so that this event can be processed in the workspace without user interaction.
- **Resources creation**: This integration requires Azure Event Grid and Azure Event Hubs instances. You can choose **Automatic** to allow the wizard to create all the required resources for you, or you can provide the name of existing Azure Event Grid and Azure Event Hubs instances. For this example, I selected **Automatic**.
- **Filter settings**: Here, you can filter files that will be processed by your Event Grid notifications by providing a file prefix or suffix (such as `.csv` or `.json`). I will not use any filters in this example, but if you did not create a new storage account for this exercise, you should specify a prefix that maps to a folder in your container where the dataset slice files will be copied to.

Your **Basics** page should look like the one shown in *Figure 5.21*.

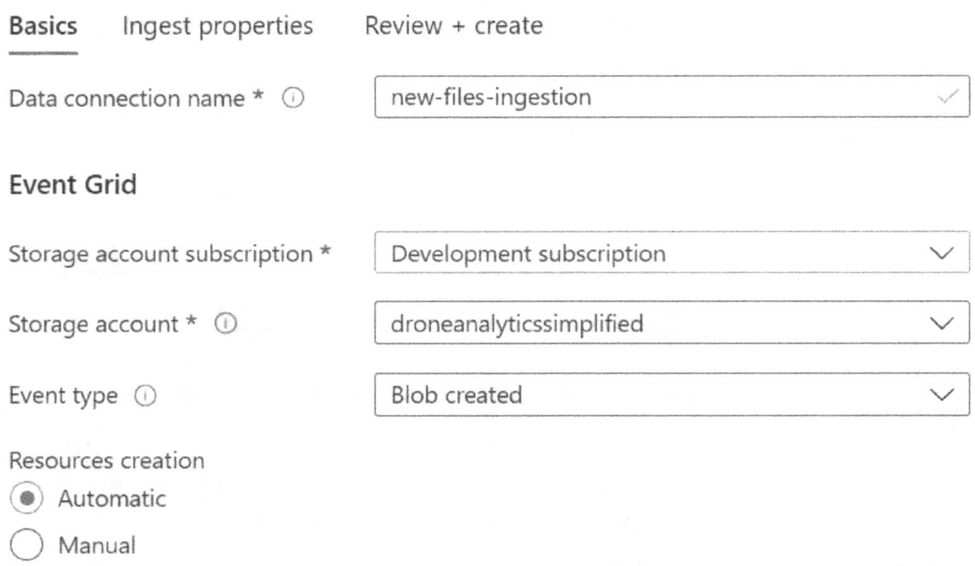

Figure 5.21 – The Basics page

7. When you are ready, select the **Next: Ingest properties** button. On the **Ingest properties** page, provide the following details:

   - **Table name**: The name of the table that will receive the data. In this example, I used `fleet data continuous`.
   - **Mapping name**: The name of the mapping we created for this table at the beginning of the process. This field is optional but recommended. I used the mapping we created previously, `fleetDataContinuousMap`.
   - **Data format**: Specify here the format of the source data. In our case, the format is **CSV**.
   - **My data has headers**: Make sure you select this option, as our files do have column headers.

   Your **Insert properties** page should look like the one shown in *Figure 5.22*.

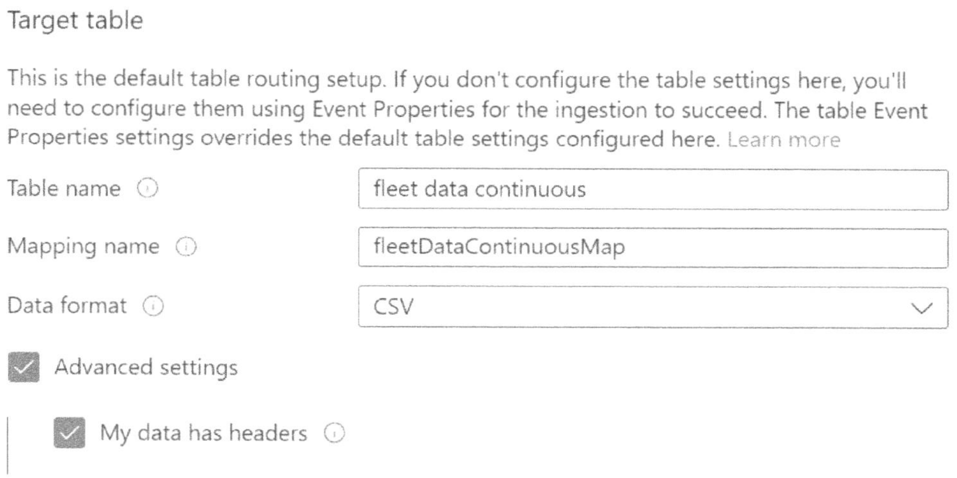

Target table

This is the default table routing setup. If you don't configure the table settings here, you'll need to configure them using Event Properties for the ingestion to succeed. The table Event Properties settings overrides the default table settings configured here. Learn more

Table name: fleet data continuous

Mapping name: fleetDataContinuousMap

Data format: CSV

Advanced settings

My data has headers

Figure 5.22 – The Ingest properties page

8. Select the **Next: Review + Create** button. On this page, you can review the details of the Event Grid subscription, Event Hub, and Event Hub namespace that will be created on your behalf. When you are ready, select **Create**.

Your data connection is now created, and any new files that you add to the ADLS Gen2 container will be ingested into your table within minutes. To test this, you can perform the following steps:

1. Go back to the home page of the Azure portal. Find the storage account that you configured with the data connection. This is the new storage account that you created at the beginning of this section and not the workspace's default storage account.
2. In the left-hand menu, select **Containers**. Then, select the `simplified-data` container that you created previously.

3. Select the **Upload** button. The **Upload blob** page will show up on the right side of the Azure portal. Select the folder button to browse your local files, and then select the `drone-telemetry-simplified - slice 1.csv` file, which you extracted during the *Technical requirements* section of this chapter. This is one of four files that contain one-fourth of the drone telemetry simplified dataset that we have been working with. It has 85,649 rows, excluding the first row. Click **Upload**.
4. After a few minutes, navigate to your Azure Synapse workspace. Select the **Develop** hub, click the + button, and pick **KQL script** to create a new KQL script window.
5. Type the following query to see whether your data was ingested:

```
['fleet data continuous']
| count
```

If your data was processed successfully, you should see a count of 84,375 rows, as shown in *Figure 5.23*.

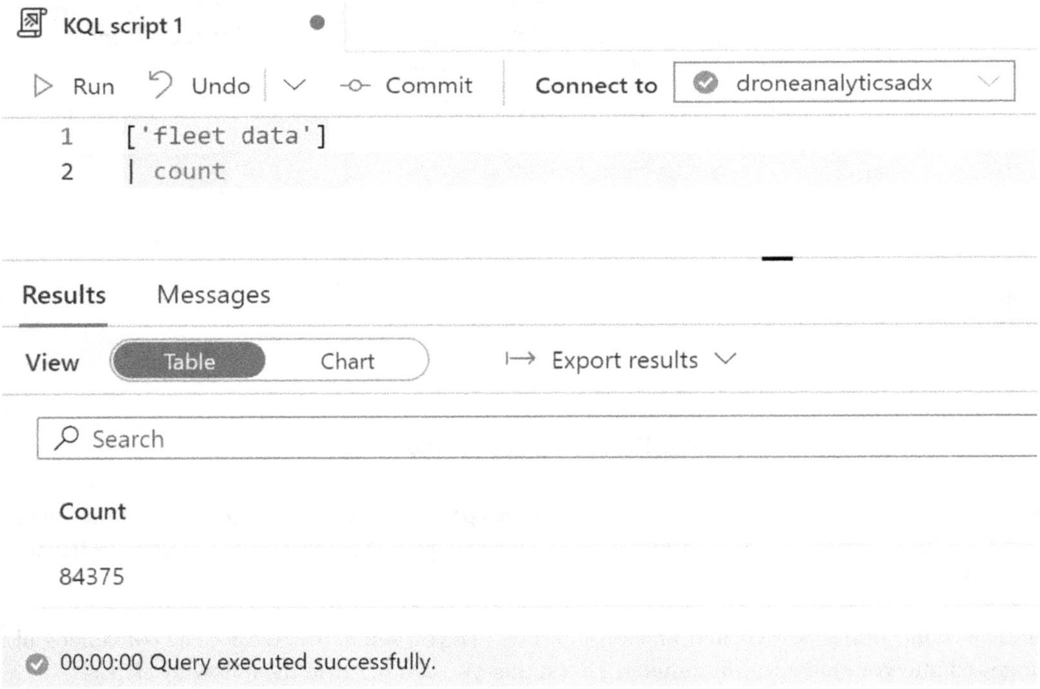

Figure 5.23 – Verifying whether data was ingested

If you would like to continue testing this approach, the `drone-telemetry-simplified - slice 2.csv` file through to the `drone-telemetry-simplified - slice 4.csv` file (which you extracted from the `Chapter 05\drone-telemetry-simplified - sliced.zip` file in the *Technical requirements* section) contain the remaining rows of the full simplified dataset. Feel free to upload them to the container and see whether data continues to get processed.

Continuous data ingestion from Event Grid notifications is a powerful scenario in data integration. New files can be added to your container by applications, devices, or individuals who curate data, and this data gets ingested into your tables automatically.

Using other data ingestion mechanisms

Before we close this chapter, let's discuss two additional approaches you can use to set up data ingestion into Azure Synapse Data Explorer: processing events in near-real time using Azure Event Hubs as a streaming platform, and building your own custom application using the Azure Data Explorer SDK.

Processing data events in real time

Azure Event Hubs is a highly scalable messaging streaming service that supports processing millions of messages per second. It integrates seamlessly with Azure data services, such as Azure Synapse Data Explorer. Some of the scenarios where Azure Event Hubs can be used to process streaming events at scale include application logging, transaction processing, and device telemetry.

Configuring data ingestion with Event Hubs requires setting up a data connection on your Azure Synapse Data Explorer Database, similar to what we did in the *Implementing continuous ingestion* section. The difference is that instead of listening to file notifications triggered from Event Grid, Event Hubs offers an interface to receive data from streaming sources and routes it to the destination table in your Data Explorer pool. Since the implementation of this scenario is quite similar to the one we just explored for continuous data ingestion, we will not discuss it in detail again. For a detailed step-by-step tutorial on configuring data ingestion from Event Hubs, refer to `https://learn.microsoft.com/azure/synapse-analytics/data-explorer/ingest-data/data-explorer-ingest-event-hub-portal`.

Building custom data ingestion applications

Finally, if you need to build your own logic for data ingestion or need to make the data ingestion process a part of your line of business applications, you can build your own solution by using the Data Explorer SDK. The SDK is available for .NET, Node.js, Go, and Java. In fact, the SDK allows you to not only ingest data but also query it and issue control commands to your Data Explorer pools. You can find the .NET SDK at `https://learn.microsoft.com/azure/data-explorer/net-sdk-ingest-data`. On this page, you can also find links to the SDK for the other languages available by navigating the menu.

In addition to the SDK on different languages, a Python library is available for the same purposes as the SDK. The Python library can be useful in any scenario where you would typically use Python, including Jupyter notebooks, Azure Synapse notebooks, or regular Python scripts. The library can easily be installed into your Python environment using your package manager (pip in the following example):

```
pip install azure-kusto-data
pip install azure-kusto-ingest
```

At the time of writing, this library required Python 3.4 or superior. For more details, including an example of how to use the Python library, visit `https://learn.microsoft.com/azure/data-explorer/python-ingest-data`.

Summary

Data ingestion is a broad topic, and there is no one way to approach this challenge. It all depends on your latency requirements, your data source types, how much control you want to have over the data ingestion process, and other factors.

First, you learned the steps in the data loading process, which we explored later in the chapter. You learned about retention policies, and how to think about the implications and benefits of keeping large volumes of data for long periods of time.

Next, you learned about the streaming and batching ingestion strategies, when to use which, and the implications of enabling streaming ingestion in your Data Explorer pool. You learned the conditions that cause batching ingestion to trigger, and how to set these conditions by using a batching policy.

Finally, you learned in detail how to implement data ingestion by using KQL control commands, an Azure Synapse pipeline, and continuously ingesting files as they are created in an ADLS container. You also learned that you can use Azure Event Hubs for streaming event processing at scale, or build your own data ingestion solutions using the Azure Data Explorer SDKs and Python library.

Data ingestion is a means to an end. In the next chapter, you will explore how to perform data analysis with Power BI, Python, and KQL.

6 Data Analysis and Exploration with KQL and Python

Now that you've learned how to ingest data into Data Explorer pools, let's look at ways to analyze this data to extract the insights you need to support a decision-making process. Part of the data analysis process is to explore your data, see its shape, and adjust it to make it more useful for you and other consumers of this data. There's no unique way to analyze data, so in this chapter, you will explore different means to achieve this task.

The chapter starts with an overview of data analysis using **Kusto Query Language** (**KQL**). You will look at examples to help retrieve data, summarize it, visualize it in simple charts, and make sense of the data by looking at its distribution. Before you move on from KQL, you will look at some quick examples to help you detect outliers in your data and use linear regression to fit the best line that represents your data.

Next, you will work with Python on Azure Synapse notebooks to create detailed web-based documents that include descriptive text, code blocks, and query results. You will create an Apache Spark pool, read data from Data Explorer pools, perform data transformation tasks, and create a lake database to offer your table to end users.

At a high level, here are the topics covered in this chapter:

- Analyzing data with KQL
- Exploring Data Explorer pool data with Python

As an analyst, data engineer, data scientist, or anyone who needs to work with data, it is helpful to get familiar with the data you are about to work with and find ways to make this data more useful to end users. We will start this chapter by making sure you are more familiar with the data you have at hand to explore and prepare it for visualizations using KQL, and next, you will take your data analysis skills further using Python.

Technical requirements

In case you haven't yet, make sure you download the book materials from our repository at `https://github.com/PacktPublishing/Learn-Azure-Synapse-Data-Explorer`. You can download the full repository by selecting **Code**, and then **Download ZIP**, or by cloning the repository by using your Git client of choice.

This chapter uses the full `fleet data` table, which you ingested in the *Running your first query* section of *Chapter 3*, *Exploring Azure Synapse Studio*. If you did not perform the tasks in this section to load data, take some time to go back and follow the instructions to ingest the full `fleet data` table.

The *Exploring Data Explorer pool data with Python* section uses basic Python syntax to load, transform, and visualize data. This book will not explain the examples in detail using the Python language. However, Python is known for its easy code readability, so if you have experience with any programming language, you should not have issues understanding the examples. A helpful tutorial on the pandas Python library, which we will use in this chapter, can be found at `https://www.w3schools.com/python/pandas/default.asp`.

Analyzing data with KQL

KQL queries are the primary tool for data exploration and analysis on Data Explorer pools. As with **Structured Query Language** (**SQL**), KQL uses a hierarchical structure to organize databases, tables, and columns and offers language elements to enable data retrieval. Unlike SQL, however, KQL supports read-only statements only, which makes sense since analytical data is meant for exploration and analysis, not for updates or deletions.

KQL gained popularity due to its support for pattern discovery, anomaly detection, statistical modeling, time series analysis, and other features. Several Azure services such as Application Insights, Log Analytics, and Azure Monitor (to name a few) offer support for the exploration of log data using KQL, which also helped increase the popularity of the language among Azure professionals.

Most KQL queries follow the pattern of tabular expression statements and have the following syntax:

```
Source
 | Operator A
 | Operator B
 | …
 | Operator N
 | RenderInstructions
```

Your queries may have no operators, or they may have as many operators as needed. Data flows from one operator to the next, with data being manipulated sequentially each time it passes an operator. For example: if `Operator A` is a filter operation (a `where` clause), then when data reaches `Operator`

B it will already be filtered, and so on. The order you choose for the operators is extremely important and can affect the result you get from your query, as well as your query performance.

To better understand how KQL works, let's look at some examples using our drone telemetry dataset. You can find all KQL queries used in the *Analyzing data with KQL* section in the `Chapter 06\KQL examples.kql` file of the book's repository.

Selecting data

Let's start with a very simple example—querying all the data in a table. To retrieve all rows and columns for a given table, simply type the name of the table and run the query:

```
['fleet data']
```

This statement produces the output seen in *Figure 6.1*. If you review the results, you will notice that our query completed with errors. This query fails because a Data Explorer pool limits the number of rows it returns to clients if it exceeds 500,000 rows or 64 MB of data:

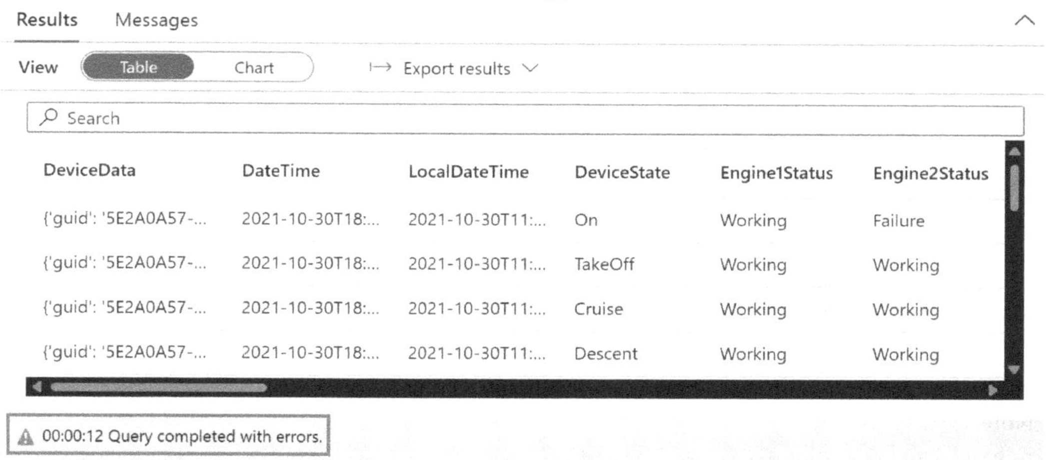

Results Messages

View Table Chart Export results

Search

DeviceData	DateTime	LocalDateTime	DeviceState	Engine1Status	Engine2Status
{'guid': '5E2A0A57-...	2021-10-30T18:...	2021-10-30T11:...	On	Working	Failure
{'guid': '5E2A0A57-...	2021-10-30T18:...	2021-10-30T11:...	TakeOff	Working	Working
{'guid': '5E2A0A57-...	2021-10-30T18:...	2021-10-30T11:...	Cruise	Working	Working
{'guid': '5E2A0A57-...	2021-10-30T18:...	2021-10-30T11:...	Descent	Working	Working

00:00:12 Query completed with errors.

Figure 6.1 – Query completed with errors due to its large data volume

> **Note**
> Since the name of our table is composed of two words separated by a space, the table name needs to be provided in between single quotation marks and square brackets. If your table does not have a space in its name, you can simply type the name of the table directly.

To overcome this limit, let's work our query a bit to reduce the amount of data sent to the client running the query. The `project` operator in KQL allows you to specify which columns you want to select from a table. As an example, the following query shows all rows for the `fleet data` table but only retrieves the columns listed after the `project` operator:

```
['fleet data']
| project DeviceData, LocalDateTime, DeviceState, CoreTemp,
BatteryTemp, Speed
```

Figure 6.2 shows the query results. However, note that it took 7 seconds to run this query:

Search

DeviceData	LocalDateTime	DeviceState	CoreTemp	BatteryTemp	Speed
{'guid': '5E2A0A57-...	2021-10-30T11:...	On	18	16	0
{'guid': '5E2A0A57-...	2021-10-30T11:...	TakeOff	2	5	15
{'guid': '5E2A0A57-...	2021-10-30T11:...	Cruise	-5	-6	90
{'guid': '5E2A0A57-...	2021-10-30T11:...	Descent	42	41	25

00:00:07 Query executed successfully.

Figure 6.2 – A simple KQL query

Selecting only those few columns vastly reduces the data amount sent to the client, so the query succeeds without errors.

> **Note**
> In SQL, when you want to execute only a block of code on a query editor that contains several queries, you highlight the code block you want to execute and then run the query. This ensures only the selected query is sent to the query processor. In KQL, to execute only one query on a query editor with several queries, all you have to do is put your cursor in any part of the query you want to execute and then run your query. Only the query where your cursor is at will be executed. You can still select a block of a query to run only the selected block.

This first example retrieved all rows for the columns specified in the `project` operator. When you are learning KQL, however, it's more useful to get a small set of rows instead of the full dataset. This prevents you from paying for the processing costs of large amounts of data and ensures you get faster query results. To limit the number of rows you get, you can use the `take` operator and specify how many rows you want to retrieve. For example, the following query retrieves the same columns as the first query but with only five rows:

```
['fleet data']
| take 5
| project DeviceData, LocalDateTime, DeviceState, CoreTemp,
BatteryTemp, Speed
```

Note that it took 7 seconds to run the previous query, while it took less than a second to run this one. You can verify this in *Figures 6.2* and *6.3*, in the footer area of the query editor:

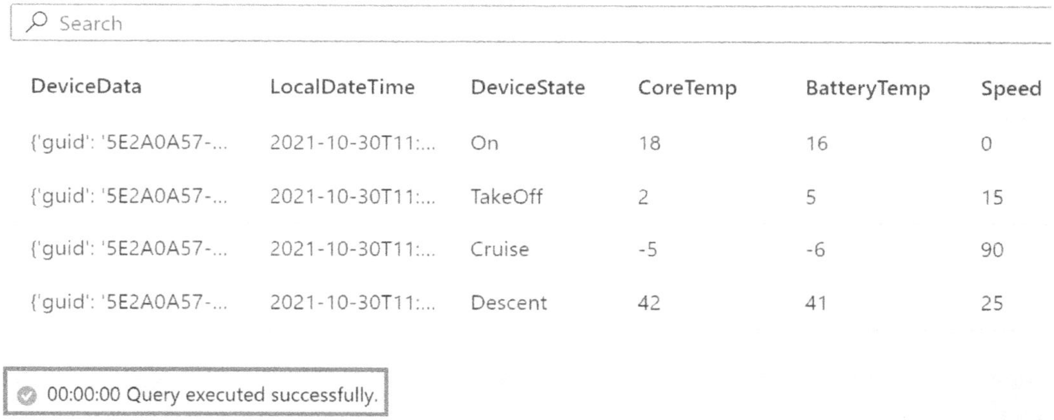

Search

DeviceData	LocalDateTime	DeviceState	CoreTemp	BatteryTemp	Speed
{'guid': '5E2A0A57-...	2021-10-30T11:...	On	18	16	0
{'guid': '5E2A0A57-...	2021-10-30T11:...	TakeOff	2	5	15
{'guid': '5E2A0A57-...	2021-10-30T11:...	Cruise	-5	-6	90
{'guid': '5E2A0A57-...	2021-10-30T11:...	Descent	42	41	25

00:00:00 Query executed successfully.

Figure 6.3 – Limiting the number of rows in your query

When you are writing queries in SQL, you would typically use the `limit` operator for this purpose (not in **Transact-SQL** (**T-SQL**), but in most SQL implementations). KQL also supports using the `limit` operator for the same function as `take`. They have the exact same behavior. The only reason the `limit` operator is available in KQL is to make it more familiar for SQL developers.

Keep in mind that using the `take` operator does not mean you are retrieving the first five rows in a table—it will retrieve the specified number of rows in no particular order. If you would like to retrieve the first few rows from a table, you should use the `top` operator, as in the following example:

```
['fleet data']
| top 5 by LocalDateTime asc
| project DeviceData, LocalDateTime, DeviceState, CoreTemp,
BatteryTemp, Speed
```

In *Figure 6.4*, you can see that only the first five rows in the table are retrieved:

DeviceData	LocalDateTime	DeviceState	CoreTemp	BatteryTemp	Speed
{'guid': 'D9FAFD7E-D3...	2021-10-30T10:...	On	19	18	0
{'guid': '40960F2F-47E...	2021-10-30T10:...	On	16	15	0
{'guid': 'F8F57534-9FF...	2021-10-30T10:...	On	22	18	0
{'guid': 'BCCDFFA3-E9...	2021-10-30T10:...	On	14	11	0
{'guid': 'D9FAFD7E-D3...	2021-10-30T10:...	TakeOff	24	20	15

00:00:00 Query executed successfully.

Figure 6.4 – Retrieving the five top rows, in ascending order

Next, let's look at filtering. Similar to the SQL syntax, to apply filters to your query, you can use the `where` operator. The following example selects only the rows where the `LocalDateTime` column has a value between `2021-10-30` and `2021-11-01` (inclusive):

```
['fleet data']
| where LocalDateTime >= datetime(2021-10-30) and LocalDateTime
<= datetime(2021-11-01)
| project DeviceData, LocalDateTime, DeviceState, CoreTemp,
BatteryTemp, Speed
```

This query output seen in *Figure 6.5* looks similar to the previous ones, but note that it retrieves all rows within the filter criteria, not only the top five:

DeviceData	LocalDateTime	DeviceState	CoreTemp	BatteryTemp	Speed
{'guid': '5E2A0A57-...	2021-10-30T11:...	On	18	16	0
{'guid': '5E2A0A57-...	2021-10-30T11:...	TakeOff	2	5	15
{'guid': '5E2A0A57-...	2021-10-30T11:...	Cruise	-5	-6	90
{'guid': '5E2A0A57-...	2021-10-30T11:...	Descent	42	41	25
{'guid': '5E2A0A57-...	2021-10-30T11:...	Delivery	42	44	0

00:00:01 Query executed successfully.

Figure 6.5 – Filtering with the where operator

As you would expect, you can sort the query results in ascending or descending order by using the `sort` operator. The following example sorts the previous query results by the `CoreTemp` column using descending order:

```
['fleet data']
| where LocalDateTime >= datetime(2021-10-30) and LocalDateTime
<= datetime(2021-11-01)
| sort by CoreTemp desc
| project DeviceData, LocalDateTime, DeviceState, CoreTemp,
BatteryTemp, Speed
```

Figure 6.6 shows the rows sorted in descending order by the `CoreTemp` column:

DeviceData	LocalDateTime	DeviceState	CoreTemp	BatteryTemp	Speed
{'guid': '7229CE93-...	2021-10-30T21:...	DeviceError	75.7899999999...	52	25
{'guid': '6FDB3D64-...	2021-10-31T08:...	DeviceError	72	49	75
{'guid': 'FCC3365F-...	2021-10-31T18:...	DeviceError	71.5500000000...	51	0
{'guid': '746C2717-...	2021-10-30T11:...	DeviceError	70.4900000000...	49	25
{'guid': 'D77FC565-...	2021-10-30T14:...	DeviceError	69.12	50	25

00:00:00 Query executed successfully.

Figure 6.6 – Sorted results

Here, we can see that the core temperature of our drones can get up to almost 76°C. That's a lot!

Working with calculated columns

Another useful feature of KQL is the `extend` operator. It allows you to create calculated columns and add them to your result set. For example, the following query retrieves the individual hour and minute values of the datetime stored in the `LocalDateTime` column using the `datetime_part` function and stores the results in the `Hour` and `Minute` calculated columns:

```
['fleet data']
| limit 5
| extend Hour = datetime_part("Hour", LocalDateTime), Minute =
datetime_part("Minute", LocalDateTime)
| project Hour, Minute, LocalDateTime
```

Figure 6.7 shows the calculated `Hour` and `Minute` columns, and the `LocalDateTime` column:

Hour	Minute	LocalDateTime
11	5	2021-10-30T11:05:23.044283Z
11	7	2021-10-30T11:07:24.244283Z
11	17	2021-10-30T11:17:51.421883Z
11	19	2021-10-30T11:19:52.621883Z
11	21	2021-10-30T11:21:26.701883Z

00:00:00 Query executed successfully.

Figure 6.7 – Using calculated columns

Creating aggregations is an easy task in KQL. The following example uses the `summarize` operator to aggregate the average drone speed in each device state reported to telemetry:

```
['fleet data']
| summarize average_speed = avg(Speed) by DeviceState
| order by average_speed desc
```

Figure 6.8 shows the results, where you can see that the average speed reported by devices when they report a `DeviceError` state is approximately 28 (km/h):

DeviceState	average_speed
Cruise	82.496119492402372
DeviceError	28.426588750913076
Descent	25
TakeOff	20.000544158458943
On	0

00:00:00 Query executed successfully.

Figure 6.8 – Average speed values by device state

You can also use the `summarize` operator to get all distinct values in a certain column:

```
['fleet data']
| summarize by DeviceState
| order by DeviceState asc
```

Figure 6.9 shows the results of this query, which displays the distinct values of the `DeviceState` column, in ascending alphabetical order:

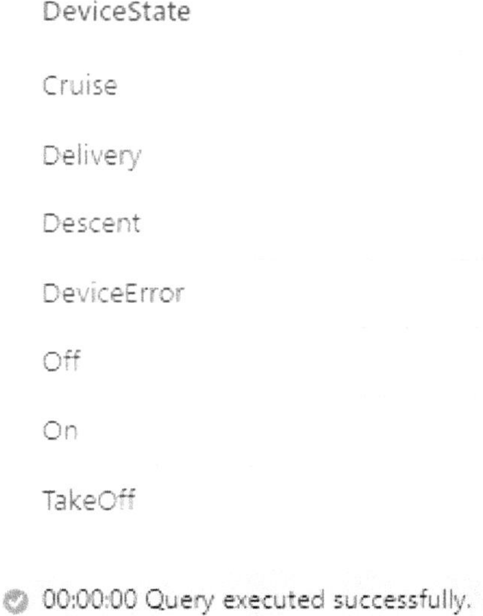

Figure 6.9 – Distinct values from the DeviceState column

As another example of the `summarize` operator, the following query shows us the days when drones performed the most deliveries by counting delivery events and aggregating the results by day:

```
['fleet data']
| where DeviceState == 'Delivery'
| summarize _count = count() by bin(LocalDateTime, 1d)
| order by _count desc
```

By the looks of *Figure 6.10*, it appears that November 10, 2021 was the busiest day so far, with a total of 1,902 delivery events reported in telemetry:

LocalDateTime	_count
2021-11-10T00:00:00Z	1902
2021-11-15T00:00:00Z	1897
2021-11-01T00:00:00Z	1893
2021-11-07T00:00:00Z	1893
2021-11-17T00:00:00Z	1891

00:00:00 Query executed successfully.

Figure 6.10 – Count of deliveries per date

As you can see from these examples, selecting the data you need with KQL is a breeze. For many more examples, I recommend the *KQL quick reference* article at `https://learn.microsoft.com/azure/data-explorer/kql-quick-reference`.

Plotting charts

Showing data in tabbed form looks nice, but a picture is worth a thousand words (that's why there are so many figures in this book!). KQL has a cool feature that allows you to render your query results as a chart right under the query editor. This is achieved with the `render` operator. As an example, the following query produces a time chart with the count of deliveries between 11/01/2021 and 11/15/2021:

```
['fleet data']
| where DeviceState == 'Delivery'
| where LocalDateTime >= datetime(2021-11-01T00:00:00.000000Z)
and LocalDateTime <= datetime(2021-11-15T23:59:59.999999Z)
| summarize _count = count() by bin(LocalDateTime, 1d)
| project LocalDateTime, _count
| order by _count desc
| render timechart
```

Figure 6.11 shows the chart plotted with the results. The comboboxes you see on the right-hand side of the figure allow you to adjust the chart options as needed:

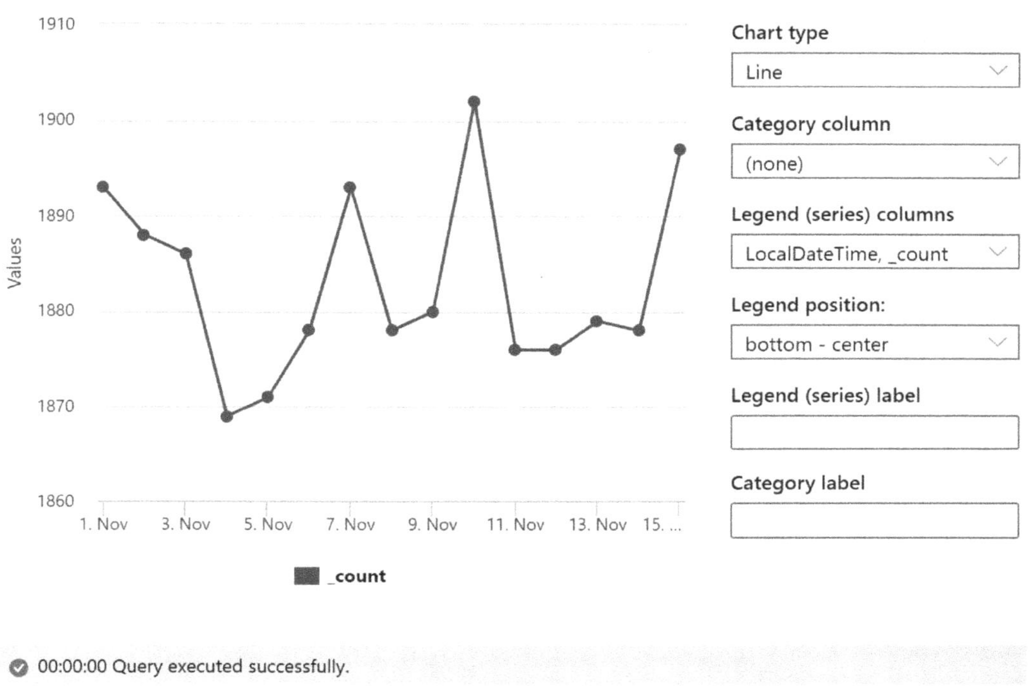

Figure 6.11 – Showing results as a chart

In *Chapter 3*, *Exploring Azure Synapse Studio*, you learned that any query executed in the query editor can be displayed as a chart by selecting the **Chart** toggle in the results pane. The `render` operator does this automatically for you, so you don't have to select the **Chart** toggle in this case to build your chart. In fact, if you run a KQL query in any supported query editor, it will still render charts for you, even if you are not using Azure Synapse Studio.

Charts become even more useful when you have several series to render and compare values. The following example renders a time chart with the average device core temperature by device state, per hour in the day:

```
['fleet data']
| where LocalDateTime >= datetime(2021-11-01T00:00:00.000000Z)
and LocalDateTime <= datetime(2021-11-01T23:59:59.999999Z)
| summarize _count = avg(CoreTemp) by bin(LocalDateTime, 1h),
DeviceState
| render timechart
```

In *Figure 6.12*, you may notice that there's a spike in device errors when its core temperature peaks or is under normal averages. With hundreds of thousands of events recorded, this can be useful information to troubleshoot problems with devices:

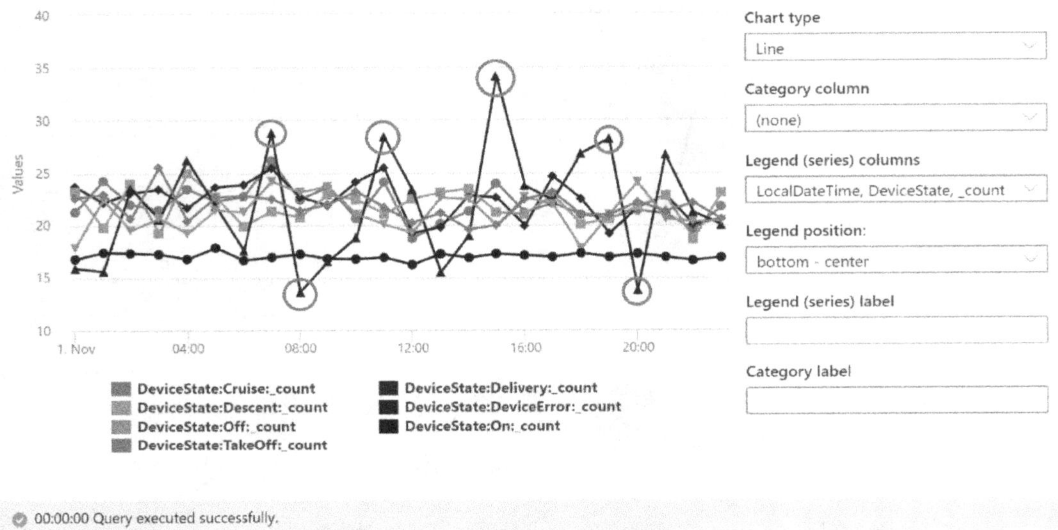

Figure 6.12 – A time chart with several different series (color version available at `https://packt.link/DQQ7A`)

You can also use other types of charts as appropriate. For example, if I wanted to see the overall average device temperature per device state, it may be more useful to visualize it as a column chart by using `render columnchart`:

```
['fleet data']
| summarize temp = avg(CoreTemp) by DeviceState
| project  DeviceState, temp
| order by temp desc
| render columnchart
```

As the chart in *Figure 6.13* shows, the average device temperature remains stable throughout all device states, except when it has the **On** state (which is when the device is turned on), and is slightly higher for `DeviceError` events:

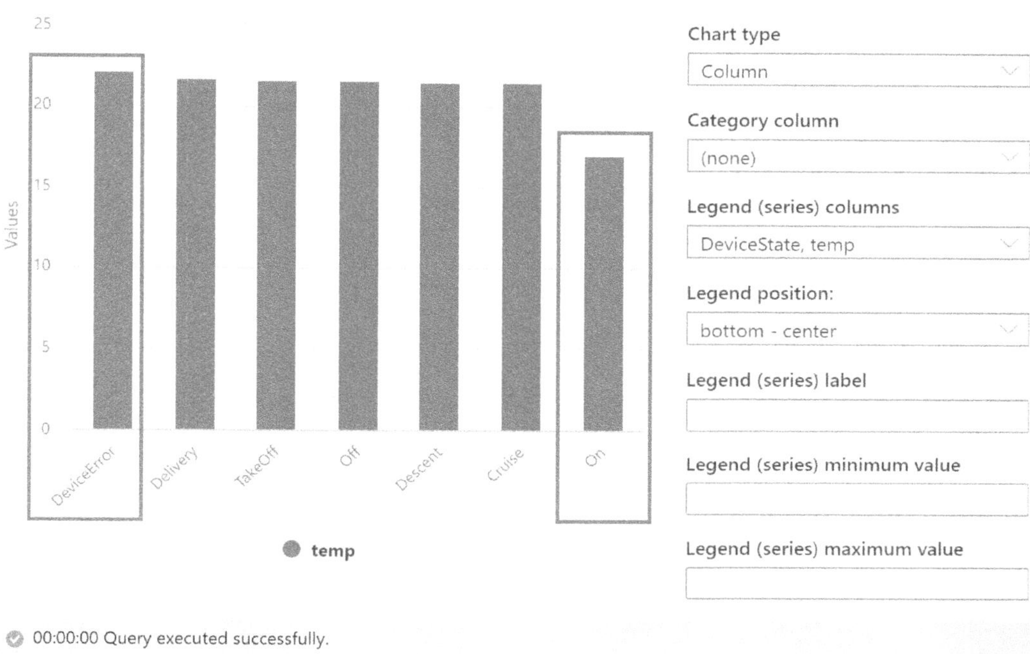

Figure 6.13 – Displaying the average device temperature as a column chart

Showing your data as charts is a great way to see the *shape* of your data without the use of any external visualization tools. It is an important aid in data preparation and exploration and helps you drive insights from your data with just a few queries.

Next, we will look at obtaining statistical measures for your data.

Obtaining percentiles

Understanding averages is helpful, but it doesn't tell the whole story. In our example dataset, more important than knowing the average temperature of drones is knowing if, most of the time, the temperature is behaving within a certain threshold. As an example, I may want to track an engineering metric where the temperature of drones should be optimal most of the time. Let's say, 98% of the times the temperature is reported, I expect it to be optimal.

To monitor metrics such as this one, let's look at obtaining percentiles for your data. In one of the products I worked on recently, we had performance goals to hit with regard to how fast certain components would show up to users in a web application. At first, we wanted the content to be available to users in 4 seconds 95% of the time, meaning our target load time was 4 seconds in the 95th percentile or, put simply, P95. After we reached this goal, we wanted the same 4-second goals, but at the 96th percentile, and so on, until we reached 4 seconds at the 98th percentile. In practice, this means users will load the page components in up to 4 seconds 98% of all times. On another product, we had a different engineering metric around device power consumption, and we set our goals at the 75th percentile because it was a much harder goal to pursue. Being able to quickly obtain and read percentiles is extremely useful when you are trying to achieve goals different from average values, maximum values, or minimum values.

Obtaining percentiles in KQL is an easy task. As an example, the following code block obtains the 25th, 50th, 75th, 90th, and 95th percentiles for values on the `CoreTemp` column:

```
['fleet data']
| summarize percentiles(CoreTemp, 25, 50, 75, 90, 95)
```

What you see in the results in *Figure 6.14* is that in 25% of all measurements, the temperature is under 7.18°C; in 50% of cases (that's the median!), it is under 19.16 °C; in 75% of cases, it's under 36°C; and so on:

percentile_CoreTemp_25	percentile_CoreTemp_50	percentile_CoreTemp_75	percentile_CoreTemp_90	percentile_CoreTemp_95
7.185638438607941	19.160154969012424	36	47	50.219024580140463

00:00:00 Query executed successfully.

Figure 6.14 – Using the percentiles function

Working with percentiles is a great way to understand how an event fits within the rest of the population in your data.

Creating a time series

Another powerful feature of KQL and Data Explorer pools is their native support for time series analysis or, put simply, the ability to organize data points in time order, typically with equally spaced points. Using time series, KQL aggregates your data across time to enable an easy data analysis through a time scale.

As an example, the following query groups the average temperature per device state, for each date provided in the `LocalDateTime` column. It also returns the `datetime` column grouped with corresponding dates for each value in the aggregated column—in this case, `AvgTemp`:

```
['fleet data']
| make-series AvgTemp=avg(CoreTemp) default=0 on LocalDateTime
from datetime(2021-11-01) to datetime(2021-11-04) step 1d by
DeviceState
```

In the results seen in *Figure 6.15*, the first value in the comma-separated values seen on the `AvgTemp` series corresponds to the average `CoreTemp` value on the day of the first value in the `LocalDateTime` series; the second value in the `AvgTemp` series corresponds to the second date in the `LocalDateTime` series; and so on. What KQL did for you here was to aggregate all the averages per day (because that's the time step that we've chosen) per `DeviceState` and put it in a simple table for you to use it:

DeviceState	AvgTemp	LocalDateTime
On	["16.998291777188328","16.9308...	["2021-11-01T00:00:00.0000000Z","2021-11-02T00:00:00.0000000Z","2021-11-03T00:00:00.0000000Z"]
TakeOff	["21.738893280632407","21.4870...	["2021-11-01T00:00:00.0000000Z","2021-11-02T00:00:00.0000000Z","2021-11-03T00:00:00.0000000Z"]
Cruise	["21.922744370860933","21.2301...	["2021-11-01T00:00:00.0000000Z","2021-11-02T00:00:00.0000000Z","2021-11-03T00:00:00.0000000Z"]
Descent	["21.790105596620908","20.9754...	["2021-11-01T00:00:00.0000000Z","2021-11-02T00:00:00.0000000Z","2021-11-03T00:00:00.0000000Z"]
Delivery	["22.323481246698357","21.3255...	["2021-11-01T00:00:00.0000000Z","2021-11-02T00:00:00.0000000Z","2021-11-03T00:00:00.0000000Z"]
Off	["21.157324840764332","21.3324...	["2021-11-01T00:00:00.0000000Z","2021-11-02T00:00:00.0000000Z","2021-11-03T00:00:00.0000000Z"]
DeviceError	["21.772012383900929","21.0736...	["2021-11-01T00:00:00.0000000Z","2021-11-02T00:00:00.0000000Z","2021-11-03T00:00:00.0000000Z"]

00:00:00 Query executed successfully.

Figure 6.15 – Generating a time series

To better understand the values produced by this query, let's look at how each column is computed:

- `DeviceState`: This was provided in the query as the expression that contains the unique values used for aggregation on distinct values. All other columns will be aggregated and grouped by this column. Therefore, we see one row for each `DeviceState` value in the dataset.
- `AvgTemp`: This column was computed based on the average `CoreTemp` value for each day in that `DeviceState`. Note that the data in this column contains values between quotation marks, separated by a comma. In *Figure 6.15*, we can observe that there are three values for each `DeviceState`—one for each day in the `LocalDateTime` column.
- `LocalDateTime`: This column contains the actual dates used to order the time series. This parameter is typically called the **axis** column. It is followed by the `step` operator, which tells the time series the time interval between each occurrence of the time series. In our case, it is 1 day. Just as with the rows for `AvgTemp`, values are provided in between quotation marks and separated by a comma.

Once we've created the time series, we can start to get more insight from our data. For example, it's easy to see statistics for values in your aggregated column such as the minimum and maximum values present in the dataset, the average, the standard deviation, and the variance using the `series_stats` function. Here's an example:

```
['fleet data']
| make-series AvgTemp=avg(CoreTemp) default=0
on LocalDateTime from datetime(2021-11-01) to
datetime(2021-11-04) step 1d by DeviceState
```

```
| extend (MIN, min_idx, MAX, max_idx, AVG, STDEV, VAR) =
series_stats(AvgTemp)
| project MIN, MAX, AVG, STDEV, VAR
```

Figure 6.16 shows the minimum (`MIN`), maximum (`MAX`), average (`AVG`), standard deviation (`STDEV`), and variance (`VAR`) for each value in the `DeviceState` column:

MIN	MAX	AVG	STDEV	VAR
16.930866596638658	17.065885774722371	16.998348049516451	0.067509606631513022	0.0045575469875416275
21.487013227513224	21.874845580404688	21.700250696183435	0.19678267323972426	0.038723420487372096
21.230168954593449	21.922744370860933	21.572021640399683	0.34637289887320744	0.11997418507382918
20.9754511775602	21.947032908704887	21.570863227628664	0.52157758516311381	0.27204317734458527
21.325545550847462	22.323481246698357	21.911867186315202	0.52140288247888111	0.27186096585728592
21.157324840764332	22.051756900212315	21.513856626714585	0.47399691291026064	0.22467307344845722
21.07368589743589	22.130428134556581	21.658708805297803	0.53740516802665461	0.28880431462175693

00:00:00 Query executed successfully.

Figure 6.16 – Obtaining series stats

In this case, we are generating the series first with the `make-series` operator, and then getting the statistics written to individual variables using the `extend` operator.

Detecting outliers

Taking a step further, your time series can help you make an advanced analysis of your data. Azure Synapse Data Explorer can detect outliers in your data by using statistical tests that provide a value indicating how significantly different that data point is from the rest of the sample. The `series_outliers` function implements a statistical test named **Tukey's test** to determine if a data point is an outlier or not. If the value returned by the function for a given data point is greater than 1.5, then the data point indicates a rise or decline anomaly. If the value is lower than -1.5, then the data point indicates a decline anomaly. Data points with a result within those ranges (greater than 1.5 or lower than -1.5) are considered outliers.

> **Note**
>
> To learn more about Tukey's test, visit `https://en.wikipedia.org/wiki/Tukey%27s_range_test`.

Let's look at an example. Say we want to see if one of the engines on our drones is behaving as expected by looking at its **rotations per minute** (**RPM**). We can use the following query:

```
['fleet data']
| make-series AvgRPM_Series=avg(Engine2RPM) default=0
on LocalDateTime from datetime(2021-11-01) to
datetime(2021-11-19) step 1d by DeviceState
| extend anomaly_score = series_outliers(AvgRPM_Series)
| mv-expand anomaly_score to typeof(double), AvgRPM = AvgRPM_
Series to typeof(double), LocalDateTime to typeof(datetime)
| where anomaly_score > 1.5 or anomaly_score < -1.5
| project LocalDateTime, DeviceState, AvgRPM, anomaly_score
| order by LocalDateTime asc
```

This query performs the following steps:

- Generates a time series on the average RPM of engine number 2, per day, by device state.
- Computes the outliers in this time series by using the `series_outliers` function. We will call this result the anomaly score.
- Expands each value in the time series (the ones separated by a comma) into new rows by using the `mv_expand` operator.
- Filters the results to show only rows with an anomaly score greater than 1.5 or lower than -1.5, as we only want to see anomalies.
- Outputs the results ordered by the `LocalDateTime` column.

What you can observe from the results in *Figure 6.17* is that on three dates, the average RPM of one of the drone engines was either too high or too low for that date and device state. From here on, an analyst would do further investigation to determine the cause of these anomalies on these days:

LocalDateTime	DeviceState	AvgRPM	anomaly_score
2021-11-04T00:00:00Z	Cruise	303.96757770632371	1.7433768257081526
2021-11-04T00:00:00Z	Off	291.7843450479233	-2.0999580117773329
2021-11-18T00:00:00Z	Descent	304.13079916192754	1.846634291436613

00:00:00 Query executed successfully.

Figure 6.17 – Performing anomaly detection

The fact that you can perform all this analysis using KQL is a very powerful trait of Data Explorer pools. You don't need to use external software or libraries to get this level of insight.

Using linear regression

As a final example, we'll look at how a device's temperature trends during a workday. The easiest way to achieve this is by using linear regression to find the line of best fit for our data. All we must do is generate our time series and use the `series_fit_line` function to compute the linear regression with the values from our time series. Here's the code to generate that:

```
['fleet data']
| make-series AvgTemp_Series=avg(CoreTemp) default=0
on LocalDateTime from datetime(2021-11-01T09:00:00.000000Z) to
datetime(2021-11-01T23:59:59.999999Z) step 1h
| extend (RSquare, Slope, Variance, RVariance, Interception,
LineFit)=series_fit_line(AvgTemp_Series)
| render timechart
```

In the chart seen in *Figure 6.18*, the oscillating line is the one with the data points from the core temperature of the drone devices on a given day. The straight line is the line we obtained by running the `series_fit_line` function. This function returns six parameters:

- **R-square (R2)**: Helps you understand how well the computed line fits your data. It is a value between 1 and 0, where 1 means the best possible fit, and 0 means the data doesn't fit any line.
- **Slope**: Linear regression is computed using the formula `y = a + bx`, where `a` is the intercept of the line and `b` is the slope. This parameter receives the values for `b` in this formula.
- **Variance**: Statistical variance of your data points.
- **Residual variance**: The difference between your data points and the expected results from the computed model.
- **Intercept**: The point of interception on the *y* axis where the slope of the line passes. This is a on `y = a + bx`.
- **Line_fit**: The values computed from linear regression for each interval in the time series. This is the value of the *y* axis of the chart that was produced:

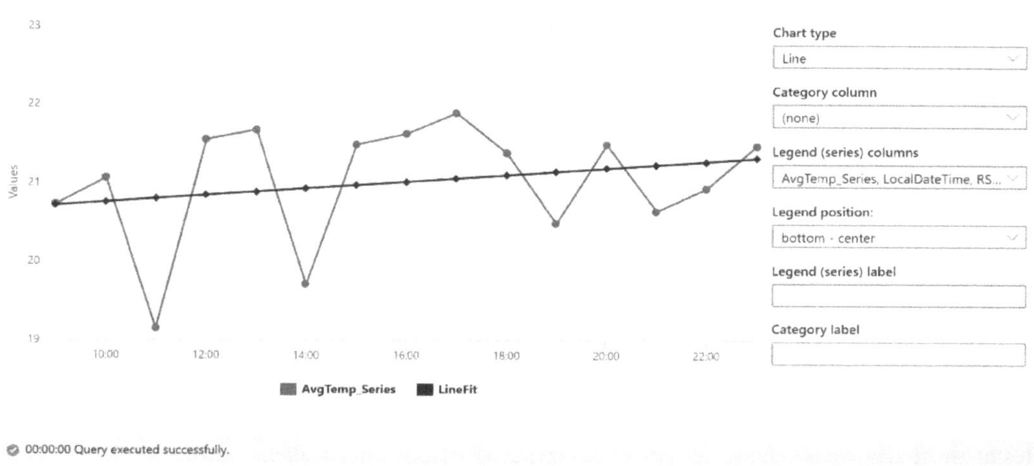

Figure 6.18 – Obtaining the line of best fit

The chart tells us that there is a rising trend in temperature throughout this particular day, raising less than 1°C—probably not a reason for concern.

When it comes to all the features and the possibilities you have in your hands when working with KQL, these examples only barely scratch the surface. Hopefully, they help you get a grasp on the language syntax and how to start writing your own queries. Make sure you check the KQL documentation at `https://learn.microsoft.com/azure/data-explorer/kusto/query/` to learn more about this powerful language.

Next, we will look at exploring data with Python using Azure Synapse notebooks.

Exploring Data Explorer pool data with Python

In the last few years, Python has gained significant popularity in the data science community. This happened for several factors, including the simplicity of the language syntax, broad community adoption, and the availability of a vast range of mathematics and statistics libraries, among other reasons. In conjunction with the R, Scala, and SQL languages, Python has become one of the primary programming languages used in data analysis, and for **artificial intelligence** (**AI**) and **machine learning** (**ML**) applications—a space that was dominated by MATLAB, especially in academia, for decades.

Azure Synapse offers support for data exploration using Python, Scala, C#, SQL, and R. Data exploration with these languages is handled by the Apache Spark analytical engine in Azure Synapse. For this chapter, we will focus on data exploration with Python.

Support for Python in Azure Synapse is achieved through PySpark: a Python **application programming interface** (**API**) for Apache Spark. You can perform data exploration, transformation, and data analysis using PySpark through Azure Synapse notebooks or by submitting jobs to Apache Spark directly using Apache Livy, the Apache Spark REST API.

> **Note**
>
> To learn more about Apache Spark in Azure Synapse, visit `https://learn.microsoft.com/azure/synapse-analytics/spark/apache-spark-overview`.

You may be wondering: why do I need PySpark, and why can't I simply use Python? The reason behind this is that Apache Spark, being an in-memory engine that runs jobs across a cluster of servers with support for parallelism and fault tolerance, needs to offer program support to scale the execution of data processing tasks across any number of nodes in an Apache Spark cluster. The programming language needs to have an awareness that data processing will happen in parallel across several machines.

PySpark enables the running of Python applications on the Apache Spark cluster, leveraging its parallelism characteristics. This is achieved mostly by implementing **resilient distributed datasets** (**RDDs**), which make data processing faster and more efficient by partitioning datasets across the nodes in an Apache Spark cluster and processing these partitions in parallel.

The easiest way to explore data with PySpark on Azure Synapse is to use Azure Synapse notebooks. Next, we will look at some examples of data exploration. To do that, however, we will need to create an Apache Spark pool in our Azure Synapse workspace.

Creating an Apache Spark pool

Creating a new Apache Spark pool is very similar to creating a Data Explorer pool. We will create our pool using Azure Synapse Studio, but this task can also be done through the Azure portal. Let's look at the steps to create an Apache Spark pool:

1. Go to your Azure Synapse workspace and select **Manage** in the **Hubs** region.
2. Under **Analytics pools**, select **Apache Spark pools**.
3. Select the + **New** button. A **New Apache Spark pool** page appears, focused on the **Basics** tab. Let's look at the parameters of this page in detail:
   - **Apache Spark pool name**: As you may have guessed, this is the name of your Apache Spark pool. Use a suggestive name that will help you identify it in the future. As an example, I used the name `sparkdronetl`, but you can choose your own.
   - **Isolated compute**: This option offers you a dedicated physical compute resource that is not shared with other users in Azure. It should be used if you need a higher degree of isolation for your data due to security requirements. For this example, I selected the **Disabled** option.
   - **Node size family**: Here, you select the family of node sizes depending on your workload demands. The **Memory Optimized** family offers you five node size options for general usage, while the **Hardware Accelerated** family offers you nodes with support for hardware acceleration that includes **graphics processing units** (**GPUs**). You should use the **Hardware Accelerated** option if your workload demands high-performance computing for complex

mathematical operations that can benefit from GPUs, such as training ML models. For our example, I picked the **Memory Optimized** option.

- **Node size**: This is the size of each compute node in your cluster. The larger your compute node, the faster your data processing will be, but also the more you will pay. The **Small (5 vCores / 32 GB)** node size is more than enough for what we will need in the examples in this book, so I chose this option.
- **Autoscale**: This allows your Apache Spark pool to automatically scale if it needs more resources, just like we've seen with Data Explorer pools. I selected the **Disabled** option for my Apache Spark pool.
- **Number of nodes**: This parameter specifies how many compute nodes you need in your Apache Spark pool. You can pick any number between 3 and 200. As you probably can tell by now, I'm running on a budget, so I selected only three nodes for my example.
- **Dynamically allocate executors**: This option allows you to set a minimum and maximum number of executors that the Apache Spark pool will use to run the different stages of a Spark job. This is useful when your data volume fluctuates over time. For this example, I chose the **Disabled** option.

4. When you are ready, select the **Next: Additional settings** > button. This takes you to the **Additional settings** tab, which exposes the following parameters:

- **Automatic pausing**: Select this option if you want your Apache Spark pool to pause after being idle for a period. Apache Spark pools can take several minutes to start, so you should disable this if you want your Apache Spark pool to be ready for use at any given time after it has started. Remember, however, that you pay for the time your Apache Spark pool is running, so this is an important setting to consider. For our example, I enabled this option.
- **Number of minutes idle**: If you enabled **Automatic pausing**, this parameter specifies how many minutes you want to wait before your pool pauses automatically. The default value is 15 minutes, but I changed this parameter to 10 minutes in my Apache Spark pool.
- **Apache Spark**: Here, you select the version of Apache Spark that will be used by your cluster nodes. At the time of this writing, the supported versions were 3.2, 3.1, and 2.4. You should use older versions if your scripts are not compatible with the newer versions of Apache Spark; otherwise, you should favor the more recent versions. Spark 3.2 introduces support for a pandas API layer on PySpark, which is something we will use in the examples in this chapter. I chose the 3.2 version for my Apache Spark pool.
- **Apache Spark configuration**: Apache Spark configurations allow you to specify properties that will be used by your pool. You can create configurations by visiting the **Apache Spark configurations** page in the **Manage** hub, under **Configurations + libraries**. Our Apache Spark pool will use the default configurations, so I did not change this for my example pool.

- **Allow session level packages**: Enabling this option allows end users to install libraries during their Apache Spark sessions, either through an Azure Synapse notebook or through a Spark job. I did not need to install any libraries beyond the ones already available by default, so I choose the **Disabled** option for this parameter.
- **Intelligent cache size**: In this parameter, you specify how much disk space you want to reserve for cache usage in your pool. Increasing the space used for the cache can significantly improve performance on your query, as the pool pre-caches the data that it needs to read from your Data Explorer pool, or from **Azure Data Lake Storage Gen2** (**ADLS Gen2**) (whichever is hosting your data). You should reduce the cache amount if your workload will require a lot of space in the local disk for RDDs. For my example pool, I left this parameter at the default of 50%.

5. When you are ready, select **Next: Tags** >. This takes you to the **Tags** tab. You learned about tags in the *Creating an Azure Synapse workspace* section of *Chapter 2, Creating Your First Data Explorer Pool*, so there's no need to discuss it again here. Feel free to add any tags you may need, and then select **Next: Review + Create** > when you are ready.
6. By now, you have seen the **Review + create** page a few times. This is your chance to review your settings before you proceed with creating your new Apache Spark pool. When you are ready, select **Create**.

It takes a few minutes to completely deploy your new Apache Spark pool. Once it completes, you will be ready to start a new Azure Synapse notebook and start exploring data using Python. Next, we will see how to do exactly that.

Working with Azure Synapse notebooks

Azure Synapse notebooks are a useful tool for data exploration, learning, presentation, and getting insights from your data. Through the Azure Synapse Studio web interface, you can create a document that mixes descriptive text with live code and execution results. Notebooks are widely used in the data science community for data exploration, learning, storytelling, and other scenarios. If you have tried a notebook experience for data exploration before, such as Jupyter notebooks, then you will feel at home with Azure Synapse notebooks.

To create a new notebook, navigate to the **Develop** hub in Azure Synapse Studio, select the ellipsis next to **Notebooks**, and select **New notebook**, as illustrated in *Figure 6.19*:

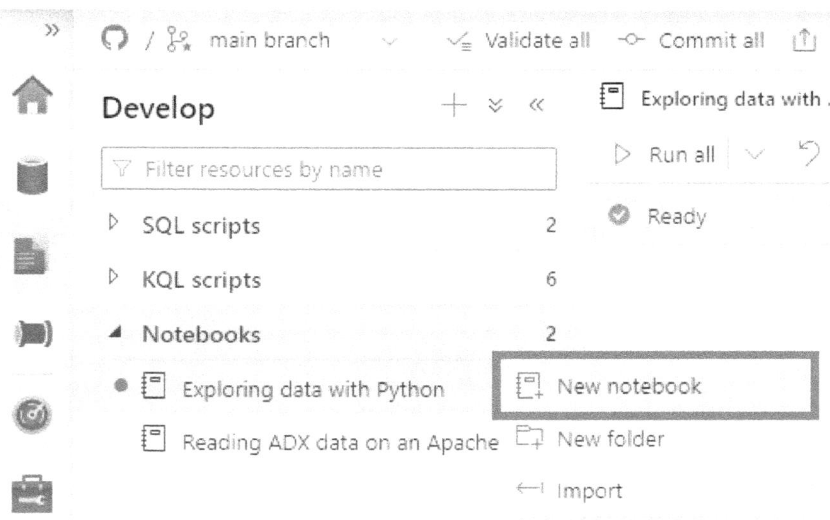

Figure 6.19 – Creating a new notebook

You will see your new notebook in the working pane of Azure Synapse Studio, with the name `Notebook 1`. Feel free to change this to a name that will help you identify this notebook later.

> **Note**
> The full notebook that will be used in this section can be found in the `Chapter 06\Exploring data with Python.ipynb` file of the book repository. It contains all the source code utilized in the examples that follow. You can import this notebook to your workspace by using the **Import** option shown in *Figure 6.19*, or you can start a new notebook and use the code provided here.

As mentioned before, notebooks allow you to mix descriptive text with live code. Descriptive text is added using the Markdown language—a markup language used for text formatting that is very easy to use. Don't worry—you don't need to know Markdown to get your work done here. Azure Synapse Studio provides formatting tools that will help you format your text and automatically generate the Markdown code required. Notebook cells that are used for descriptive text are called Markdown cells.

Code cells, on the other hand, are the place to write your data exploration code. Code cells will use the language configured in the notebook by default, but you can have different cells running different languages while interacting with the same data by using what are called **magic commands**. For example, if your notebook is set to work with PySpark (Python), you can type `%%sql` at the beginning of a new code cell and simply write your code in SQL instead of Python, and then use Python again in the next code cell.

> **Note**
>
> To learn more about magic commands and development with Azure Synapse notebooks, visit `https://learn.microsoft.com/azure/synapse-analytics/spark/apache-spark-development-using-notebooks`.

You can create Markdown or code cells throughout your notebook by selecting the + **Markdown** and + **Code** buttons respectively. If you cannot see these buttons, simply hover your mouse pointer over the cell above where you want to insert a new one in the notebook and the buttons will reveal themselves, as shown in *Figure 6.20*:

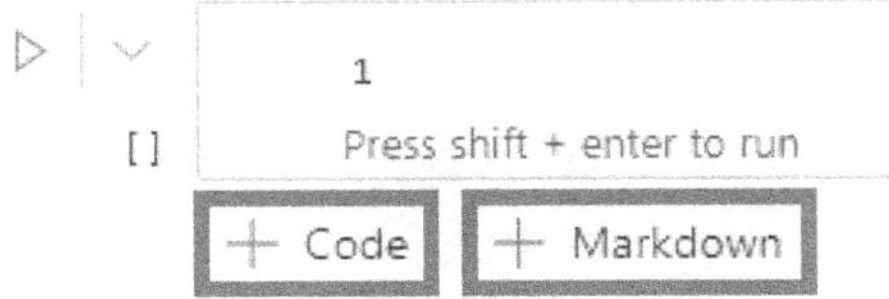

Figure 6.20 – The Code and Markdown buttons

For simplicity, this section walks you through the steps to use the code cells where we will perform the data analysis and will skip the instructions for the Markdown cells. Do not worry—they are not functionally required. Feel free to explore Markdown cells and build a descriptive notebook that helps you tell the story of your data, or simply import the notebook provided in the `Chapter 06\Exploring data with Python.ipynb` file of the book repository.

To create a new Markdown cell, select the + **Code** button in your new notebook. As we did in the *Analyzing data with KQL* section, let's look at some examples of working with PySpark (Python) in a notebook. You will perform a few data exploration tasks, and then prepare the data for *Chapter 7, Data Visualization with Power BI*, when we will learn about the Power BI integration with Azure Synapse.

Reading data from Data Explorer pools

Azure Synapse Studio helps you get started quickly with data exploration of Data Explorer pool databases using notebooks. With the click of a mouse, it creates a new notebook with pre-generated code that reads data from a table in your Data Explorer pool database. Let's see how this works:

1. In Azure Synapse Studio, navigate to the **Data** hub.
2. Expand **Data Explorer Databases (Preview)** to reveal your Data Explorer pool.
3. Expand your pool by selecting the arrow next to the Data Explorer pool name. In your database, expand **Tables** and right-click the `fleet data` table. Select **New Notebook**, and then **Load to DataFrame**.

4. This creates a new notebook with a few lines of Python code. This code block creates a new Spark DataFrame named `kustoDf` that contains 10 rows from the `fleet data` table. I've made a few changes to this code block:

   - Imported a few Python libraries that we will use throughout this notebook. It's generally a good practice to have your `import` statements right at the beginning of your script to make it easy to find all the libraries you will need, even though Python allows you to have import statements anywhere in your code (as long as it's before a code line that uses that library).
   - Changed the name of the Spark DataFrame from `kustoDf` to `df`. This name can be any valid Python variable name, and I chose `df` simply because that's what I call DataFrames when I work in Python.
   - Removed the `take 10` sentence from the line that sets the `kustoQuery` option parameter of the Spark DataFrame. We will need all our data, not only 10 rows.

The resulting code is shown here:

```
import matplotlib as mpl
import matplotlib.pyplot as plt
import numpy as np
import pandas as pd
from pyspark.sql.types import DateType
import seaborn

mpl.rcParams['agg.path.chunksize'] = 10000

df = spark.read \
  .format("com.microsoft.kusto.spark.synapse.datasource") \
  .option("spark.synapse.linkedService", "kustoPool") \
  .option("kustoCluster", \
  "https://droneanalyticsadx.drone-analytics.kusto.
azuresynapse.net") \
    .option("kustoDatabase", "drone-telemetry") \
    .option("kustoQuery", "['fleet data']") \
    .load()

display(df)
```

Note

Before you run this code cell, if you copied this code from the book's sample notebook, make sure you adjust the `kustoCluster`, `kustoDatabase`, and `kustoQuery` parameters with the name of your Data Explorer pool, database, and table, respectively.

You can run this code cell by selecting the **Run cell** button on the left side of the code cell, or by pressing *Ctrl* + *Enter* on your keyboard (*Command* + *Return* if you are using macOS). When you run the first code cell, your Apache Spark pool will start. This operation may take several minutes to complete.

As a result, you will see your query results displayed, as in *Figure 6.21*, on a grid, under the code cell:

DeviceData	DateTime	LocalDateTime	DeviceState
{'guid': '5E2A0A57-5FDD-42D2-8...	2021-10-30 18:05:23.044283	2021-10-30 11:05:23.044283	On
{'guid': '5E2A0A57-5FDD-42D2-8...	2021-10-30 18:07:24.244283	2021-10-30 11:07:24.244283	TakeOff
{'guid': '5E2A0A57-5FDD-42D2-8...	2021-10-30 18:17:51.421883	2021-10-30 11:17:51.421883	Cruise
{'guid': '5E2A0A57-5FDD-42D2-8...	2021-10-30 18:19:52.621883	2021-10-30 11:19:52.621883	Descent
{'guid': '5E2A0A57-5FDD-42D2-8...	2021-10-30 18:21:26.701883	2021-10-30 11:21:26.701883	Delivery
{'guid': '5E2A0A57-5FDD-42D2-8...	2021-10-30 18:23:27.901883	2021-10-30 11:23:27.901883	TakeOff
{'guid': '5E2A0A57-5FDD-42D2-8...	2021-10-30 18:33:55.079483	2021-10-30 11:33:55.079483	Cruise
{'guid': '5E2A0A57-5FDD-42D2-8...	2021-10-30 18:35:56.279483	2021-10-30 11:35:56.279483	Descent
{'guid': '5E2A0A57-5FDD-42D2-8...	2021-10-30 18:37:30.359483	2021-10-30 11:37:30.359483	Off
{'guid': '5E2A0A57-5FDD-42D2-8...	2021-10-30 19:37:30.359483	2021-10-30 12:37:30.359483	On

Figure 6.21 – Data retrieved from your Data Explorer pool database

Once we have a dataset in our hands, it's useful to peek at its structure to see which data we have in hand. We can do that by using the `display` function again to show our Spark DataFrame, combined with the `summary` parameter set to `True`:

```
display(df, summary=True)
```

This displays a summary of your DataFrame, listing all columns, their type, a count of unique values, and a count of missing values. *Figure 6.22* shows this output:

name	type	unique	missing
DeviceData	string	150	0
DateTime	timestamp	337500	0
LocalDateTime	timestamp	337500	0
DeviceState	string	7	0
Engine1Status	string	2	0
Engine2Status	string	2	0
Engine3Status	string	2	0
Engine4Status	string	2	0
Engine1RPM	bigint	599	0
Engine2RPM	bigint	599	0
Engine3RPM	bigint	599	0

Figure 6.22 – Showing the DataFrame's structure

Clicking any of the rows highlighted in blue will give you column statistics that include how many occurrences you have of each value present in that column. This is something useful to do immediately after you read data from some source (including, but not limited to, data sourced from Data Explorer pools) so that you can better understand the schema of the data you will work with, its data types, how much data you have, and more.

Plotting charts

Azure Synapse notebooks offer robust support for data visualization. There are three main mechanisms you can use to visualize data with charts using notebooks:

- The **Chart** feature of the query editor
- Using the `displayHTML` command, combined with Java's **Data-Driven Documents** (D3) library
- Using external libraries for data visualization

The **Chart** feature of the query editor works similarly to how we explored it in the *Analyzing data with KQL* section of this chapter. To plot charts using this feature in an Azure Synapse notebook, all you need to do is use the `display` function to show the results of a Spark DataFrame, and then select the **Chart** toggle next to **View**. As an example, if we wanted to plot a line chart showing the trend of the `CoreTemp` column over a span of 10 minutes, we could use the following code:

```
display(df.select('CoreTemp','LocalDateTime') \
    .filter(df['LocalDateTime'] >= '2021-11-01 12:00:00') \
    .filter(df['LocalDateTime'] <= '2021-11-01 12:10:00'))
```

This code block produces the result in a table format. To view it as a line chart, simply select the **Chart** toggle and set the **Chart type** option to **Line chart**, as highlighted in *Figure 6.23*:

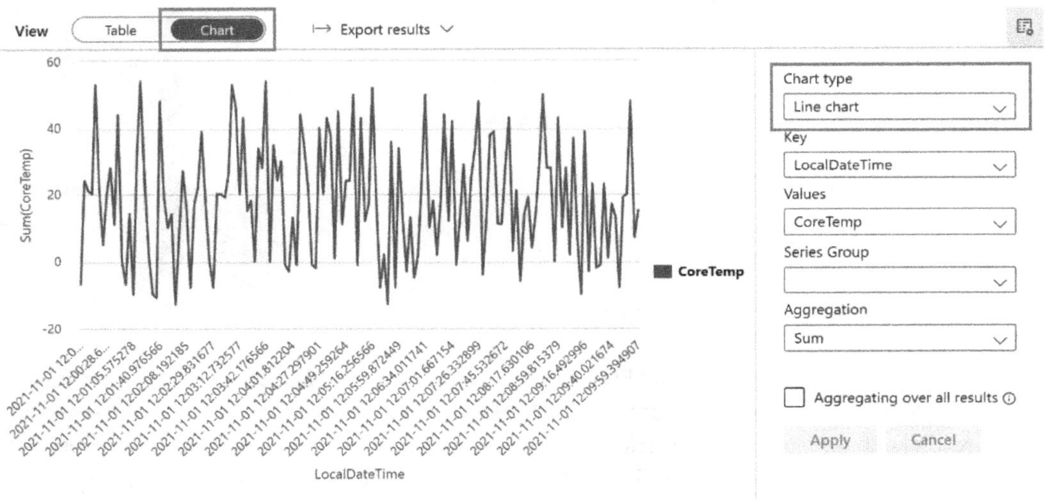

Figure 6.23 – Plotting the results of a DataFrame as a chart

This feature is very simple to use and doesn't require any additional libraries to work. However, the chart options are quite limited: you can't format chart series, change colors, or change legend settings or other settings. For richer charts, you have to choose a different technique.

An interesting option if you are an HTML person is to use the D3.js library (also known simply as D3). D3 is a JavaScript library used to display data on web pages. It allows binding data to **scalable vector graphics** (**SVG**) that show data as charts dynamically as web pages are resized or as users interact with the charts. Users can write their D3 charts using HTML and call the `displayHTML` function of PySpark to display the charts. The downside of this approach is that it requires a lot of HTML coding, so it may not be as productive as the other options.

> **Note**
> To learn more about D3.js, visit `https://d3js.org/`.

My favorite way to plot charts in notebooks is to use Python libraries such as Matplotlib, Seaborn, Plotly, and others. These libraries are already preloaded into the Apache Spark environment provided by Azure Synapse, so all you have to do is import them before you use them. Let's look at a few examples.

One of the most popular libraries for data manipulation in Python is the pandas library. It offers data structures for data manipulation and a long list of features that make it easy to read and modify data in memory using Python. Spark DataFrames offer a convenient `toPandas` method, which converts your Spark DataFrame to a Pandas DataFrame. Before we start using the Python libraries for data visualization, we have to convert our Spark DataFrame to a pandas DataFrame, as most of these libraries were built to work with pandas and not PySpark.

Using the `toPandas` function is simple:

```
pandas_df = df.toPandas()
```

Now that we have our new pandas DataFrame, we can start building the charts. We will start by plotting a bell curve, which shows the normal distribution of your data. For our example, we will use the `CoreTemp` column. The bell curve helps understand the data distribution in your table, with the mean value at the top of the curve and all the other possible values symmetrically distributed around the mean. Plotting the bell curve requires us to compute what's called the **probability density function** (known as **PDF**, illustrated in *Figure 6.24*) for the values in the `CoreTemp` column. The PDF calculates, for a value of any given sample, the relative likelihood that a variable would be close to a value in the dataset:

$$f(x) = \frac{1}{\sqrt{2\pi\sigma^2}} e^{-\frac{(x-\mu)^2}{2\sigma^2}}$$

Figure 6.24 – The probability density function

Some Python packages, such as SciPy, provide libraries to compute the PDF for you. In this example, we will simply build the expression and plug in the values to avoid importing a package only for this need. After we compute the PDF, we will use the Matplotlib library in Python to plot the chart.

The code to compute the PDF and generate a bell curve is as follows:

```
# Compute the PDF
meanTemp = pandas_df['CoreTemp'].mean()
stdTemp = pandas_df['CoreTemp'].std()
y =  1/(stdTemp * np.sqrt(2 * np.pi)) * np.exp( - (pandas_
df['CoreTemp'] - meanTemp)**2 / (2 * stdTemp**2))
```

```
# Plot the bell curve
plt.style.use('seaborn')
plt.figure(figsize = (5, 5))
plt.scatter(pandas_df['CoreTemp'], y, marker = '.', s = 10,
color = 'blue')
```

This code produces an output similar to the one seen in *Figure 6.25*:

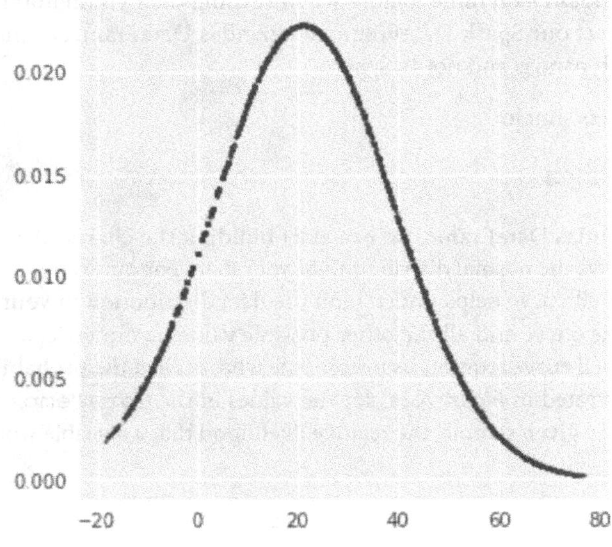

Figure 6.25 – Bell curve

Reviewing this chart, we can observe that the data is symmetrical, meaning that for new values of the `CoreTemp` value, you can generally expect values with a probability to be closer to the center of the bell, where most values are. There is little tendency that this column will produce outliers or many extreme values.

Another way to visualize how values are distributed in a certain column is to use a histogram. They are much simpler to produce and give you an easy way to visualize the most common value ranges for your column, as well as the data distribution across the full range of your data. You can produce a histogram easily using Matplotlib:

```
plt.hist(pandas_df['CoreTemp'], bins=7, alpha=0.5)
plt.show()
```

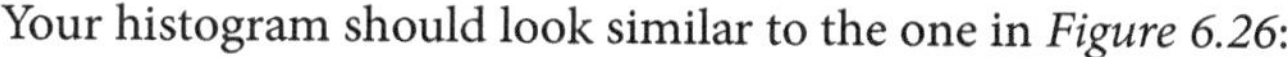

Your histogram should look similar to the one in *Figure 6.26*:

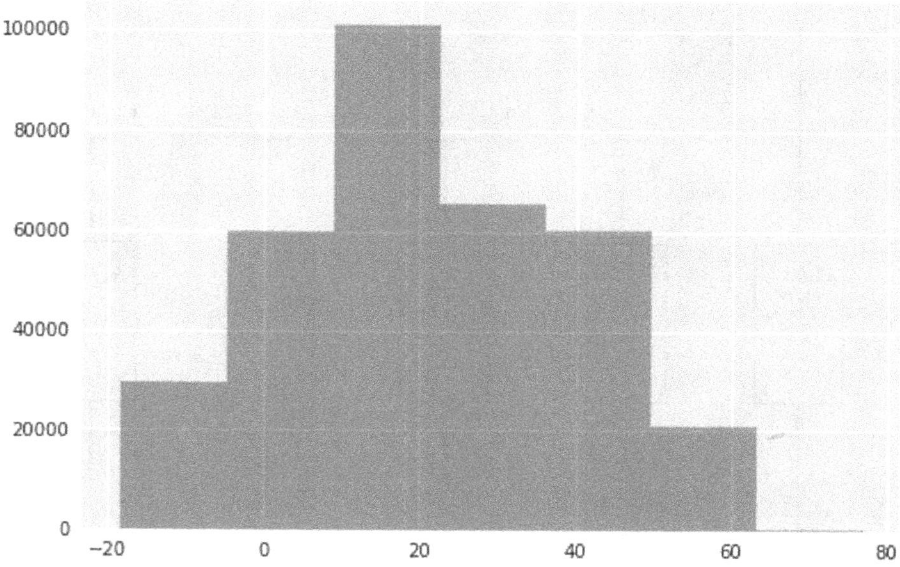

Figure 6.26 – Histogram showing how values are distributed in your data

In this case, we can see that there are approximately 30,000 occurrences of `CoreTemp` with values somewhere between about -19 and -3, close to 60,000 values between about -3 and 9, and so on.

Another useful way to get a good sense of how your data is spread out is to use **box plots**. These allow you to easily see the spread and skewness of the values in your dataset by plotting their quartiles. Box plots are also useful to identify outliers as they plot these data points individually beyond the chart area, making them obvious and easy to spot.

Producing box plots with Matplotlib in Python is a breeze. Here is an example:

```
plt.boxplot(pandas_df[['CoreTemp','Engine1Temp',
    'Engine2Temp','Engine3Temp','Engine4Temp']])
plt.xticks([1,2,3,4,5], ['CoreTemp','Engine 1',
    'Engine 2','Engine 3','Engine 4'])
plt.show()
```

You should see a chart similar to the one in *Figure 6.27*:

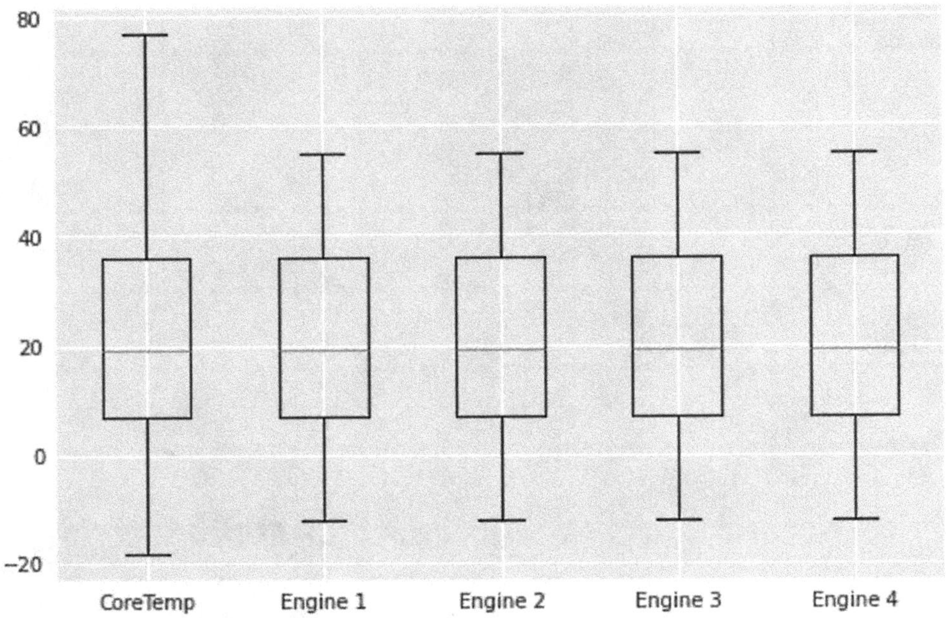

Figure 6.27 – A box plot

In this chart, we observe the following:

- The areas in the rectangle provide a range from the first quartile (which is the 25th percentile in a normal distribution) to the third quartile (75th percentile)
- The maximum and the minimum values are the ranges expressed beyond the rectangles
- The median value is expressed by the horizontal line inside the rectangles

We can also notice that the data distribution is consistent across the temperature values from **Engine 1** through **Engine 4**, which is what we would expect. Additionally, the `CoreTemp` values observe a broader temperature range when compared to that of the individual engines, which could indicate an issue with the drone's core temperature.

Let's look at one final example of plots. A common task when analyzing data is to try to understand how variables (or columns) in your dataset relate to each other, and how much correlation there is between them. This can be achieved by using a **correlation matrix**. pandas provides a convenient `corr` function, which, given a certain DataFrame, computes the correlation of columns in such a dataset excluding null values. For example, the following code produces a correlation matrix for a subset of our DataFrame:

```
correlation_matrix =
pandas_df[['Engine1RPM','Engine1Temp','CoreTemp',
'BatteryTemp','Altitude','Speed','DistanceFromBase','RFSignal',
'PayloadWeight']].corr()
```

Now that we have the correlation matrix stored in the `correlation_matrix` variable, we can use the Seaborn library to produce a heatmap that visually represents our matrix:

```
seaborn.set(rc = {'figure.figsize':(14, 8)})
seaborn.heatmap(correlation_matrix, annot=True)
```

This produces a plot similar to the one in *Figure 6.28*:

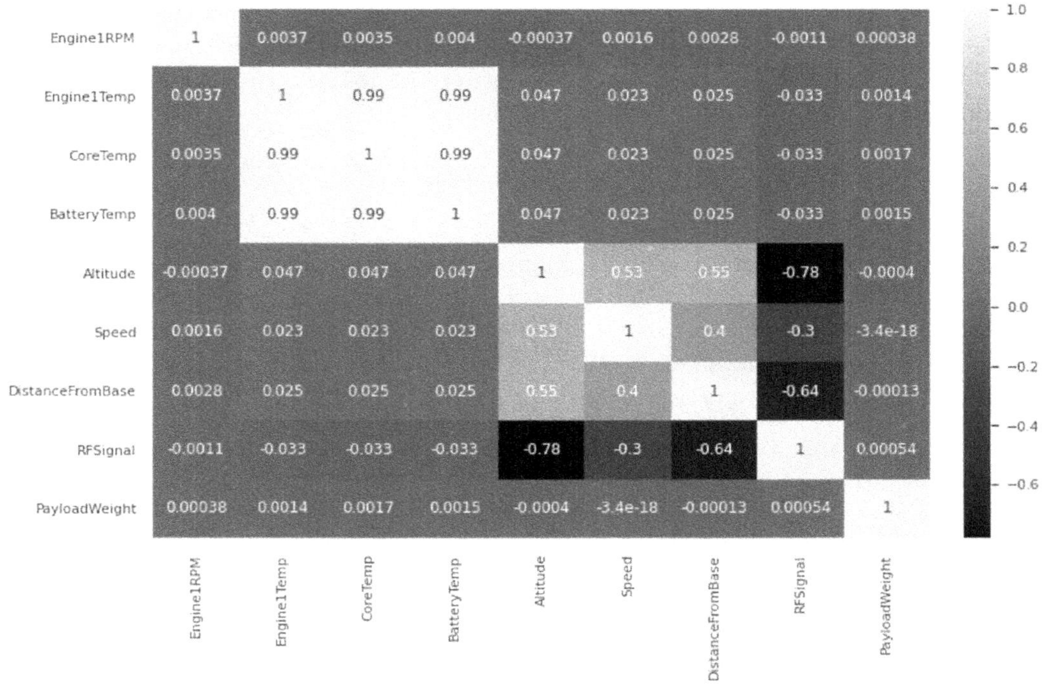

Figure 6.28 – The correlation matrix plotted as a heatmap

The values of a correlation matrix range between `-1` and `1`. Values that are closer to `-1` are called negative correlations, while values closer to 1 are called positive correlations. Negative correlations indicate that the variables are related but tend to move in opposite directions, while positive correlations indicate that the variables are also related but tend to move in the same direction. Values closer to 0 represent a weak or inexistent linear relationship between the variables.

In the example illustrated in *Figure 6.28*, we notice that probably there is collinearity between `RFSignal` and `Altitude`, with a `-0.78` correlation coefficient.

Performing data transformation tasks

So far, we've loaded the data from the Data Explorer pool into a Spark DataFrame and explored the data to understand its schema and the distribution of some columns. We will now look at ways to make this data easier to read and consume for an end user.

For the first data transformation task, we will assume a scenario where users will need to build a report and need a date column to easily filter data, and maybe produce some charts. The user will need the date only, not the specific time the telemetry event was captured. We notice that the table we are working with has a column named `DateTime` and a column named `LocalDateTime`. The `DateTime` column stores the date and time in the **Coordinated Universal Time** (**UTC**) format, while the `LocalDateTime` column stores the same date and time, but in the local time zone. We will produce a new column that contains the date only, without the time, derived from the `LocalDateTime` column. The new column will be named `LocalDate`. We can achieve this by creating a new code cell and using the following code snippet:

```
df = df.withColumn('LocalDate', df['LocalDateTime'].
cast(DateType()))
```

The `withColumn` function of the Spark DataFrame object returns a new DataFrame with a new column or replaces an existing column of the same name with a new one. The name of the new column (or the one being replaced) is the first parameter of the function (`LocalDate` in this example), and the new value that the column will receive is contained in the second parameter (in our case, the `LocalDateTime` column cast to `DateType`).

We can then verify that our data transformation worked by displaying a few rows from our DataFrame using the following code:

```
display(df.select('LocalDateTime','LocalDate'))
```

This command will produce an output similar to what is illustrated in *Figure 6.29*:

LocalDateTime	LocalDate
2021-10-30 11:05:23.044283	2021-10-30
2021-10-30 11:07:24.244283	2021-10-30
2021-10-30 11:17:51.421883	2021-10-30
2021-10-30 11:19:52.621883	2021-10-30
2021-10-30 11:21:26.701883	2021-10-30
2021-10-30 11:23:27.901883	2021-10-30
2021-10-30 11:33:55.079483	2021-10-30

Figure 6.29 – The new LocalDate column with dates only

You can also use data transformation tasks to produce new columns based on calculations. As an example, our dataset contains a `CoreTemp` column that stores the core temperature of the drone at the time the telemetry event was generated. This column contains temperature values using the Celsius scale. Let's say we want to produce a new column, which displays the equivalent temperature in the Fahrenheit scale. We can perform this task by using the same `withColumn` function we used before but by applying a simple formula as the column expression for the new column.

For simplicity, you can create a new code cell that contains the following code to produce the new column and visualize a few rows with the results at once:

```
from pyspark.sql.functions import round
df = df.withColumn('CoreTemp-F', round(df['CoreTemp'] * 1.8 +
32,2))
display(df.select('CoreTemp','CoreTemp-F'))
```

This code produces an output similar to the one seen in *Figure 6.30*:

CoreTemp	CoreTemp-F
18.0	64.4
2.0	35.6
-5.0	23.0
42.0	107.6
42.0	107.6
20.0	68.0
32.0	89.6
-4.0	24.8

Figure 6.30 – The new CoreTemp-F column with temperatures in the Fahrenheit scale

In this case, we see that the row that contains a `CoreTemp` value of `11` now also contains a `CoreTemp-F` column with a value of `51.8`. The same applies to each row in the DataFrame, each row with its respective values.

As a final example of data transformation, let's see how we can extract data from the JSON columns in the dataset to make data analysis easier for our end user. There are two JSON columns in our source data: `DeviceData` and `EventData`. These columns contain a JSON document, which may provide different values depending on the telemetry event. An example of the `DeviceData` column has the following data:

```
{
    'guid': '90C5BED1-232B-432A-B559-2DAB2F59A97F',
    'deviceID': 'drn00001',
    'deviceName': 'DT-F-725C'
}
```

This is a simple JSON document with a collection of name/value pairs that provide some attributes about the drone that generated the telemetry event. Now, let's look at an example of the `EventData` column:

```
{
    'averageSpeed': 75,
    'stageDistance': 2938,
```

```
    'accumulatedDistance': 6977
}
```

This is also a simple JSON document with value/pair combinations. These values provide additional attributes about the drone at the time the telemetry event was generated.

Let's think of an example where an analyst will want to know the drones in the fleet with the highest accumulated distance, as they may require maintenance after reaching a certain mileage. For this purpose, we will create two new columns: one that makes it easier to see what the name of the device in the telemetry event is, and one with the accumulated distance of that drone. We can achieve this by using the following code:

```
from pyspark.sql.functions import col, get_json_object
df = df.withColumn('deviceName', get_json_object(col
('DeviceData'),'$.deviceName').alias('deviceName'))
df = df.withColumn('accumulatedDistance', get_json_
object(col('EventData'),'$.accumulatedDistance').
alias('accumulatedDistance').cast('int'))
display(df.select('DeviceData','EventData','deviceName',
'accumulatedDistance'))
```

Your code should produce a result similar to the one in *Figure 6.31*:

DeviceData	EventData	deviceName	accumulatedDistance
{'guid': '5E2A0A57-5FDD-42D2-8...	{'averageSpeed': 0, 'stageDistanc...	DT-F-312A	0
{'guid': '5E2A0A57-5FDD-42D2-8...	{'averageSpeed': 15, 'stageDistan...	DT-F-312A	640
{'guid': '5E2A0A57-5FDD-42D2-8...	{'averageSpeed': 90, 'stageDistan...	DT-F-312A	5764
{'guid': '5E2A0A57-5FDD-42D2-8...	{'averageSpeed': 25, 'stageDistan...	DT-F-312A	6404
{'guid': '5E2A0A57-5FDD-42D2-8...	{'averageSpeed': 0, 'stageDistanc...	DT-F-312A	6404
{'guid': '5E2A0A57-5FDD-42D2-8...	{'averageSpeed': 25, 'stageDistan...	DT-F-312A	7044
{'guid': '5E2A0A57-5FDD-42D2-8...	{'averageSpeed': 75, 'stageDistan...	DT-F-312A	12168
{'guid': '5E2A0A57-5FDD-42D2-8...	{'averageSpeed': 25, 'stageDistan...	DT-F-312A	12808
{'guid': '5E2A0A57-5FDD-42D2-8...	{'averageSpeed': 0, 'stageDistanc...	DT-F-312A	12808

Figure 6.31 – Extracted deviceName and accumulatedDistance values from the JSON columns

It's important to note that none of these data transformation tasks resulted in data loss. We are not replacing the values in the source columns with the *transformed* ones; we are always creating new columns with the new, desired values. Additionally, these changes are not written back to the source table schema in the Data Explorer pool database.

However, if you would like to perform further analysis using the new columns we just created as part of the data transformation process, there is a way to persist this data and offer a dataset for users to consume. This can be done by implementing a lake database.

Creating a lake database

A lake database allows you to store data in ADLS Gen2 while maintaining database and table metadata in Azure Synapse. All your data will be physically written in Parquet, Delta, or CSV files but exposed to users as if they were regular SQL tables and databases. Users can then perform queries normally using the serverless SQL pools, or Apache Spark pools, using the SQL and Spark APIs.

Using the lake database, we can persist the Spark DataFrame we have been working on to disk, with all the data transformations, and use it as if it were a SQL table. This SQL table can be used for user queries, Power BI reports, or any other purpose that a SQL table could satisfy.

To persist our Spark DataFrame on the lake database, we should first create a new database, like so:

```
spark.sql('CREATE DATABASE IF NOT EXISTS drone_telemetry')
```

Next, we will switch the context to the new database with the `USE drone_telemetry` command, and then we will use the `saveAsTable` Spark DataFrame method to write the contents of our Spark DataFrame to this new managed table:

```
spark.sql('USE drone_telemetry')
df.write.mode('overwrite').saveAsTable('fleet_data')
```

The data will be written to your workspace's ADLS Gen2 account as a series of Parquet files, and Azure Synapse will maintain the metadata needed to allow you to query the new table seamlessly across the serverless SQL and Apache Spark compute engines. You can use the new table normally, as you would query any other SQL table.

To verify your table has been created successfully, you can perform the following steps:

1. Navigate to the **Data** hub and expand **Lake database**.
2. You should see your new `drone_telemetry` database listed in the database list under **Lake database**, as seen in *Figure 6.32*. Expand your database's name, and then expand **Tables**. Click on the ellipsis next to your table's name (in this example, the `fleet_data` table) and select **New SQL script** > **Select TOP 100 rows**:

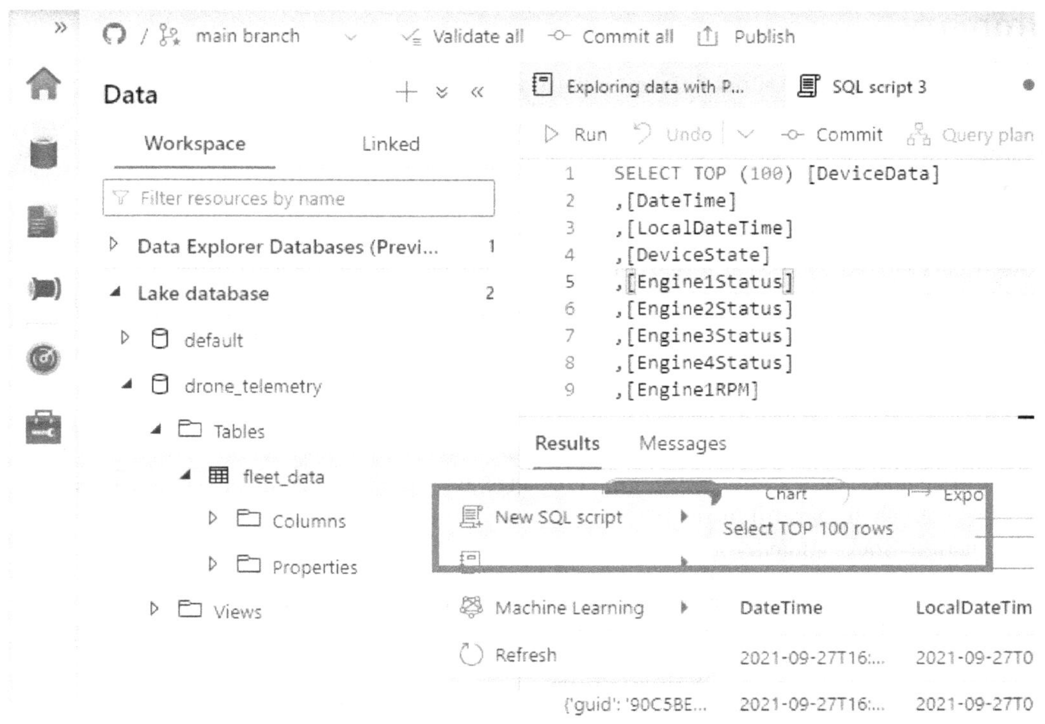

Figure 6.32 – Querying from the lake database

3. This will populate a new SQL query window with the T-SQL statement to select 100 rows from the `fleet_data` table. Select **Run** to execute the query. The serverless SQL pool engine retrieves data from the lake database. Note that this table contains the new columns we created in the *Performing data transformation tasks* section.

The lake database can be queried using the serverless SQL pool, which is always available for you. You don't need to start, pause, or resume your serverless SQL pool. It works for you on demand and, differently from Data Explorer pools, dedicated SQL pools, or Apache Spark pools, you are not charged for the time it is running. You are charged by the amount of data processed in queries. To learn more about how pricing works for serverless SQL pools, visit `https://learn.microsoft.com/azure/synapse-analytics/sql/data-processed`.

Summary

In this chapter, you learned about different ways to work with data in Azure Synapse Data Explorer. The Data Explorer engine stores and manages large volumes of data efficiently, and this chapter helped you understand how to make sense of the data you have on Data Explorer pools using the different tools available for data exploration with Azure Synapse workspaces.

First, you used different KQL queries to navigate through your data and get familiar with the drone telemetry dataset. You created calculated columns, plotted information on charts using only the query editor, and then explored your data by looking at percentiles. You then created a time series and used the native features of KQL to detect outliers in your data, and even analyzed the trends in your data using linear regression.

Next, you used Azure Synapse notebooks to explore data using Python. You created your first Apache Spark pool, used it to read data from our Data Explorer pool, and used different Python (PySpark) libraries to visualize data and transform data for later use. Finally, you created a lake database to persist the data changes you'd made to allow users to connect to it and consume the data with the transformations you'd applied.

The possibilities within each of the topics that were covered are endless. In fact, a whole book could be written for each one of these topics. My hope is that the examples provided here help you scratch the surface and experiment with the different flavors for data exploration using Azure Synapse Data Explorer. I encourage you to spend more time on each of the topics and build upon the examples provided here to reach your own learning goals.

In the next chapter, you will learn how to use the data we worked on during this chapter to build Power BI reports that are integrated into your Azure Synapse workspace.

7

Data Visualization with Power BI

Data visualization is key to data analysis and exploration. It helps you present complex data more simply and makes data easier to understand by users at all technical levels, alongside individuals without any technical background at all. By visually showing the trends, patterns, and distribution of your data, you empower individuals in their decision-making process with a single source of truth: the data behind the story.

Azure Synapse offers a native integration with the Power BI service, allowing you to author Power BI reports and publish them to the Power BI service directly from within Azure Synapse Studio. In this chapter, you will create Power BI reports that visualize data in your Data Explorer pool as well as the data you've transformed using Azure Synapse notebooks. You will learn how to configure the Power BI integration with Azure Synapse and use Azure Synapse Studio to author reports using your datasets.

At a high level, we will cover the following topics in this chapter:

- Creating a Power BI report
- Adding data sources to your Power BI report
- Connecting Power BI with your Azure Synapse workspace
- Authoring Power BI reports from Azure Synapse Studio

The skills you will learn in this chapter go beyond the work you are doing with Data Explorer pools. In fact, you will learn that Data Explorer pools will offer you the data for the Power BI report, but the skills you will learn here apply to any data source you connect to your Power BI reports. So, enjoy the learning, and let's go!

Technical requirements

In case you haven't already, make sure you download the book materials from our repository at `https://github.com/PacktPublishing/Learn-Azure-Synapse-Data-Explorer`. You can download the full repository by selecting **Code**, and then **Download ZIP**, or by cloning the repository by using your git client of choice.

This chapter uses the full `fleet data` table, which you ingested in the *Running your first query* section of *Chapter 3, Exploring Azure Synapse Studio*. If you did not perform the tasks in this section to load data, take some time to go back and follow the instructions to ingest the full `fleet data` table.

If you haven't read and completed the examples from *Chapter 6, Data Analysis and Exploration with KQL and Python*, make sure you do that before you attempt to reproduce the examples within this chapter. The *Adding data sources to your Power BI report* section of this chapter connects to the lake database that we created in that chapter, so if you are hoping to reproduce the examples in this chapter, first, you will need to complete the previous chapter.

Finally, to create reports with Power BI, you will need a Windows PC with Power BI Desktop. You can edit and publish reports in Azure Synapse Studio, but the first report needs to be created on your desktop PC. To download Power BI Desktop, visit `https://powerbi.microsoft.com/downloads/`. If you don't have previous experience with Power BI, a great resource is a tutorial from `https://learn.microsoft.com/power-bi/fundamentals/desktop-getting-started`. Finally, to connect your Azure Synapse workspace with the Power BI service, you need a Power BI Pro or Power BI Premium license. A free trial is available at `https://powerbi.microsoft.com/pricing`.

Introduction to the Power BI integration

Azure Synapse offers integration with Power BI to allow you to visualize, create, and modify Power BI reports from Azure Synapse Studio. This is a powerful capability that increases team productivity by allowing team collaboration at all stages in the analytics process from within a single tool: Azure Synapse Studio. In this section, to better understand how this collaboration works in real life, we will create a new report using Power BI Desktop, connect a Power BI workspace with your Azure Synapse workspace, and then work on this report in Azure Synapse Studio.

Power BI reports need a data source definition to connect to data. At the time of writing, the Power BI data source for Azure Synapse was only available using Power BI Desktop. We will use Power BI Desktop to create the data source, author an initial report, and publish this initial report and data source to the Power BI service. After this data source has been published, we can reuse it in Azure Synapse Studio.

> **Note**
>
> To view the latest details about the Azure Synapse data connector for Power BI, refer to the documentation at `https://learn.microsoft.com/power-bi/connect-data/service-azure-sql-data-warehouse-with-direct-connect`.

Creating a Power BI report

To make this flow easier to understand, let's create a new report and see how we can work with it in Azure Synapse Studio. Proceed with the following steps:

1. Launch Power BI Desktop. If you haven't installed it already, please refer to the *Technical requirements* section at the beginning of this chapter.
2. You will see a window with resources to help you get started with Power BI. Select **Get data**.
3. In the **Get Data** window, pick **Azure** from the left-hand side and then **Azure Data Explorer (Kusto)**, as illustrated in *Figure 7.1*.

   Even though this connector refers to the standalone service Azure Data Explorer, the same connector can be used to access Data Explorer pools in Azure Synapse. Later, we will create a different connection using the Azure Synapse Analytics SQL connector.

When you are ready, click on the **Connect** button:

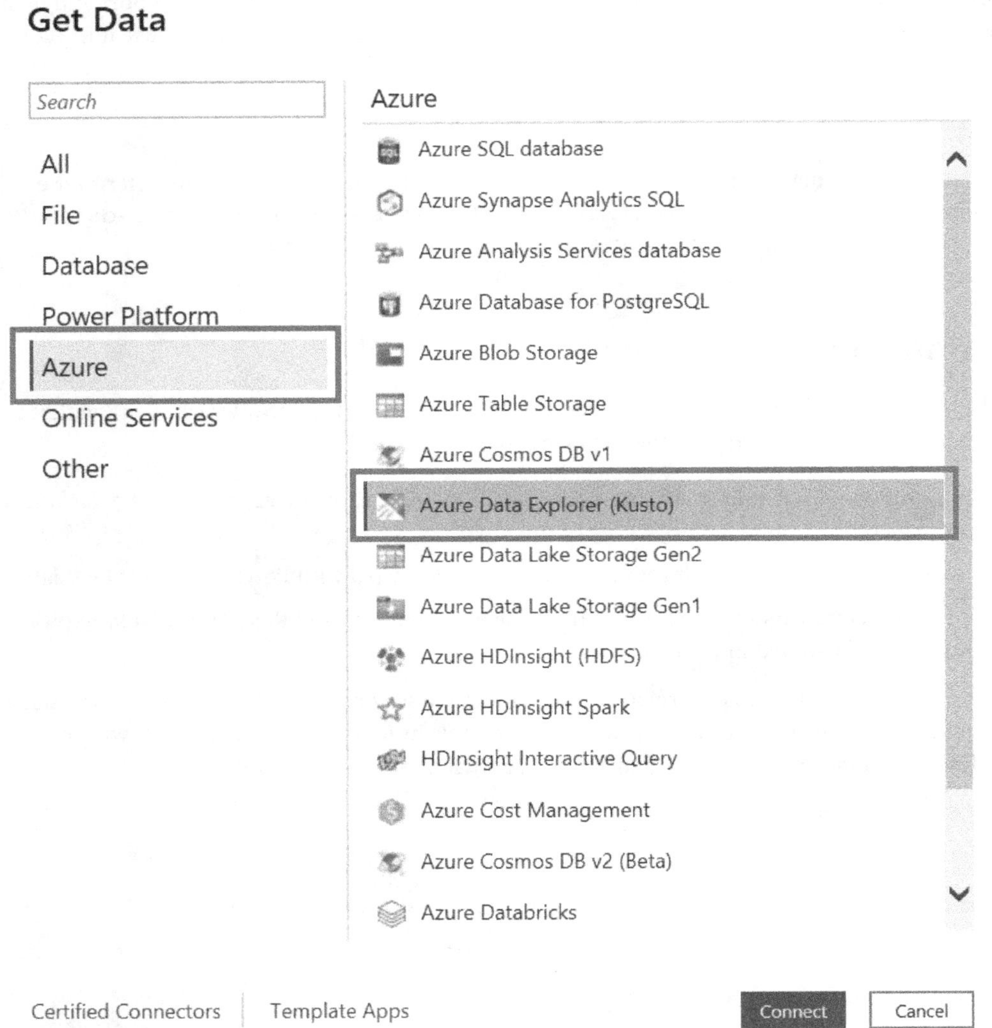

Figure 7.1 – Selecting the data source type

4. In the **Azure Data Explorer (Kusto)** window, you should provide the details required to connect to your Data Explorer pool database. The connector needs the following information:

   - **Cluster**: This is the URI of your Data Explorer pool. The URI uses the https://<pool>.<workspace>.kusto.azuresynapse.net format, where <pool> is the name of your Data Explorer pool, and <workspace> is the name of your Azure Synapse workspace name. You can find this information in the Azure portal by navigating to the Data Explorer pool resource. Here, you should use the URI of your Data Explorer pool.
   - **Database**: This is the name of the database you will connect to. In our example, I'm using the name drone-telemetry.
   - **Table name of Azure Data Explorer query**: Here, you can specify a KQL query or the name of a table to use all the rows. For this example, I'm using a KQL query that retrieves the telemetry data in a specific week, as the telemetry data in this example dataset is stale:

```
['fleet data']
| where LocalDateTime >= datetime(2021-11-01) and
LocalDateTime <= datetime(2021-11-07)
```

   In a real-time telemetry environment, you could use a query that gets the most recent data – for example, all rows from the last seven days.

   - **Advanced options**: Here, you can specify whether you would like to limit your query to a certain number of records or data volume in bytes. For this example, I'm leaving these fields blank.
   - **Data connectivity mode**: Here, you can specify whether Power BI will import the result of your query into Power BI (and then refresh data as needed), or whether Power BI will connect to the data source directly to update the charts. **Import** is the most common option, and the preferred one for Power BI as it offers the best performance for the end user and reduces stress on the data source. You should use **DirectQuery** if you need to work with real-time data and to always have the latest information in a report or dashboard – which is common for the telemetry and log scenarios described in this book. For this example, since we are working with stale data, I'll use the **Import** option.

5. Your Azure Data Explorer (Kusto) window should look similar to the one shown in *Figure 7.2* (don't forget to use the name of *your* Data Explorer pool in the **Cluster** field). Select **OK** when you are ready:

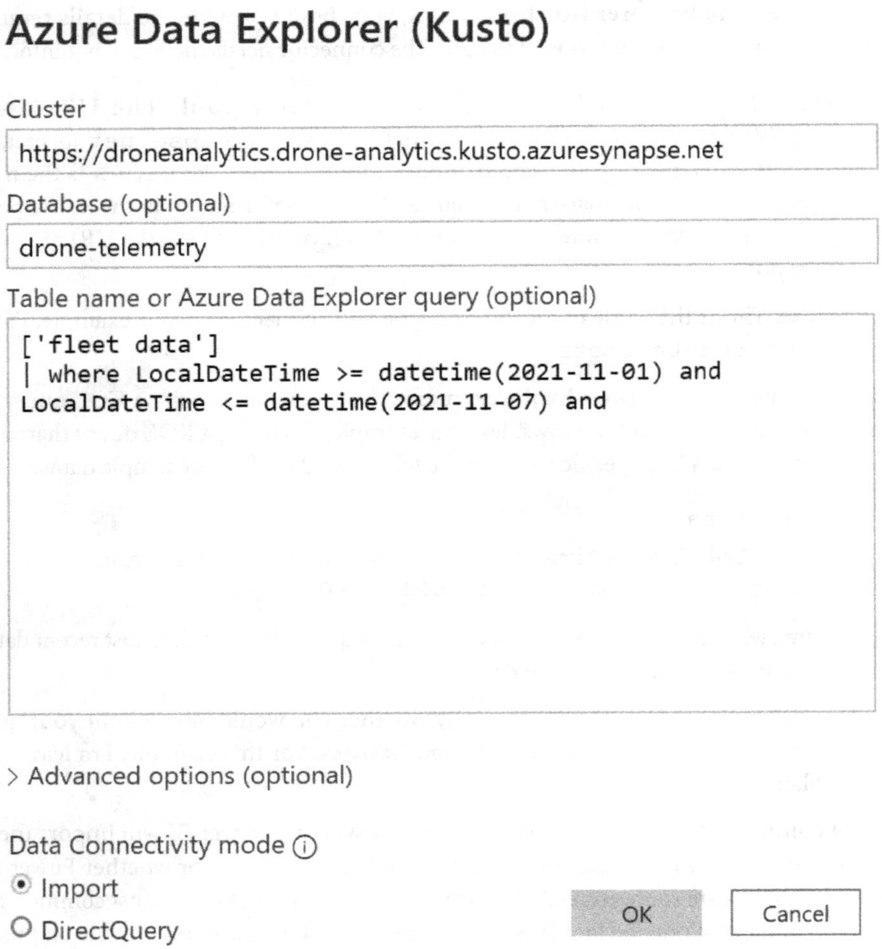

Azure Data Explorer (Kusto)

Cluster

https://droneanalytics.drone-analytics.kusto.azuresynapse.net

Database (optional)

drone-telemetry

Table name or Azure Data Explorer query (optional)

```
['fleet data']
| where LocalDateTime >= datetime(2021-11-01) and
LocalDateTime <= datetime(2021-11-07) and
```

> Advanced options (optional)

Data Connectivity mode ⓘ

◉ Import

○ DirectQuery

OK Cancel

Figure 7.2 – Connecting to your Data Explorer pool

6. Power BI will need your credentials to connect to your Data Explorer pool. When you see the Azure Data Explorer (Kusto) window with the **You are not signed in** notice, select **Sign in** and provide your credentials. Select **Connect** when you are ready.
7. Next, Power BI shows you a preview of your data based on your query. Select the **Load** button.
8. In the **Fields** pane, you will see a new object called **Query1**. This is your new data source. To make it easier to identify, right-click on **Query1** and select **Rename**. Change the name of your data source to a friendly name, such as `fleet-data (ADX)`.

You will build a report similar to the one shown in *Figure 7.3*, which when used with real-time telemetry data, will allow you to monitor failure rates for drone engines and the average payload weight for the past few days:

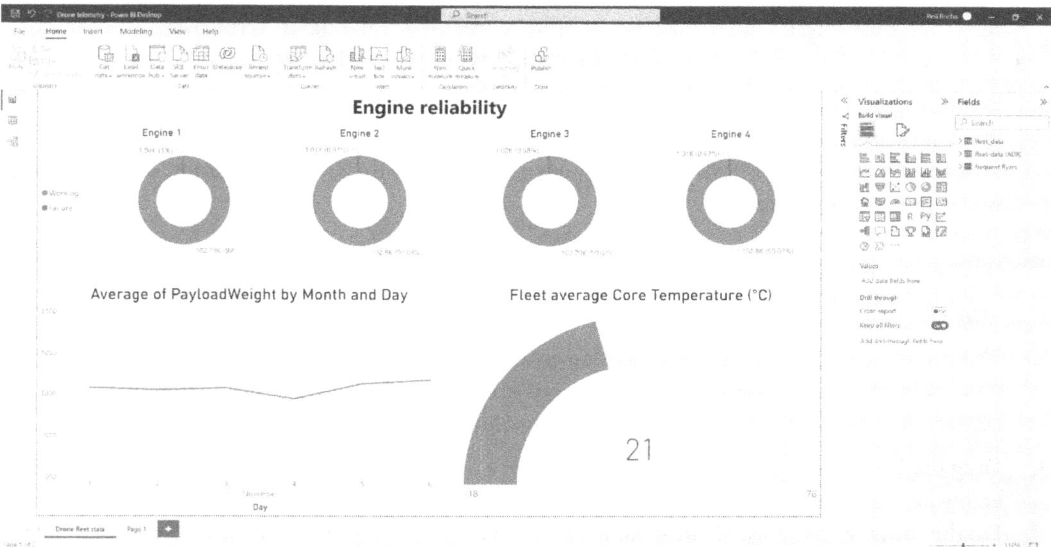

Figure 7.3 – A Power BI report connected to the Data Explorer pool

To build this report, follow these steps:

1. On the right-hand side of Power BI Desktop, expand your dataset and select the `Engine1Status` column. This will create a table in the canvas area showing the values for the `Engine1Status` column.
2. Go to the **Visualizations** pane and, making sure your table with the `Engine1Status` column has been selected in the canvas area, select the **Donut** chart option.
3. Under the **Visualizations** area, you should see `Engine1Status` in the **Legend** field. Drag `Engine1Status` from the **Fields** area again, but this time to the **Values** field. You should now see `Count of Engine1Status` in the **Values** field and the donut chart showing you the split between working and failed counts, as shown in *Figure 7.3*.
4. Repeat steps 1–3 for the `Engine2Status`, `Engine3Status`, and `Engine4Status` columns. Adjust the position of the charts to your preference or follow the example in *Figure 7.3*.
5. In the Power BI app menu, select **Insert**, and then select the textbox option to insert a caption over the four donut charts. Type in `Engine reliability` and format the text style as desired.

Next, you will create the line chart in the lower-left corner of the report:

1. Select a blank part of the canvas and check the `PayloadWeight` column under the **Fields** section.
2. You will see a bar chart with the sum of `PayloadWeight.` In the **Visualizations** pane, select the **Line** chart. Under the **Visualizations** area, select the down arrow next to `Sum of PayloadWeight` and change the aggregation from **Sum** to **Average**.
3. With the line chart selected, drag the `LocalDateTime` column from the **Fields** area into the **X-axis** field of the line chart. Select the **X** symbol next to **Year** and **Quarter** to leave only **Month** and **Day** as part of the *x* axis of your chart. Adjust the size and position of your chart in the canvas as needed.

Finally, you will create a gauge that shows the average core temperature:

1. Select a blank area in the canvas and check the `CoreTemp` column in the **Fields** area.
2. This produces a table with all the `CoreTemp` values present in your dataset. In the **Visualizations** pane, select **Gauge**. Under the **Visualizations** area, select the down arrow next to `Count of CoreTemp` and change the aggregation from **Count** to **Average**.
3. Drag the `CoreTemp` column from the **Fields** area to the **Minimum** value field. Select the down arrow next to `Count of CoreTemp` and change the aggregation from **Count** to **Minimum**. Do the same with the **Maximum** value field, replacing the aggregation with **Maximum**.

Adjust the chart title and sizes to your preference, or in accordance with the example in *Figure 7.3*, and you are done. Next, we will create additional data sources in the Power BI report for future use and publish the report and the data sources to your Power BI workspace.

Adding data sources to your Power BI report

Before you publish your new report, let's create additional data sources for it. This will allow us to build new reports later in Azure Synapse Studio.

The first report connected directly to your Data Explorer pool, and if configured to use the **DirectQuery** connectivity mode, could be used as a real-time report for telemetry. For the new data sources, we will connect to the lake database that we created in the *Exploring Data Explorer pool data with Python* section of *Chapter 6, Data Analysis and Exploration with KQL and Python*, and leverage the new columns we created as part of the data transformation process.

To create the new data sources, execute the following steps:

1. In Power BI Desktop, click on **Get data** in the toolbar.
2. In the **Get Data** window, pick **Azure** from the left-hand side of the window, and then select **Azure Synapse Analytics SQL**. Select **Connect**.
3. You will see a dialog with the title **SQL Server database**. Don't panic – you are in the right place! The same drivers used to connect to SQL Server can be used to connect to Azure Synapse SQL. Within this dialog, you should provide the following details required to connect to your serverless SQL pool:

   - **Server**: This is the URI of your serverless SQL pool. You can find this information in Azure Synapse Studio, by navigating to the **Manage** hub, selecting **SQL pools**, and clicking on the **Built-in** pool. This value is listed in the **Workspace SQL endpoint** textbox. Additionally, you can infer the name of your serverless SQL pool from the name of the workspace, as this endpoint follows the `<workspace>-ondemand.sql.azuresynapse.net` format, where `<workspace>` is the name of your Azure Synapse workspace. In my example, the URI is `drone-analytics-ondemand.sql.azuresynapse.net`, but you should insert the URI of your serverless SQL pool instead.
   - **Database**: This is the name of the database you used when you created the lake database in the *Exploring Data Explorer pool data with Python* section of *Chapter 6, Data Analysis and Exploration with KQL and Python*. In this example, I used the name `drone_telemetry`.
   - **Data Connectivity mode**: This is the same as explained in the *Creating a Power BI report* section. For this example, I will use **Import**.
   - **Advanced options**: For the first of two data sources we create, we will use a SQL query that fetches some specific information that we'll use in our report. This query returns the top 10 distinct combinations of device names and accumulated distances, within a date range. As a reminder, we created the column that computes the accumulated distances for each drone in the *Performing data transformation tasks* section of *Chapter 6, Data Analysis and Exploration with KQL and Python*. Use the following code in the **SQL statement (optional, requires database)** box:

     ```
     SELECT DISTINCT TOP 10 DeviceName, accumulatedDistance
     FROM [drone_telemetry].[dbo].[fleet_data]
     WHERE LocalDate >= '2021-11-01' AND LocalDate <= '2021-
     11-07'
     ORDER BY accumulatedDistance DESC
     ```

4. When you are ready, select **OK**.
5. You will be presented with a dialog that shows the result of your query (or some rows of your table when you don't specify the SQL statement). Select **Load**.
6. In the **Fields** pane, you will see a new object called **Query1**. This is your new data source. To make it easier to identify, right-click on **Query1** and select **Rename**. Change the name of your data source to a friendly name, such as `frequent flyers`.
7. Repeat steps 1–4 to create an additional data source, but this time, do not provide the name of the database as you did in step 3. Additionally, don't provide a predefined SQL statement as you did in step 3. Instead, only provide the **Server** parameter (which has the same value as before) and select **OK**.
8. You will see the **Navigator** dialog, which allows you to pick database objects that will be used in this data source. Expand the **drone_telemetry** database and check the box next to the **fleet_data** table, as shown in *Figure 7.4*. Review the data in the grid to make sure it contains the columns you created in the data transformation task. When you are ready, click on **Load**:

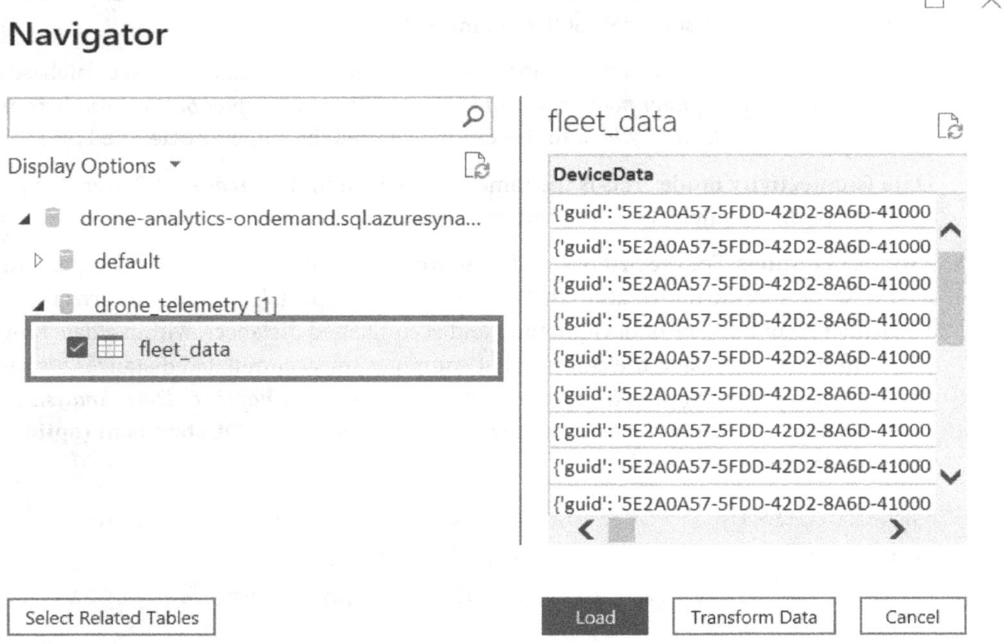

Figure 7.4 – Picking tables in the Navigator dialog

9. In the **Fields** pane, just like before, rename the new `fleet_data` data source object to `fleet-data (Lake DB)` to differentiate it from the previous connection to the `fleet-data` table in the Data Explorer pool.

Your new data sources are now created locally in your Power BI Desktop document. To publish your report and the new data sources to your Power BI workspace, click on the **Publish** button in the toolbar and follow the prompts. If prompted to save your changes, make sure you save your Power BI Desktop document. When done, your report and data sources should now be accessible from Azure Synapse Studio.

However, to work with Power BI in Azure Synapse Studio, you need to connect Power BI with your Azure Synapse workspace. Let's explore that next.

Connecting Power BI with your Azure Synapse workspace

To connect a Power BI workspace with your Azure Synapse workspace, perform the following steps:

1. Navigate to the **Home** hub of Azure Synapse Studio by selecting the **Home** button of the hub area.
2. In the center area of the workspace's landing page, select the **Visualize** tile. This opens the **Connect to Power BI** page, which prompts for values of the following parameters:

    - **Name**: This is the name of the linked service you will create in your workspace for Power BI. Give it a familiar name that will help you easily identify it in the future. For this example, I'm using the name `Drone telemetry Power BI workspace`.
    - **Description**: You can add any descriptive text that will help you describe this integration. This field is not required.
    - **Tenant**: This is the Power BI tenant name and ID. You should select from the drop-down control the tenant where your Power BI workspace is being hosted.
    - **Workspace name**: This is the name of the Power BI workspace you want to integrate with. Pick the desired one from the drop-down box.

Your **Connect to Power BI** page should look similar to the one illustrated in *Figure 7.5*:

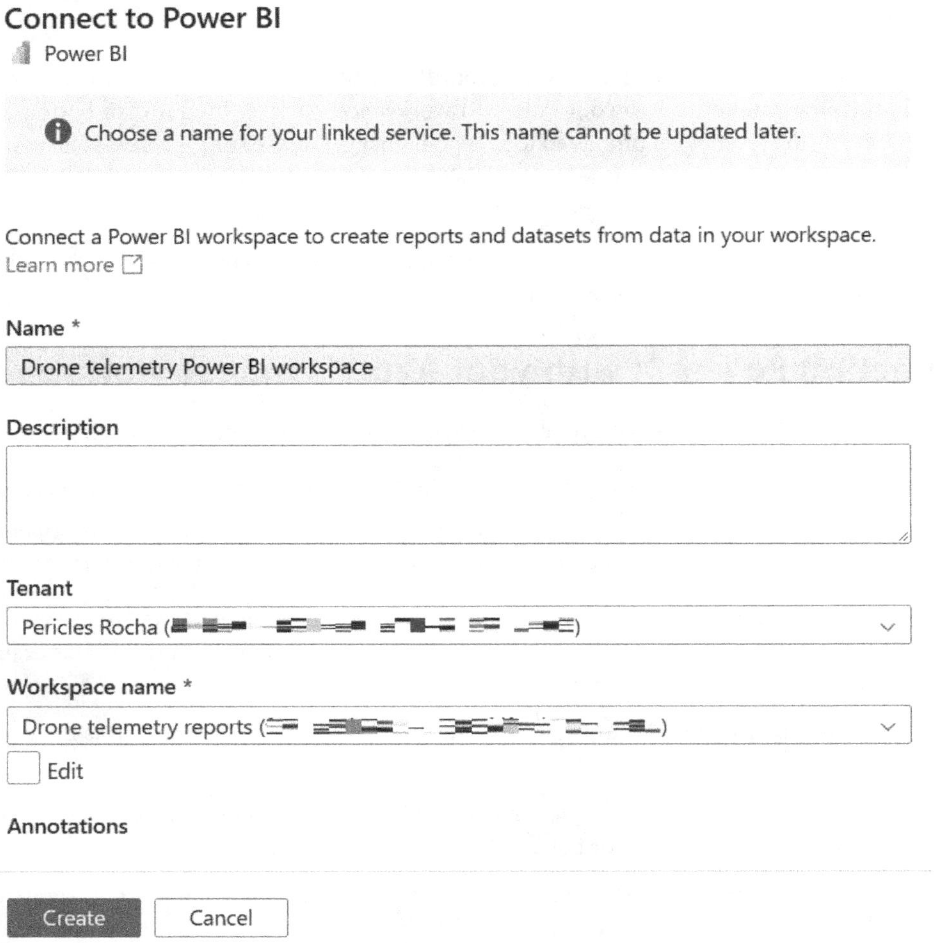

Connect to Power BI

Power BI

Choose a name for your linked service. This name cannot be updated later.

Connect a Power BI workspace to create reports and datasets from data in your workspace.
Learn more

Name *

Drone telemetry Power BI workspace

Description

Tenant

Pericles Rocha ([illegible])

Workspace name *

Drone telemetry reports ([illegible])

Edit

Annotations

Create Cancel

Figure 7.5 – Configuring the Power BI workspace integration (with the Tenant and Workspace names obfuscated)

3. When you are ready, select **Commit**.

To confirm the connection to your Power BI workspace, go to the **Develop** hub. You should now see Power BI listed under **Notebooks**, as shown in *Figure 7.6*. Expand Power BI to see your workspace, as well as the datasets and reports we published in the *Adding data sources to your Power BI report* section:

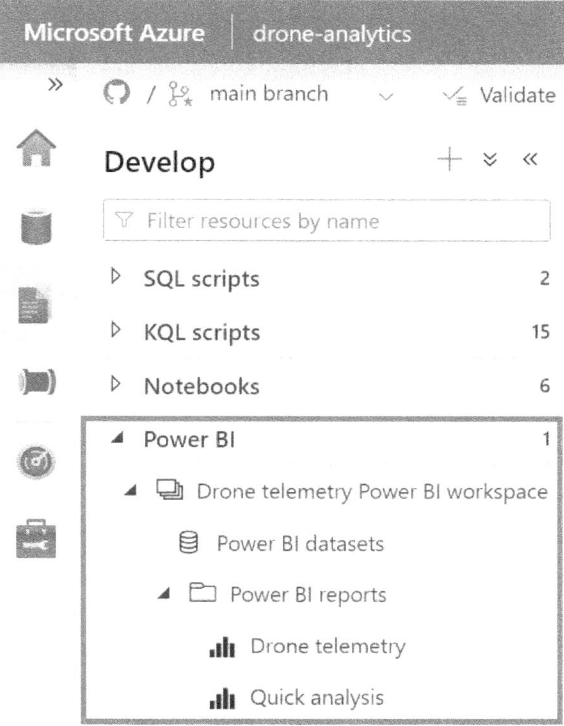

Figure 7.6 – The new Power BI group in the Develop hub

From now on, you can collaborate on Power BI reports with your colleagues in your Azure Synapse workspace. Azure Synapse Studio offers an interface to not only view but also create and edit existing reports that then get published directly into the Power BI service. We'll investigate this capability next.

Authoring Power BI reports from Azure Synapse Studio

One of the benefits of Azure Synapse workspaces is that you get to collaborate with colleagues working on the same project, sharing all the workspace assets. When you connect your workspace with Power BI, you get the benefit of collaborating on Power BI reports too with a report authoring experience embedded into Azure Synapse Studio.

In the previous section, you created a report using Power BI Desktop and published it to your Power BI workspace. Next, we will create a new Power BI report using Azure Synapse Studio, consuming the data sources that we deployed with our report.

To create a new Power BI report in Azure Synapse Studio, perform the following steps:

1. Navigate to the **Develop** hub, click on the + button, and select **Power BI report**.
2. From the **New Power BI report** page, select your dataset. The dataset you deployed with Power BI Desktop should have the name of the Power BI Desktop file that you saved locally to your PC. Select the dataset and click on **Create**.
3. You will be presented with a Power BI canvas just like the one you used in Power BI Desktop. In the following steps, you will create a report that is similar to the one shown in *Figure 7.7*, which shows the key influencers that affect drone battery life, and a table with the drones with the most miles in the period specified in our query:

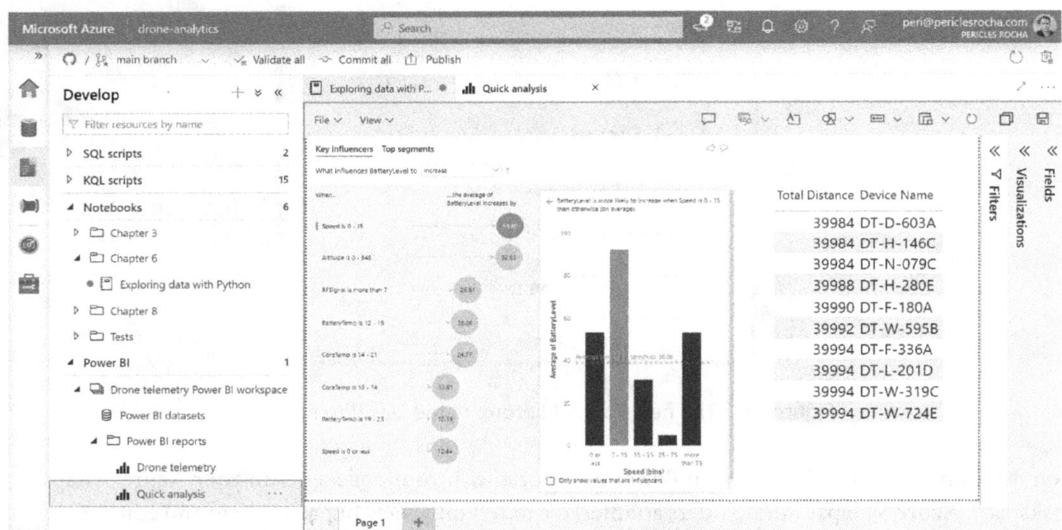

Figure 7.7 – Creating a new Power BI report in Azure Synapse Studio

To create this report, follow these steps:

1. Select a blank part in the report canvas area. In the **Fields** pane, select the `BatteryLevel` column under the `fleet_data` data source (not the one that references the Data Explorer pool). This action produces a bar chart with the sum of the `BatteryLevel` column.
2. Keep this bar chart selected and change it to a key influencers chart by going to the **Visualizations** pane and selecting the **Key influencers** chart type.
3. Drag the `BatteryTemp` column from the **Fields** pane into the **Explain by** field under the **Visualizations** area.

4. Select the down arrow next to `Sum of BatteryTemp` and select **Don't aggregate**. This causes this field to show actual values and not an aggregation of all the values in this column.
5. Repeat steps 3–4 with the `PayloadWeight`, `RFSignal`, `Speed`, `DistanceFromBase`, `CoreTemp`, and `Altitude` columns. Your **Explain by** field of the key influencers chart should look similar to *Figure 7.8*:

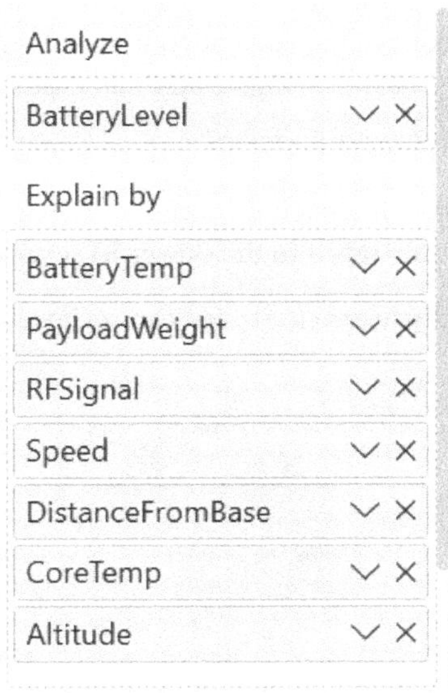

Figure 7.8 – Columns used in the key influencers chart

6. Resize and position the key influencers chart to your preference.

Next, we will create a table showing the top 10 drones with the most miles. To create this chart, follow these steps:

1. Select a blank part in the report canvas area. In the **Fields** pane, select the `accumulatedDistance` column under the `frequent flyers` data source. This action produces a bar chart with the sum of the `accumulatedDistance` column.
2. Keep this bar chart selected and change it to a table by going to the **Visualizations** pane and selecting the **Table** chart type.
3. Drag the `DeviceName` column from the **Fields** pane into the **Columns** field under the **Visualizations** area.
4. Select the down arrow next to `Sum of accumulatedDistance` and select **Don't aggregate**. This causes this field to show actual values and not an aggregation of all the values in this column.

5. Click on the `accumulatedDistance` column header twice to order it in decreasing order, showing the top drones by accumulated distance on the top.
6. Optionally, double-click on the `accumulatedDistance` column in the **Columns** area of the **Visualizations** pane to rename it to `Total Distance`. Do the same with the `deviceName` column to rename it to `Device Name`.
7. Additionally, you can format your table by making the font larger and removing the total row. These customizations can be done in the **Format your visual** section of the **Visualizations** pane. Feel free to browse the options that are available here and customize the table to your preference.
8. Resize and position the chart to your preference.

Your report is now ready. You can give it a friendly name by selecting **File** and **Save As**. When you save your report, it is also automatically saved in the Power BI service.

Let's take a better look at the key influencers chart. You can review it closely in *Figure 7.9*:

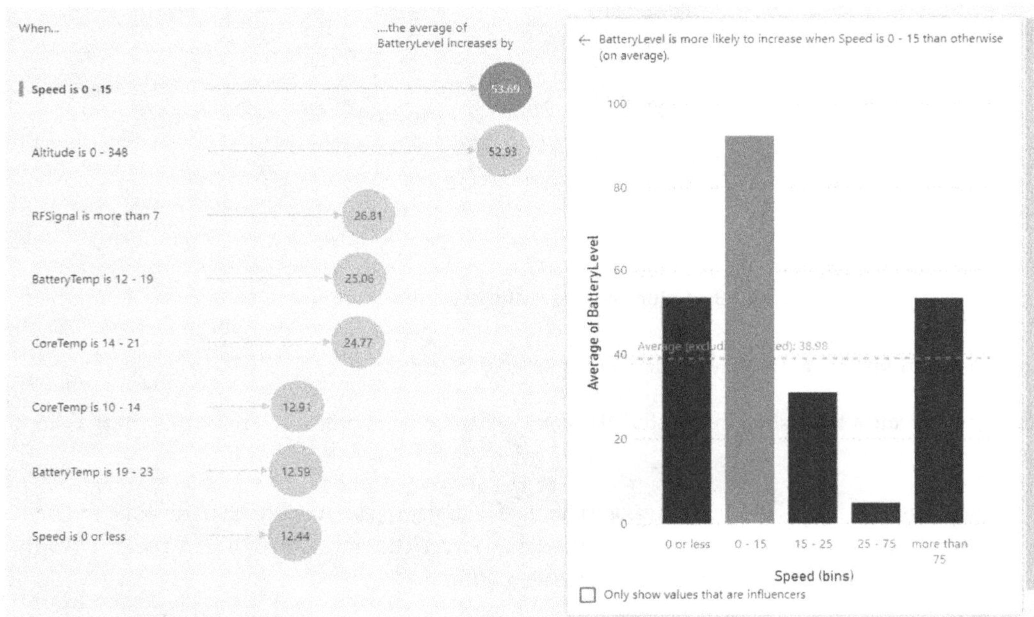

Figure 7.9 – The key influencers chart

By analyzing the key influencers chart, we conclude that battery consumption is optimal when the following applies:

- The drone's speed is from 0 to 15 km/h
- The drone's altitude is from 0 to 348 meters
- The drone's **radio frequency** (**RF**) signal is more than 26.81
- The drone's battery temperature is between 12 and 19 degrees
- The drone's core temperature is between 14 and 21 degrees

Other factors also influence battery life, but with less importance. By selecting the circles on the left-hand side of the chart, you can see a breakdown of the values in this column that influence the battery life column as a bar chart.

The table on the right-hand side of our report shows the drones with the most accumulated miles in the period of our query. This can be used to determine which drones may need maintenance based on mileage.

Summary

In this chapter, you learned how to configure and leverage Power BI integration with Azure Synapse. You started by creating your first Power BI report using Power BI Desktop, which you connected to your Data Explorer pool for a live view of your telemetry data.

Next, you added a data source to connect to the lake database that we created in *Chapter 6, Data Analysis and Exploration with KQL and Python*, and created a new report that uses that data source. You connected the Power BI service to your Azure Synapse workspace and opened your new report in Azure Synapse Studio.

Finally, made changes to your report directly in Azure Synapse Studio, without using Power BI Desktop, and even created a new report directly from the web interface. This makes it easy for anyone to author new Power BI reports, regardless of the operating system they use.

Data analysis goes beyond writing queries and reports. In the next chapter, we will look at how to use other Azure Synapse Analytics and Azure services to enhance your data using machine learning.

8

Building Machine Learning Experiments

By now, you've learned how to load data into your Data Explorer pools and perform analysis using the **Kusto Query Language** (**KQL**), Python/PySpark, and Power BI. The next boundary to explore in analytics is how to do a more comprehensive analysis of your data, and even predict future behavior. **Machine Learning** (**ML**) helps us cross this new boundary, and Azure Synapse offers a vast array of options to apply ML algorithms to your data. In this chapter, you will learn how to build ML experiments with your Data Explorer pool.

First, you will learn how ML is applied to day-to-day tasks by discussing the most common algorithms that are used today. Note that this will not be an extensive list of all the ML algorithms, but it will help you understand the core tasks that it helps resolve.

Next, you will train a new model using **Automated Machine Learning** (**AutoML**). You will create a new Azure Machine Learning workspace and link it to your Azure Synapse workspace. Then, you will walk, in detail, through the process of configuring AutoML and working with it through an Azure Synapse notebook. You will use the best model obtained from AutoML to predict some test values and evaluate the model's performance.

Finally, you will learn other ways to use ML with Azure Synapse. You will explore options such as using pre-trained models to leverage the Azure Cognitive Services integration and using Apache Spark's MLlib in Azure Synapse. Additionally, we'll look at a quick example of how you can use some simple functions in KQL to score values in real time using KQL queries. To wrap things up, we will briefly discuss building applications using the SynapseML library.

In this chapter, we will cover the following topics:

- Understanding the application of ML
- Introducing ML into your projects with AutoML
- Exploring additional ML capabilities in Azure Synapse

Technical requirements

In case you haven't already, make sure you download the book materials from our repository at `https://github.com/PacktPublishing/Learn-Azure-Synapse-Data-Explorer`. You can download the full repository by selecting **Code**, and then **Download ZIP**, or by cloning the repository using your Git client of choice.

This chapter uses the full `fleet data` table, which you ingested in the *Running your first query* section of *Chapter 3, Exploring Azure Synapse Studio*. If you did not perform the tasks in this section to load data, take some time to go back and follow the instructions to ingest the full `fleet data` table.

Just as with *Chapter 6, Data Analysis and Exploration with KQL and Python*, this chapter also uses basic Python syntax in the code examples. If you need a refresher on Python and Pandas, a helpful tutorial can be found at `https://www.w3schools.com/python/pandas/default.asp`.

ML can be an overwhelming topic. Even though we will not go too deep into ML, you might be curious to learn more. If you need a good book to get started on ML, I recommend *Machine Learning for Absolute Beginners* by Oliver Theobald.

Finally, this chapter discusses Azure services outside of Azure Synapse, such as Azure Machine Learning and Azure Cognitive Services, without covering them in detail. Microsoft Learn is a useful resource to learn more about these services if you need to become familiar with them. You can access it at `https://learn.microsoft.com/`.

Understanding the application of ML

Today, ML is widely used in every industry to help companies and non-profit organizations gain deeper insights into their business through data. As a branch of artificial intelligence, it allows you to apply statistics with historical data to predict outcomes, forecast future values, detect anomalies in patterns, and more.

Some of the most popular techniques or algorithms used in ML are listed as follows:

- **Regression**: This helps you respond to questions such as how much or how many. For example, how many units will I sell next quarter? How much crime will a region have next month? What is the projected fuel consumption given certain variables?
- **Classification**: This helps you categorize events or data points. For example, will this customer churn (yes or no)? What animal is this?
- **Clustering**: This helps you group data points together. For example, which customers tend to purchase this product? Which demographics are likely to engage in my campaign?
- **Dimensionality reduction**: This helps you choose the variables that matter the most for your analysis and remove any redundancy identified by collinearity between them. For example, which variables (also known as features or columns) should I use to train my ML model?

- **Anomaly detection**: This helps you understand whether something is out of the ordinary. For example, is this a fraudulent transaction? Is that drone failing? Is my network being attacked?

Note that many other approaches and techniques are used in ML and, for each of the approaches described, there are several ways to solve these problems.

ML is not something new. The term was coined by Arthur Lee Samuel, who is known for his work teaching computers to play checkers in 1959, which is the year he helped make the term popular. It gained popularity in the last decade due to the decreasing costs of storage, allowing you to use greater volumes of data to train models, and the broad availability of compute resources, especially in the cloud.

To mention a real-life experience, allow me to share an experience I had in the mid-2000s when I briefly worked on a project for an oil company. In this project, they used over 100 servers on a cluster that processed data, with an application that implemented an embarrassingly parallel approach to processing massive amounts of data. Building and managing this large cluster was an expensive and time-consuming task, which is not something most companies could do without some serious capital and operational investment. However, nowadays, with the availability of compute platforms for ML in the cloud, companies can now train models and analyze large volumes of data and pay only for the time they are using the compute resources and storage. There's no need for the expenditure of hardware and software, only to leave the environment idle most of the time. Cloud ML platforms offer huge value to customers in that regard.

When it comes to Azure Synapse, there are a few different ways you can use ML with Data Explorer pools in your projects. Next, we will explore some of the different approaches that you might want to take depending on your project requirements. If you currently don't have experience with ML but would like to get started, a great way to do that is to use AutoML. Let's look at what AutoML is and how you can use it.

Introducing ML into your projects with AutoML

One of the challenges for data scientists and analysts when working with ML is that they need to perform a long series of tests before they can find the optimal parameters to train a model that maximizes its results. From finding and transforming the right features to identifying the algorithms that perform best and tuning their parameters just the right way, this is an extensive task that requires many hours of testing and iteration.

In 2018, Microsoft announced the availability of its AutoML feature in Azure ML during a preview. AutoML works like a recommendation system for ML algorithms: you provide your data, set AutoML to the task you want to achieve (e.g. classification, forecasting, or regression), provide your task parameters (e.g. model metrics or cost constraints), and AutoML will test several different algorithms and parameters until it finds the best model for your data. Then, you can export this model or register it in your Azure Machine Learning workspace for later use.

> **Note**
> To learn more about AutoML, visit `https://learn.microsoft.com/azure/machine-learning/concept-automated-ml`.

As mentioned earlier, AutoML is a capability of the Azure Machine Learning service. The good news is that Azure Synapse integrates with Azure Machine Learning through a linked service, and even offers some in-product integrations in Azure Synapse Studio to make your job easier. Let's look at an example of how we could use it to train a model that predicts what a drone's core temperature would be in certain conditions.

Creating an Azure Machine Learning workspace

To configure the integration between your Azure Synapse workspace and Azure Machine Learning, you need to have an Azure Machine Learning workspace created beforehand. If you don't have one already, you can create one by following these steps:

1. Navigate to `https://ml.azure.com` and log in with the same account you use for your Azure Synapse workspace.
2. Select **Workspaces** from the left-hand menu, and then click on the **+ New** button.
3. You will see the **Create new workspace** dialog that's illustrated in *Figure 8.1*. It requires the following input parameters:
   - **Workspace name**: This is the name you want to give to your workspace. Use a name that will help you identify it in the future. For this example, I named it `telemetry_experiments`.
   - **Subscription**: This is the Azure subscription that you will use to host this service. Pick the desired one from the drop-down menu.
   - **Resource group**: This is the resource group where you will store this service. You can pick an existing one or create a new one. For this example, I'm creating a new resource group named `rg-AzureML`. This will make it easier to delete this and any dependent resources after I no longer need the Azure Machine Learning integration.
   - **Region**: This is the Azure region where your resources will be deployed to. It's important to choose the same region where your Data Explorer pools reside (that is, the region of your Azure Synapse workspace) to avoid any data transfer between regions. For this example, I picked **East US**:

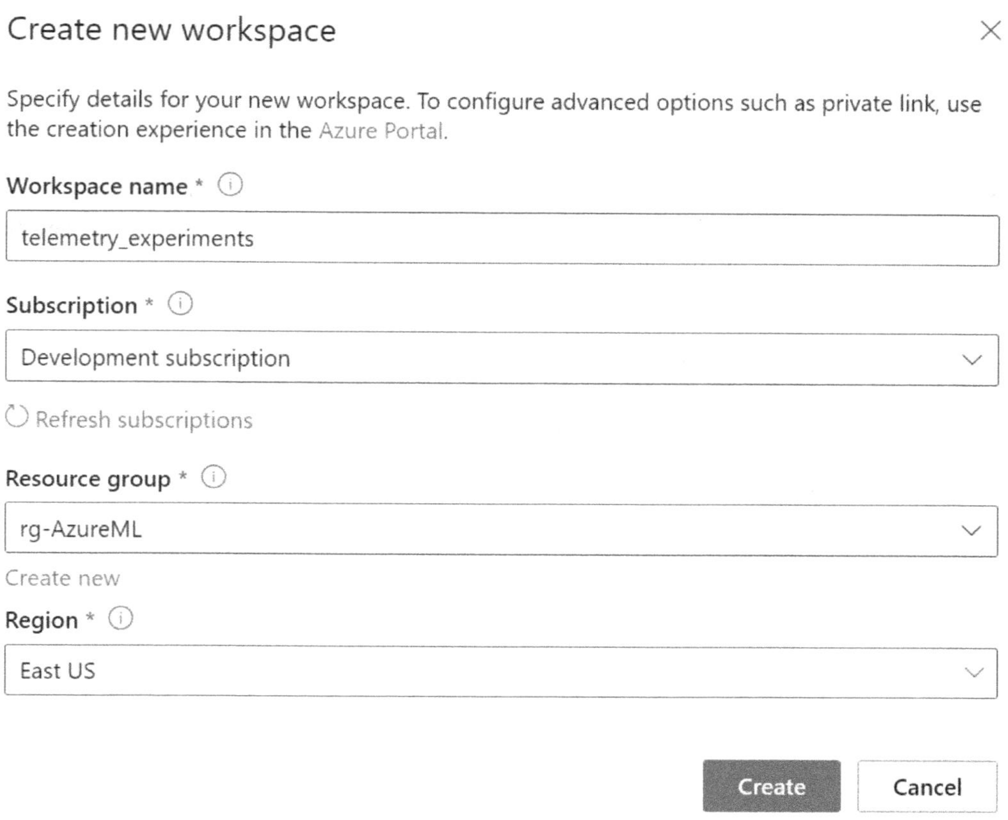

Create new workspace

Specify details for your new workspace. To configure advanced options such as private link, use the creation experience in the Azure Portal.

Workspace name *

telemetry_experiments

Subscription *

Development subscription

Refresh subscriptions

Resource group *

rg-AzureML

Create new

Region *

East US

Create Cancel

Figure 8.1 – Creating your Azure Machine Learning workspace

When you are ready, click on **Create**. It takes about a minute for your new workspace to be provisioned.

Before we are done, we need to grant the proper permissions to allow our Azure Synapse workspace to interact with the new Azure Machine Learning workspace. We do that by adding our Azure Synapse workspace's managed identity to the **Contributor** role of the Azure Machine Learning workspace. Here's how to do it:

1. Navigate to the Azure portal at `https://portal.azure.com` and find your new Azure Machine Learning workspace. The easiest way to find it is to simply type `telemetry_experiments` (or the name of your workspace if you chose a different one) into the global search box.
2. From your Azure Machine Learning workspace page, select **Access control (IAM)**. Click on the **Add role assignment** button.
3. Select the **Contributor** role and click on **Next**.

4. Next to **Assign access to**, select **Managed identity**.
5. Click on + **Select members** to reveal the **Select managed identities** page. In the **Managed identity** combo box, select **Synapse workspace**, and then select your workspace name under the **Select** textbox. The workspace identity will move to the **Selected members** group. When you are ready, click on the **Select** button.

Now that you have your Azure Machine Learning workspace created and the proper permissions assigned, we are ready to configure the linked service between your Azure Synapse workspace and Azure Machine Learning.

Configuring the Azure Machine Learning integration

To leverage the AutoML feature of Azure Machine Learning in Azure Synapse, we need to configure a new linked service between the two services. This configuration is similar to the one we did for Power BI in *Chapter 7, Data Visualization with Power BI*, only with slightly different parameters.

To configure the linked service, run the following steps:

1. Navigate to Azure Synapse Studio at `https://web.azuresynapse.net` and log in to your workspace.
2. In the hubs region, select the **Manage** hub.
3. Under **External connections**, select **Linked services**.
4. From the **Linked services** page, select + **New** to reveal the **New linked service** page.
5. Select the **Compute** option, and then pick **Azure Machine Learning**. Click on **Continue**.
6. The **New linked service** page appears. It requires the following parameters:

    - **Name**: This is the name of the linked service. You will need this name when we use AutoML, so make sure you use a name that's easy to remember. For this example, I picked `AzureML_telemetryexperiments`.
    - **Connect via integration runtime**: This is the compute infrastructure that executes the data integration tasks. By default, Azure Synapse uses **AutoResolveIntegrationRuntime**. This integration runtime tries to detect the Azure region where you will copy data and allocates the compute infrastructure for performing the data integration job in that same region. Optionally, you can host an integration runtime within your own premises. For this example, I used the default **AutoResolveIntegrationRuntime** option.
    - **Authentication method**: This is how your workspace will authenticate against the Azure Machine Learning service. Since we added the workspace's managed identity to the contributor role of the Azure Machine Learning workspace, we will use **System Assigned Managed Identity** here.

- **Azure Machine Learning workspace selection method**: Select this if you wish to browse your Azure subscription to find the Azure Machine Learning workspace, or if you want to type in the workspace details manually. If you don't have many Azure Machine Learning workspaces in your subscription, it's easier to just choose the **From Azure subscription** option. Just pick the Azure subscription where you created your workspace, and then pick the workspace from the **Azure Machine Learning workspace name** combo box.

7. Click on **Test connection** to make sure everything has been configured correctly. If you get an error with an HTTP status code of `Forbidden`, make sure you review the workspace permissions, as per the *Creating an Azure Machine Learning workspace* section. After your connection test succeeds, click on **Commit**.

Your linked service has now been created, and you are ready to start using AutoML integrated with Azure Machine Learning. Next, we will run an experiment that finds the best model to predict a drone's core temperature given a telemetry event.

Finding the best model with AutoML

The process of using AutoML to find the best model for your ML task consists of four steps: loading your data, configuring your Azure Machine Learning experiment, running the experiment, and exploring the results. You will perform these tasks using your Apache Spark pool and PySpark.

> **Note**
>
> A notebook with all the code used in this section can be found in the `Chapter 08\Regression example.ipynb` file of the book repository.

Azure Synapse Studio offers a convenient wizard to help set up your experiment, which produces almost all the PySpark code for you. All you have to do is make any adjustments you would like to do with your data, and then write the logic that explores the results of the experiment. To build your first AutoML experiment, follow these steps:

1. Go to Azure Synapse Studio at `https://web.azuresynapse.net` and log in to your workspace.
2. From the **Data** hub, expand **Lake Database** and find the `drone_telemetry` database you created in *Chapter 7*, *Data Visualization with Power BI*. Don't worry! You will change your notebook code to retrieve data from your Data Explorer pool instead of the lake database.
3. Expand **Tables** and find the `fleet_data` table. Click on the ellipsis to reveal the context menu. Hover over **Machine Learning** and select **Train a new model**.
4. The **Train a new model** wizard page appears, as shown in *Figure 8.2*. This page describes the model types you can use with AutoML depending on the question you want to answer with your data. For our example, we want to predict a drone's core temperature given a telemetry event, so we will choose **Regression**. Click on **Continue** when you are ready:

Train a new model

fleet_data

This wizard will help you to train a machine learning model using Automated Machine Learning.

Choose a model type

Select the machine learning model type for the experiment based on the question you are trying to answer. Once you have selected the model type, you will be prompted with a few settings before the experiment run is created. Learn more

Classification

Determine the likelihood of a specific outcome being achieved (binary classification) or identify the category an attribute belongs to (multiclass classification).

Example: Predict if a customer will renew or cancel their subscription.

Regression

Estimate a numeric value based on input variables.

Example: Predict housing prices based on house size.

Time series forecasting

Estimate values and trends based on historical data.

Example: Predict stock market trends over the next year.

Continue | Cancel

Figure 8.2 – The Train a new model wizard

5. Next, you will see the **Configure experiment** page, as illustrated in *Figure 8.3*. This page of the wizard requires the following input parameters:

   - **Azure Machine Learning workspace**: This is the name of your Azure Machine Learning workspace and linked service name, as named in the *Configuring the Azure Machine Learning integration* section. Pick the correct one from the list.
   - **Experiment name**: Here, you can specify the name of the experiment that will be submitted to Azure Machine Learning. In this example, I used `Predict_Core_Temp`.

- **Best model name**: When AutoML runs, it will experiment with dozens of models and parameters until it reaches the optimal model based on your data. Here, you specify the name you will give to this optimal model. I named my model `Predict_Core_Temp - Best Model`.
- **Target column**: This field lists all the columns available in your table. You should choose the column that you want to predict values for using regression. In this case, you should pick the `CoreTemp` column.
- **Apache Spark Pool**: Here, you specify the Apache Spark pool that will be used to process your AutoML task:

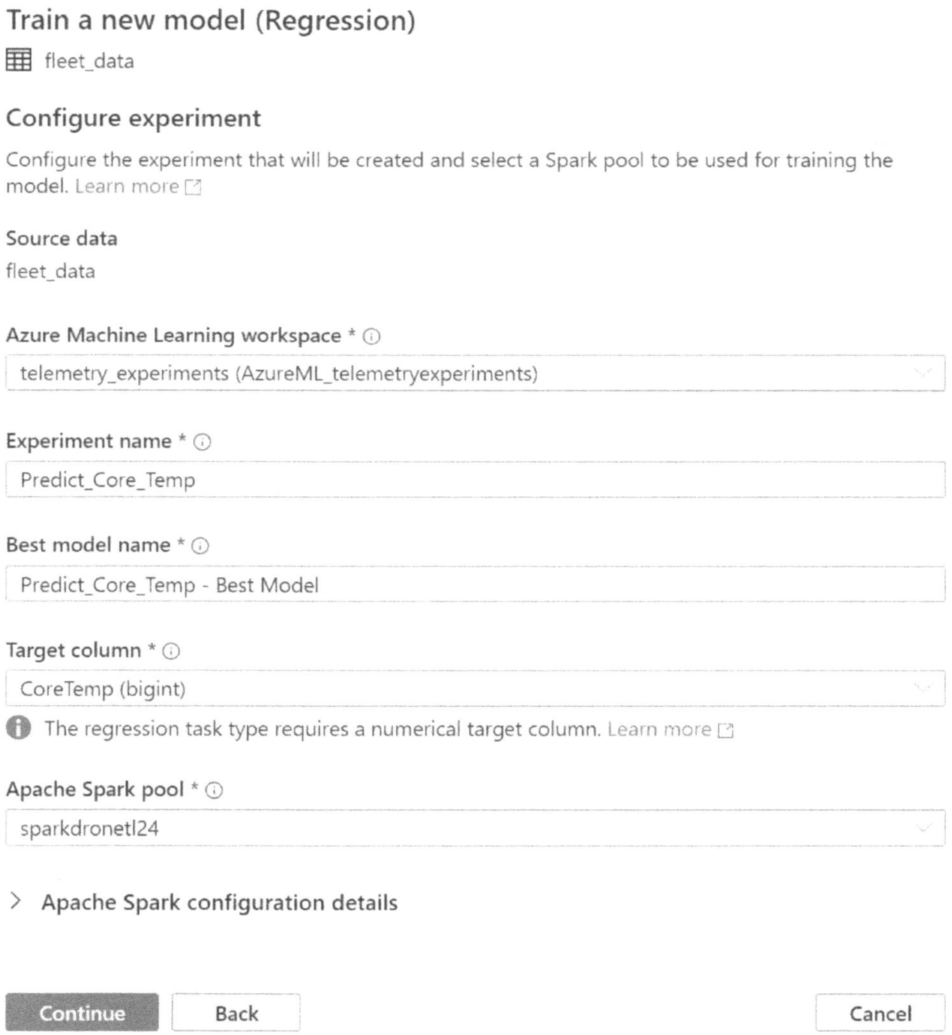

Train a new model (Regression)

fleet_data

Configure experiment

Configure the experiment that will be created and select a Spark pool to be used for training the model. Learn more

Source data

fleet_data

Azure Machine Learning workspace *

telemetry_experiments (AzureML_telemetryexperiments)

Experiment name *

Predict_Core_Temp

Best model name *

Predict_Core_Temp - Best Model

Target column *

CoreTemp (bigint)

The regression task type requires a numerical target column. Learn more

Apache Spark pool *

sparkdronetl24

> Apache Spark configuration details

Continue Back Cancel

Figure 8.3 – Configuring your AutoML experiment

6. When you are ready, click on **Continue**. You will be directed to the **Configure the regression model** page, which is the last one in the wizard. This page, as shown in *Figure 8.4*, requires the following input parameters:

   - **Primary metric**: This is the metric that AutoML uses to evaluate the performance of the models as it tests them. AutoML uses this metric to determine which is the best model for your experiment. For this example, I picked **R2 score**, which is a standard way to measure the performance of regression models. R2 (pronounced R squared) measures the variance in the predictions obtained, as explained by your dataset.
   - **Maximum training job time (hours)**: This parameter determines how long your job should run before it is stopped automatically. Training ML models can take hours, depending on how many features (or columns) your dataset has, the data volume, and your compute size options. Use this parameter to set a time boundary to control costs. For this example, I am limiting the job to run for 1 hour at most.
   - **Max concurrent iterations**: AutoML starts a series of child jobs in parallel to minimize the total execution time of the experiment. This parameter sets how many child jobs can run concurrently at any given time. The recommendation is to set this to the same number of nodes you have in your Apache Spark pool, so each child job runs in one node at a time. Since the Apache Spark Pool that I'll use for this experiment has two nodes, I will set this parameter to `2`.
   - **ONNX model compatibility**: This option allows you to export your best model in the **Open Neural Network eXchange** (**ONNX**) format. ONNX is an open source format for ML models that can be used on many different platforms. In Azure Synapse, you can import an ONNX model to a table on a dedicated SQL pool and use the `PREDICT` **Transact-SQL** (**T-SQL**) statement to consume ML models directly from a T-SQL query, an immensely powerful capability. Since this is not the focus of our experiment, I will choose **Disable**:

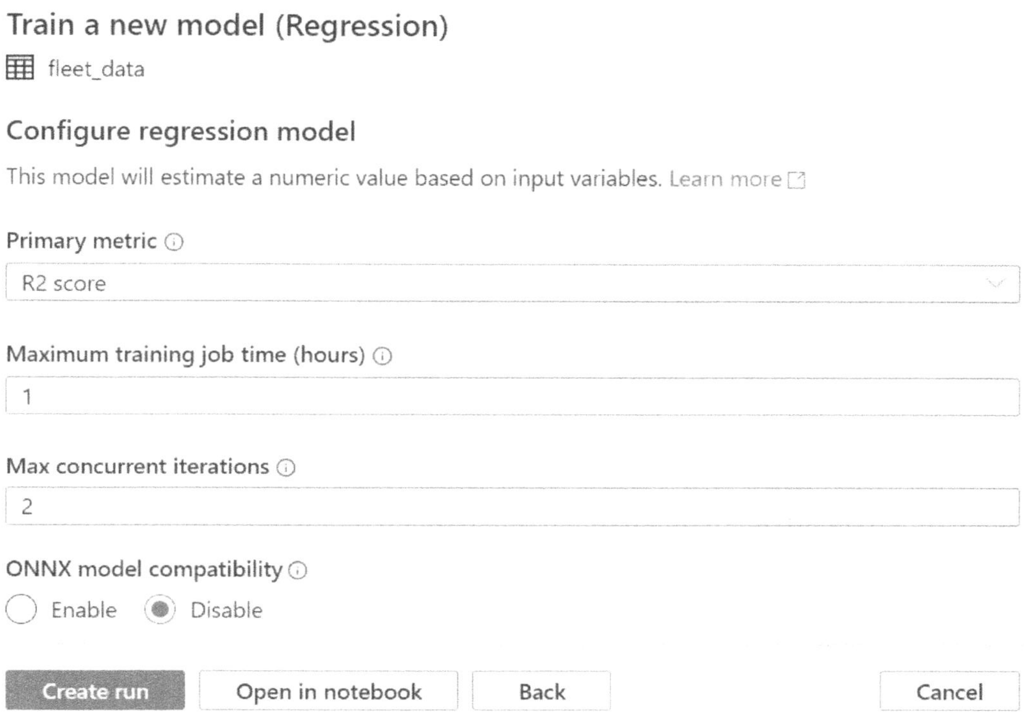

Train a new model (Regression)

fleet_data

Configure regression model

This model will estimate a numeric value based on input variables. Learn more

Primary metric

R2 score

Maximum training job time (hours)

1

Max concurrent iterations

2

ONNX model compatibility

Enable Disable

Create run Open in notebook Back Cancel

Figure 8.4 – Configuring job parameters

7. When you are ready, click on the **Open in notebook** button.

This opens your new AutoML experiment as an Azure Synapse notebook. Let's review each of the code cells generated by the wizard and make a few minor changes.

The first code cell imports the libraries needed for this notebook:

```
import azureml.core
from azureml.core import Experiment, Workspace, Dataset,
Datastore
from azureml.train.automl import AutoMLConfig
from notebookutils import mssparkutils
from azureml.data.dataset_factory import TabularDatasetFactory
```

Next, we define variables with the name of the linked service object that we created in the *Configuring the Azure Machine Learning integration* section, the name of the experiment that we created in the *Finding the best model with AutoML* section, and then we create a new experiment using the connection to the linked service:

```
linkedService_name = "AzureML_telemetryexperiments"
experiment_name = "Predict_Core_Temp"
ws = mssparkutils.azureML.getWorkspace(linkedService_name)
experiment = Experiment(ws, experiment_name)
```

In the next code cell, you will see that the notebook that was automatically generated by the **Train a new model** wizard produces a SQL query that selects everything from the `fleet_data` table in the lake database. We want to change this to read from the table in the Data Explorer pool. Additionally, we will perform a few actions with our data once we have loaded it into a Spark DataFrame:

- **Downsampling the data**: Our original table has over 337,500 rows. To test this notebook and make sure everything works, it's a good idea to work with a smaller subset of your data to avoid the increased costs of data retrieval and the compute costs. This code reduces the Spark DataFrame to 1% of its size. Once you are done with testing and ready to test this model at full scale, feel free to remove this line of code.
- **Selecting a few columns**: Not all the columns in your table will be useful to train a model. For example, the `DeviceData` column, which is a JSON column that contains the device's ID and name, is not relevant to determining the device's core temperature. Here, you'll only select columns that you think might help train the regression model. Additionally, in this example, you will also produce three new columns that store the hour, the day of the month, and the day of the week of the telemetry event. Having this type of data separated into new columns can sometimes help find insights related to a certain fixed time or day an event happens.
- **Splitting the data into training and validation sets**: When you train ML models, you want to separate a portion of your data to train the model and use the other portion to validate the model's accuracy (by comparing the predicted data against actual values). In this example, we will set 80% of the available data to the training set and 20% to the validation set.
- **Creating a tabular dataset from the Spark DataFrame**: This is already in the notebook that was generated by the wizard, but we want to make sure we change this function to use our training set stored in the `training_set` variable and not the original full Spark DataFrame.

The new code for this cell is the following:

```
# Changing the query to the Data Explorer pool table
# Make sure you change this code to use YOUR Data Explorer
pool,
# database, and table.
```

```
df  = spark.read \
    .format("com.microsoft.kusto.spark.synapse.datasource") \
    .option("spark.synapse.linkedService", "kustoPool") \
    .option("kustoCluster", "https://<YourDataExplorerPoolName-
Here>.<YourAzureSynapseWorkspaceNameHere>.kusto.azuresynapse.
net") \
    .option("kustoDatabase", "drone-telemetry") \
    .option("kustoQuery", "['fleet data']") \
    .load()

# Downsampling data
df = df.sample(True, 0.01, seed=1234)

# Select specific columns
from pyspark.sql.functions import *
df = df.select(
      'DeviceState',
      'Engine1Status','Engine2Status','Engine3Status','En-
gine4Status',
      'Engine1RPM','Engine2RPM','Engine3RPM','Engine4RPM',
      'Engine1Temp','Engine2Temp','Engine3Temp','Engine4Temp',
      'CoreTemp','BatteryTemp','CoreStatus','MemoryAvaila-
ble','BatteryLevel',
      'Altitude','Speed','DistanceFromBase','RFSignal','Pay-
loadWeight',
      date_format('LocalDateTime', 'hh').alias('HourOfDay'),
      dayofmonth('LocalDateTime').alias('DayOfMonth'),
      dayofweek('LocalDateTime').alias('DayOfWeek')
      )

# Create the training and validation data frames
training_data, validation_data = df.randomSplit([0.8,0.2],
seed=1234)

# Create a dataset from the training data frame,
# using the default workspace data store
datastore = Datastore.get_default(ws)
```

```
dataset = TabularDatasetFactory.register_spark_dataframe(train-
ing_data, datastore, name = experiment_name + "-dataset")
```

Next, we will add a code cell to see the structure of our new Spark DataFrame and make sure everything looks good:

```
display(df, summary=True)
```

As you can see, in *Figure 8.5*, the new columns we created to store the hour, the day of the month, and the day of the week values are present in the Spark DataFrame:

name	type	unique	missing
CoreStatus	string	2	0
MemoryAvailable	bigint	3348	0
BatteryLevel	double	43	0
Altitude	bigint	256	0
Speed	bigint	5	0
DistanceFromBase	bigint	1848	0
RFSignal	bigint	9	0
PayloadWeight	bigint	1601	0
HourOfDay	string	12	0
DayOfMonth	int	21	0
DayOfWeek	int	7	0

Figure 8.5 – Reviewing the structure of your Spark DataFrame

The next step is to configure the parameters of the AutoML task. We will make a slight change to the code that was generated by the wizard: create a new Python dictionary with a series of parameters that we will then use when we create the actual AutoML configuration object. Our new settings limit the experiment's timeout to 15 minutes and define other useful parameters. Here's the code:

```
import logging

automl_settings = {
    "iteration_timeout_minutes": 10,
    "experiment_timeout_minutes": 15,
    "enable_early_stopping": True,
    "featurization": 'auto',
```

```
    "verbosity": logging.INFO,
    "n_cross_validations": 2}

automl_config = AutoMLConfig(spark_context = sc,
                             task = 'regression',
                             training_data = dataset,
                             label_column_name= 'CoreTemp',
                             primary_metric = 'r2_score',
                             max_concurrent_iterations = 2,
                             enable_onnx_compatible_models =
False, **automl_settings)
```

Now, we're finally ready to submit our AutoML experiment to Azure Machine Learning. Here, you can reuse the code generated by the wizard without any changes. The code to submit the experiment is as follows:

```
run = experiment.submit(automl_config, show_output=True)
```

The preceding command generates a table as output, providing details about the experiment. The **Details Page** column, as highlighted in *Figure 8.6*, offers a link to this experiment in Azure Machine Learning Studio, where you can see the status of the experiment as it runs:

```
Submitting spark run.
No run_configuration provided, running on local with default configuration
```

Experiment	Id	Type	Status	Details Page	Docs Page
Predict_Core_Temp	AutoML_4cfa1ed6-47fc-4154-8979-f326d720424b	automl	NotStarted	Link to Azure Machine Learning studio	Link to Documentation

Figure 8.6 – Monitoring the experiment's execution

The AutoML experiment will build a series of models with different algorithms parameters using your data. Next, we will retrieve the run that provided the best result according to the metric we specified (that is, R2) and the model that produced this run. Once we have retrieved this model, we will register it in the Azure Machine Learning model registry for future use. This code block is the same, as it was created automatically for you:

```
run.wait_for_completion()

import mlflow
```

```
# Get best model from automl run
best_run, fitted_model = run.get_output()

artifact_path = experiment_name + "_artifact"

mlflow.set_tracking_uri(ws.get_mlflow_tracking_uri())
mlflow.set_experiment(experiment_name)

with mlflow.start_run() as run:
    # Save the model to the outputs directory for capture
    mlflow.sklearn.log_model(fitted_model, artifact_path)

    # Register the model to AML model registry
    mlflow.register_model("runs:/" + run.info.run_id + "/" +
artifact_path, "drone-analytics-fleet_data-20221026113234-
Best")
```

But how good is this model, really? Let's perform a few tasks to see its accuracy. The easiest way to do this is to use real data and see whether the model that we got can predict values close to the real ones. We can do this with the following code block:

```
import pandas as pd

# Create a Pandas DataFrame from the validation set
validation_data_pd = validation_data.toPandas()

# Create a new Pandas DataFrame with the REAL CoreTemp data
df_compare = validation_data_pd.pop('CoreTemp').to_frame()

# Predict the CoreTemp column using the fitted model
y_predict = fitted_model.predict(validation_data_pd)

# Add the predicted values to the new Pandas DataFrame
# for comparison
df_compare['Predicted'] = y_predict
display(df_compare)
```

This produces the table shown in *Figure 8.7* as a result. Note that, in most cases, the predicted value is pretty close to the real value from our validation set. Those are really good results:

CoreTemp	Predicted
-6	-2.9683907018972504
-1	1.1794805423380597
39	40.60617681624501
50	47.75019920626331
17	16.787901774350072
49	44.849166973169744
10	12.58377476890226
7	6.150294167784525

Figure 8.7 – A sample of real values versus predicted

To finalize the validation process, let's compute the R2 score of our model since that is the metric that we chose to evaluate it. The *y-axis* will show all the values that were predicted by our model, and the *x-axis* will show all the actual values in the dataset. An optimal model should see the data points meet at the same *y-* and *x-*coordinates, or as close as possible. You can use this code in a new code cell to perform this task:

```
import matplotlib.pyplot as plt
import numpy as np
from sklearn.metrics import mean_squared_error, r2_score

# Compute the R2 score by using the predicted and actual
CoreTemp
y_test_actual = df_compare['CoreTemp']
r2 = r2_score(y_test_actual, y_predict)

# Plot the actual versus predicted CoreTemp values
plt.style.use('ggplot')
plt.figure(figsize=(10, 7))
```

```
plt.scatter(y_test_actual,y_predict)
plt.plot([np.min(y_test_actual), np.max(y_test_actual)], [np.
min(y_test_actual), np.max(y_test_actual)], color='lightblue')
plt.xlabel("Actual")
plt.ylabel("Predicted")
plt.title("Actual vs Predicted CoreTemp R^2={}".format(r2))
plt.show()
```

As you can see, in *Figure 8.8*, our model was very efficient at predicting values for our dataset, with an R2 score of 0.98:

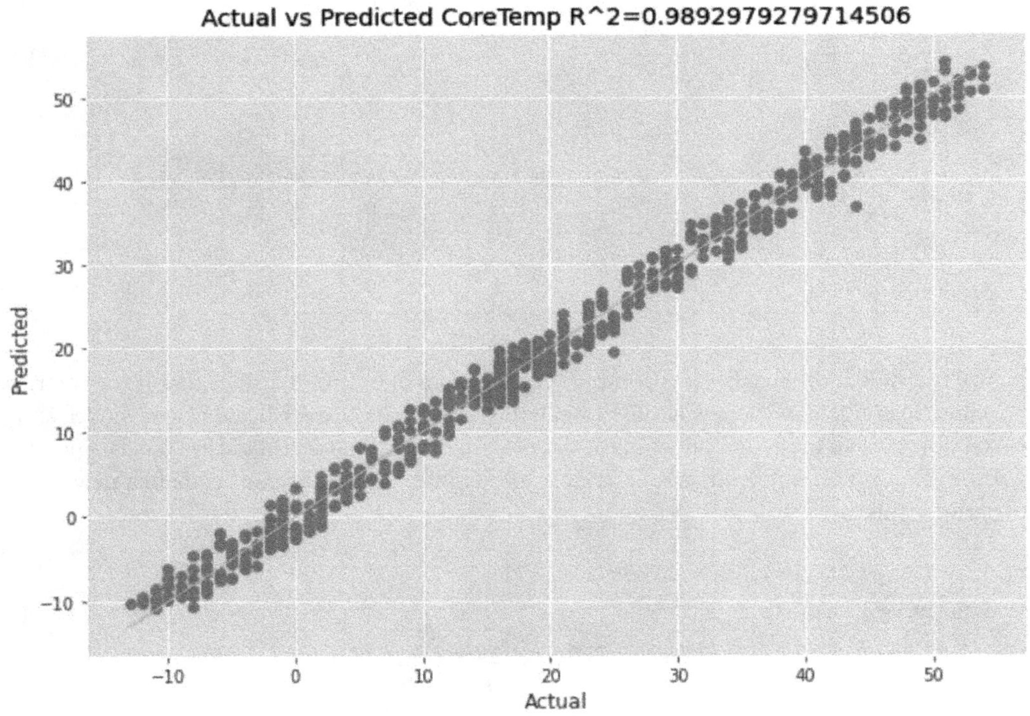

Figure 8.8 – Comparing all predicted values with actual values

An R2 score of 1 represents a perfect model that can predict all occurrences perfectly, while a model that has an R2 score of 0 represents a model that is not effective. For a regression model, we want to pursue an R2 score that's as close to 1 as possible.

Using AutoML is one way to easily introduce ML models into your Azure Synapse workspace and get deeper insights into your Data Explorer pool data. However, it is not the only way to work with ML in Azure Synapse. Let's briefly review a few other alternatives you have available.

Exploring additional ML capabilities in Azure Synapse

As we have seen, AutoML accelerates the adoption of ML in your existing analytics environment, especially if you are getting started with ML. It tests your data with different combinations of algorithms and parameters to find the model that offers the best result possible. However, if you're an experienced ML engineer, there are other options available in Azure Synapse that will help you build your own projects while leveraging the parallel compute capabilities in Apache Spark and proximity with your Data Explorer pool data. Let's briefly describe these options.

Using pre-trained models with Cognitive Services

Azure Cognitive Services is a cloud offering from Microsoft that provides APIs that developers can use to consume pre-trained artificial intelligence models in their applications. These pre-trained models facilitate building applications for computer vision, speech, language, and decision-making tasks. Since these models are pre-trained, you can consume them directly to perform predictions with your data without having to train any new models.

Azure Synapse integrates with Cognitive Services in a very similar way to how it integrates with Azure Machine Learning, so we will not cover it in detail here to avoid redundancy. In short, you need to create an instance of the service using the Azure portal, provide the rights for your workspace to allow access to your Cognitive Services instance, and create a linked service using Azure Synapse Studio. Once you've done that, you can replicate the steps from the *Finding the best model with AutoML* section of this chapter, but then pick the **Predict with a model** option, instead of picking the **Train a new model** option, when you select the ellipsis next to the table name, as shown in *Figure 8.9*:

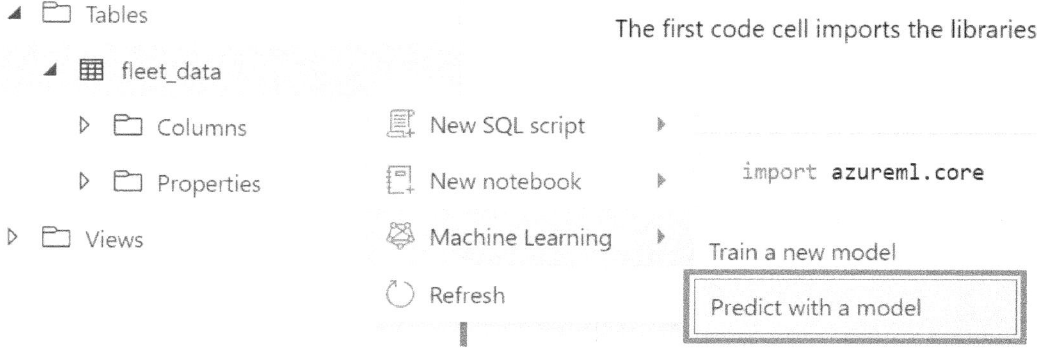

Figure 8.9 – Using pre-trained models with Azure Cognitive Services

This action invokes the **Predict with a model** wizard, as illustrated in *Figure 8.10*. The wizard allows you to select one of two pre-trained models: **Anomaly Detector**, which helps identify anomalies or rare events in your data, and **Sentiment Analysis**, which evaluates the sentiment of certain text (for example, customer feedback) indicating whether the sentiment of what's written in this text block is positive, neutral, or negative:

Predict with a model

fleet_data

Choose a pre-trained model

Azure Cognitive Services

This experience allows you to enrich the selected dataset with pre-trained Azure Cognitive Services models.

Anomaly Detector

Anomaly detection is the identification of rare items, events or observations which raise suspicions by differing significantly from the majority of the data. Learn more

Sentiment Analysis

Evaluates the sentiment (positive/negative/neutral) of a text and also returns the probability (score) of the sentiment. Learn more

Continue Cancel

Figure 8.10 – Choices of pre-trained models

Similar to the **Train a new model** wizard, the **Predict with a model** wizard configures your Cognitive Services task with your linked service and dataset details and produces a notebook ready for usage.

To learn more about the Cognitive Services integration in Azure Synapse, visit `https://learn.microsoft.com/azure/synapse-analytics/machine-learning/overview-cognitive-services`.

Finding patterns using KQL

Azure Synapse Data Explorer supports ML algorithms directly in KQL through three plugins: `autocluster`, `basket`, and `diffpatterns`. All three plugins work with clustering algorithms. In addition to these plugins, you can use time series analysis to perform anomaly detection, as you saw in the *Detecting outliers* section of *Chapter 6, Data Analysis and Exploration with KQL and Python*, and even forecasting.

However, remember that the most common usage scenarios for Data Explorer pools are to store and analyze telemetry and log data. Being able to identify patterns in your data helps not only to understand common patterns by themselves but most importantly also helps analyze failures – occurrences that don't fit the common patterns.

> **Note**
>
> The code example in this section can be found in the `Chapter 08\ Finding patterns with KQL.kql` file of the book's repository.

As an example, let's go back to the drone telemetry database in your Data Explorer pool. We will use the `autocluster` plugin to look at a small sample of the data (all the telemetry from 1 day) and retrieve the common patterns. Simply start a new KQL query in the develop hub and run the following query:

```
let date_start=datetime(2021-11-01 00:00);
let date_end=datetime(2021-11-01 23:59);
['fleet data']
| where LocalDateTime between(date_start..date_end)
| evaluate autocluster()
```

This query produces the result shown in *Figure 8.11* (some columns have been hidden for brevity):

SegmentId	Count	Percent	...	DeviceState	Engine1Status	Engine2Status	Engine3Status	Engine4Status
0	3666	21.149186569747318	...		Working	Working	Working	Working
1	3599	20.76266297450098	...	Descent	Working	Working	Working	Working
2	1809	10.436137071651091	...	TakeOff	Working	Working	Working	Working
3	1800	10.384215991692628	...	TakeOff	Working	Working	Working	Working
4	1791	10.332294911734165	...	Delivery	Working	Working	Working	Working
5	1803	10.401523018345449	...	Cruise	Working	Working	Working	Working
6	1786	10.303449867312796	...	Cruise	Working	Working	Working	Working
7	16485	95.102111457251638	...		Working	Working	Working	Working

Figure 8.11 – Results obtained with the autocluster plugin

We can make a few observations from the preceding results:

- Eight different segments were identified with the data. These are the most common patterns identified by the algorithm in your dataset. The `Count` and `Percent` columns in the result set denote how many rows are captured by this pattern, along with the percentage of occurrences such as this one over the total number of rows. Note that these patterns may overlap, that they are not distinct, and that they don't necessarily cover all the rows in your dataset.
- For the `SegmentID 1` row, we notice that 20% of the rows in the dataset fit in a pattern where the `DeviceState` column equals `Descent`, and the value for the `Engine1Status` through `Engine4Status` columns equals `Working`. The same logic applies to each of the rows in the result set.
- The most common pattern by far is listed in the last row, where 95% of the values in the dataset will fit it. This pattern overlaps with all the other patterns, where the value for the `Engine1Status` through `Engine4Status` columns equals `Working`. As you saw with the Power BI report that you built in *Chapter 7, Data Visualization with Power BI*, the drone engines are quite reliable, and 1 of 4 engines fails only about 1% of the time. Therefore, seeing that in 95% of instances all engines are in a `Working` state would make sense.

You can assume that every time you get a telemetry event that doesn't fit your regular patterns, it is possible that this event is an anomaly, which at the very least would be worth investigating.

As you have seen, in this example, KQL can help you apply ML algorithms to get quick insights, such as this one. To learn more about the ML capabilities in KQL, make sure you visit the *Machine Learning capability in Azure Data Explorer* article at `https://learn.microsoft.com/azure/data-explorer/machine-learning-clustering`. However, if you need to train ML models and apply your models as needed to large volumes of data, KQL will not satisfy your needs. Let's look at how we can use Apache Spark to train robust models using traditional Python libraries.

Training models with Apache Spark MLlib

Apache Spark pools in Azure Synapse offer support for Spark's ML library, known simply as MLlib. This library brings several tools to Apache Spark needed by data scientists:

- **ML algorithms**: As you would expect, MLlib supports several ML algorithms for classification, regression, and clustering tasks, among others.
- **Featurization**: When preparing data for modeling, it's common to transform it to make it more useful for ML algorithms. This might involve transforming non-numeric data into a numerical representation, performing dimensionality reduction, feature selection, and more. MLlib offers a series of tools for featurization tasks.
- **Pipelines**: Training models consists of several tasks from construction and evaluation to model tuning. MLlib's API helps users create and optimize pipelines.

- **Persistence**: MLlib includes APIs to persist models and pipelines for reuse and sharing.
- **Common utilities**: Finally, MLlib offers a series of libraries for data handling, algebra, and statistics.

MLlib can be used directly in Azure Synapse notebooks with an Apache Spark pool, without the need to install any custom packages in your Apache Spark cluster. Simply import the packages you need, directly into your code cell, and they will be ready for you to use.

To learn more about MLlib in Azure Synapse, visit `https://learn.microsoft.com/azure/synapse-analytics/spark/apache-spark-machine-learning-mllib-notebook`.

Building applications with SynapseML

Many times, putting together all the pieces that make up a data science project can be an overwhelming task. You need to track all the libraries that you use, the ML frameworks, the programming language, the package versions, and more. It feels like you are building a puzzle with 1,000 pieces.

Microsoft built SynapseML as a library that brings together multiple components to make it easier to build end-to-end ML pipelines. It unifies several algorithms and frameworks in a single API, leveraging Apache Spark for processing. SynapseML abstracts over different data source types and supports several programming languages.

What's even more interesting is that, even though using SynapseML in Azure Synapse notebooks is an easy task, you don't necessarily need to use SynapseML inside Azure Synapse. You can install it on other services such as Databricks, run your own instance on a Docker container, or use it from a Jupyter notebook connected to an Apache Spark instance. You'll still leverage the full API, bringing together a full set of algorithms, data handling, and ML pipeline tools.

To learn more about SynapseML, visit `https://learn.microsoft.com/azure/synapse-analytics/machine-learning/synapse-machine-learning-library`.

Summary

Historically, application log and telemetry data were used merely for support, diagnostics, and live site monitoring. With the reduced costs in storage and compute resources, this data can now be retained for longer for richer analysis. ML brings a huge opportunity to use log and telemetry data to gain deep, meaningful insights with real business impact.

In this chapter, you learned how to introduce ML models to your Data Explorer pool data using AutoML. We explored how to create Azure Machine Learning workspaces, how to configure the linked service to connect your Azure Synapse workspace to Azure Machine Learning, and finally, how to retrieve the best model from an AutoML experiment.

Next, you learned about other means to bring ML into your Azure Synapse projects. We looked at using pre-trained models to make predictions using Azure Cognitive Services, using KQL plugins to find patterns in data, and training ML models using Apache Spark MLlib.

Lastly, we briefly discussed SynapseML as an end-to-end API you can use to build ML applications.

In the next chapter, we will look at data exports: why they are important, what some of the strategies you can use are, and how to perform data extraction from Data Explorer pools.

9
Exporting Data from Data Explorer Pools

Now that you've learned how to ingest, analyze, and enrich data, let's look at exporting data, which is a key part of the data management process. This chapter closes *Part 2, Working with Data*, by looking at the reasoning behind data exports and the common approaches you can take when working with Azure Synapse Data Explorer.

Azure Synapse Data Explorer offers a fully managed data management solution for log and telemetry data, with blazing-fast performance for queries. If that's the case, why would you want to move data off of Data Explorer pools?

To start the chapter, we will address that question by looking at some of the scenarios where it can be useful to export data from Data Explorer pools. Even though it is a common task that doesn't need that much explanation, we will explore some real-life scenarios that may be applicable to your reality.

Next, you will learn about common mechanisms you can use, from saving the query results in your query editor to robust server-side export techniques. These methods will help you export data to local files in your computer, cloud storage, SQL databases, or Data Explorer external tables.

Lastly, you will learn how to configure continuous data export to keep a job running that exports any new data at a certain recurrence. This approach is helpful if you need to maintain copies of your table in sync without manual intervention.

Here are the topics covered in this chapter:

- Understanding data export scenarios
- Exporting data with client tools
- Using server-side export to pull data
- Performing robust exports with server-side data push
- Configuring continuous data export

As you can see, there's a lot to learn about data export. So, let's dive right into it.

Technical requirements

If you haven't yet, make sure you download the book materials from the repository at `https://github.com/PacktPublishing/Learn-Azure-Synapse-Data-Explorer`. You can download the full repository by selecting **Code** and then **Download ZIP**, or by cloning the repository by using your git client of choice.

This chapter uses the full `fleet data` table that you ingested in the *Running your first query* section of *Chapter 3, Exploring Azure Synapse Studio*. If you did not perform the tasks in this section to load data, take some time to go back and follow the instructions to ingest the full `fleet data` table.

Some of the examples in this chapter involve creating **shared access signatures** (**SAS**) from your **Azure Data Lake Storage** (**ADLS**) **Gen2** account to allow your Data Explorer pool to export data to ADLS and to create external tables. We will not cover creating SAS tokens in detail in this chapter. So, if you need a refresher on how to create SAS tokens for your storage accounts, please refer to the documentation at `https://learn.microsoft.com/azure/storage/common/storage-sas-overview`.

The *Exporting to SQL tables* section of this chapter uses an Azure SQL Database instance to receive data exported from your Data Explorer pool. This example requires any Azure SQL Database instance that is accessible from the internet that you have access to, in order to create the destination table and export data. We will not cover the process of creating a Azure SQL Database instance. So, if you need some help, refer to the tutorial at `https://learn.microsoft.com/azure/azure-sql/database/single-database-create-quickstart`. Any basic instance with minimal compute resources will be sufficient for the examples in this chapter. If you have a Microsoft SQL Server instance that is accessible via the internet, then that would also work as long as you configure access accordingly.

Understanding data export scenarios

As much as possible, when you are thinking about building your analytical environment, you should work with the premise of keeping a single point of truth in your data, meaning, you have one repository that holds the true version of your data. This is a structural principle of data management, especially in analytics, to ensure everyone makes decisions based on the same data. If someone needs to know how much product the company sold the previous month or the failure rate of IoT sensors, this one repository holds the truth that helps answer these questions. The moment you start copying data to different places, it becomes susceptible to user changes that affect a table's layout, or even the data itself, creating several different versions of the data and removing that single point of truth.

However, not every analytical problem is solved in one single platform. The ability to export data to a different location is an essential tool in any database engine, allowing analysts to collaborate, experiment, share data with external parties, and more. Some common scenarios that may require exporting data include the following:

- **Experimentation**: Performing constant queries against Data Explorer pools can become an expensive task quickly. An analyst may choose to use a slice of the full data to perform experimentation, without the expectation to enjoy the performance benefits offered by the source database engine, and later scale the experiment to the full dataset. This experiment may run on the user's machine or on a cheaper SQL database hosted on the user's premises.
- **Development**: Building applications, reports, dashboards, or any assets that consume data can also cause frequent queries to the underlying database engine that holds that data. For development purposes, you should consider an environment different from that of your production data, such as a simple copy of your original tables obfuscating any sensitive data, such as **personally identifiable information** (**PII**). You could then work with this copy of the data without risks.
- **Data sharing**: Providing external partners with access to some of your data. Instead of providing access to your data directly, you may provide data exports to partners that contain only the relevant data for that organization. For example, you could export the occurrences of failures detected by sensors in a certain plant to individuals on that plant only, without sharing data from other plants.

As you can see, even though you should always think about keeping that single point of truth, there are benefits to data export scenarios. Quite frankly, data export scenarios are unavoidable. You should think about your Data Explorer pools as the one place that holds that truth but you should enable data exporting to enhance productivity in your organization.

We will look next into a few different approaches for exporting data with Azure Synapse Data Explorer.

Exporting data with client tools

Data export with client tools is the ability to perform client-side exports through the tool you are using to query your Data Explorer pools. In this book, our tool of choice has been Azure Synapse Studio, so let's use it in our example. Whenever you perform any queries in Azure Synapse Studio, regardless of whether it is a SQL query or a KQL query, you will see the **Export results** button above your query results, as illustrated in *Figure 9.1*. This allows you to immediately export the results of this query to a file that will be downloaded to your local machine. The supported file formats here are **comma-separated values** (**CSV**) files, **JavaScript Object Notation** (**JSON**) files, or **Extensible Markup Language** (**XML**) files.

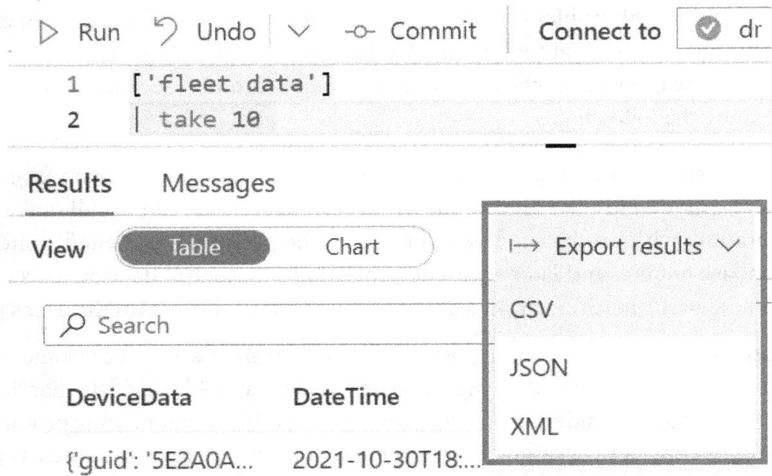

Figure 9.1 – Exporting query results

Most business intelligence and data management client tools in the industry, if not all, offer a similar capability to this, allowing you to save the results of a query to a local file.

This is the simplest and easiest way to export data from Data Explorer pools but it doesn't scale well if you're exporting larger volumes of data, or if you need to export data to a more reliable data store, such as a database server. Let's look at more robust ways to export data using Data Explorer pools.

Using server-side export to pull data

A very common pattern used in databases is to perform a query that writes its results to another table. In KQL, this pattern is called **Ingest from Query**. There are four KQL control commands that support Ingest from Query in Data Explorer pools:

- `.set`: This command can be used only if the destination table, that is, the table that will receive the data from a query result, doesn't yet exist. If it does exist, the command fails.
- `.append`: Produces the opposite of the `.set` command. If the destination table doesn't yet exist, the command fails. If the table already exists, it appends the query results to the destination table.
- `.set-or-append`: This command combines the `.set` and `.append` commands. If the destination table exists, data is appended to the table. If it doesn't, the table is created and data is ingested.
- `.set-or-replace`: Similar to `.set-or-append` except that if the destination table already exists, the operation replaces any data in the destination table.

Let's look at a quick example: if we wanted to create a table (or update it if it already existed) that contains all the telemetry events where the `DeviceState` column reported by a drone was equal to `DeviceError`, this could be achieved with the following code block:

```
.set-or-append ['fleet data - failures'] <|
   ['fleet data']
   | where  DeviceState == 'DeviceError'
```

> **Note**
> This code example can be found in the `Chapter 09\server-side export - pull.kql` file of the book repository.

This command creates a table named `fleet data - failures` with the rows specified in the query. As you can see in *Figure 9.2*, the command's output reports out the ID of the extent where the data was stored, the original data size, the size of the extent, the size of the compressed data, the size of the index, and the row count of the new table.

Search

ExtentId	OriginalSize	ExtentSize	CompressedSize	IndexSize	RowCount
4453981f-bbf2-47d2-b350-80f28...	3204978	1304106	1231793	72313	6845

00:00:02 Query executed successfully.

Figure 9.2 – Results from the data export operation

This data export mechanism is ideal when you need a copy of your data inside the Data Explorer pool but it doesn't allow you to export the data to a different data store or to files. You can still query the new table and export it within Azure Synapse Studio, as described in the *Exporting data export with client tools* section of this chapter, but you will meet the same scale challenges as before. In addition to that, these commands open a single network connection between the process that produces the query and the process that writes the results to the new table, limiting throughput. So, while it scales better than exporting data through the query results in the client tool, we can do a little better in terms of scale.

> **Note**
> Microsoft recommends that you limit data ingestion tasks such as this one to less than 1 GB per operation. Also, note that these data ingestion tasks are resource intensive, as we mentioned in *Chapter 5, Ingesting Data into Data Explorer Pools*, so it's better to avoid running this at peak usage times as it may interfere with the performance of user queries.

There is a better way to export data that is both more scalable and allows exporting data to different destinations. We will explore this option next.

Performing robust exports with server-side data push

Server-side exports with the push model are the most scalable and flexible way to export data from Data Explorer pools. They allow you to export data to a cloud storage container (such as ADLS Gen2 or Amazon S3), to a SQL table, or to a Data Explorer external table. The mechanism that allows the push method is exposed through the `.export` commands that perform a query on a Data Explorer pool table and writes the query results to the destination store.

Data exports with server-side data push have the following advantages:

- **Support for async exports**: You can run your export task and continue doing your work while the task runs in the background. The `.show operations` command can be used to track the progress of all data export tasks. The `.show operation details` command retrieves completion results specific to a task in particular.
- **Allows scalable writes**: When writing to cloud storage, you can specify multiple connection strings to allow scalable writes to the destination storage.

> **Note**
>
> All code examples for the *Performing robust exports with server-side data push* section can be found in the `Chapter 09\server-side export - push.kql` file of the book repository.

Let's look at each of the supported destinations for this method and explore some examples.

Exporting to cloud storage

Exporting data to cloud storage allows easy experimentation and gives flexibility for analysts to work with data in any format, without any table schema constraints, and with the possibility to combine data from multiple sources in one data lake.

To export data to cloud storage, use the following syntax:

```
.export [async] [compressed] to OutputDataFormat (
StorageConnectionString [, ...] ) [with ( PropertyName =
PropertyValue [, ...] )] <| Query
```

Let's break down the syntax:

- `async`: Use this option to indicate whether this will be an async export operation.
- `compressed`: Compresses the files generated in the destination cloud storage.
- `OutputDataFormat`: Specifies the file format used to store data in the destination cloud storage. The supported file formats are CSV files, **tab-separated values** (**TSV**) files, JSON files, and Parquet files.
- `StorageConnectionString`: Specifies the connection string that contains the details needed to connect to the destination cloud storage. It includes the path and authentication details.
- `PropertyName` and `PropertyValue`: Lists any optional export parameters used by the export operation, such as the need to include headers in the destination files, to use a name prefix, file size limits, and others.

> **Note**
>
> To see all available optional parameters, refer to `https://learn.microsoft.com/azure/data-explorer/kusto/management/data-export/export-data-to-storage`.

As an example, the following code snippet exports the `DeviceData` and `CoreTemp` columns from the `fleet data` table, where `DeviceState` is equal to `DeviceError`, and writes the results to a file in the `data-exports` container of an ADLS Gen2 account. The destination file will have a prefix of `DeviceError-`, include column headers, and use the `UTF8NoBOM` encoding. Make sure you replace `<storageaccountname>` with the name of your storage account and `<SASKeyHere>` with the SAS key that you create for this purpose:

```
.export async compressed to csv (
        h@"https://<storageaccountname>.blob.core.windows.net/
data-exports?<SASKeyHere>"
    ) with (
        namePrefix="DeviceError-",
        includeHeaders="all",
        encoding ="UTF8NoBOM"
    )
  <| ['fleet data'] | where DeviceState == 'DeviceError' |
project DeviceData, CoreTemp
```

This command produces the result seen in *Figure 9.3*:

Figure 9.3 – Exporting data to cloud storage

> **Note**
> Make sure your SAS key includes read, write, and list blobs permissions. If you need a refresher on creating SAS tokens, refer to the *Ingesting data* section of *Chapter 5, Ingesting Data into Data Explorer Pools*.

By navigating to your ADLS Gen2 container in the Azure Portal, you will find the new file generated by the export operation. This is illustrated in *Figure 9.4*.

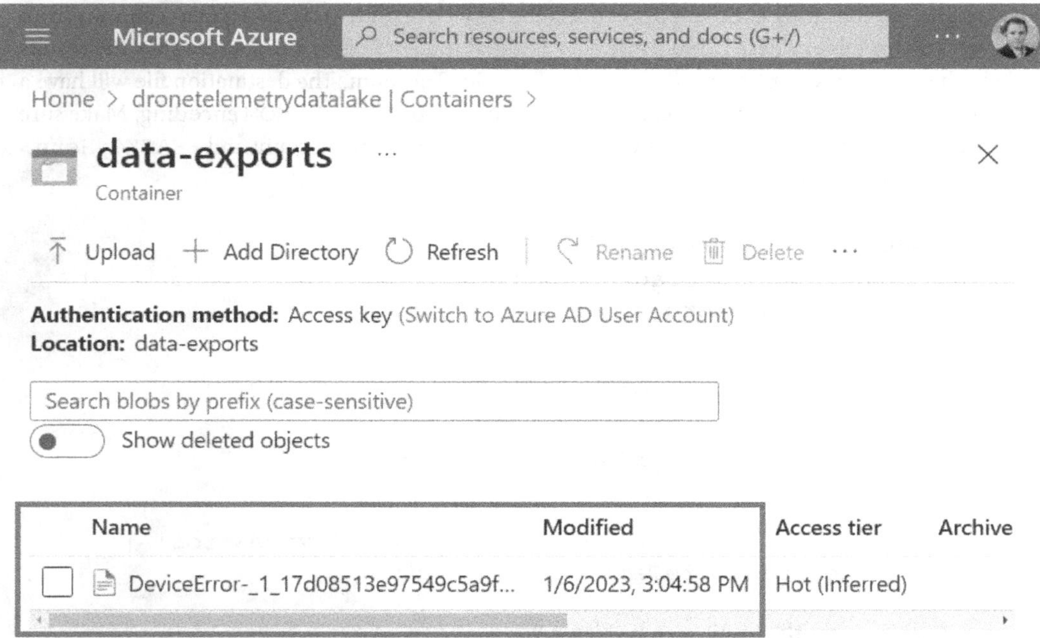

Figure 9.4 – The file generated by the export operation

In this case, the export operation generated a single file. If you are working with larger data volumes or specify a size limit to the destination file, the operation may generate several output files. If for some reason the operation fails, the `.export` command does not roll back the operation: any files that were written to cloud storage will remain there and you must delete them manually. This `.export` operation does not perform retries.

Exporting to SQL tables

Exporting data from a Data Explorer pool to a SQL table works similarly to exporting to cloud storage: it performs a query in the desired table and writes the results to a table residing on a SQL database. The destination database must be accessible from the public Azure cloud.

To export data to a SQL table, use the following syntax:

```
.export [async] to sql SqlTableName SqlConnectionString [with
(PropertyName = PropertyValue,...)] <| Query
```

Let's break down the syntax:

- `async`: Use this option to indicate whether this will be an `async` export operation.
- `SqlTableName`: The name of the table in the SQL database that will receive the data. To protect this command from SQL injection attack attempts, the table name can only use letters, digits, spaces, underscores, dots, and hyphens.
- `SqlConnectionString`: The connection string that contains the details needed to connect and authenticate to the destination database server.
- `PropertyName` and `PropertyValue`: List any optional export parameters used by the export operation, such as to create the destination table if one doesn't exist.

> **Note**
> To see all available optional parameters, refer to `https://learn.microsoft.com/azure/data-explorer/kusto/management/data-export/export-data-to-sql`.

As an example, the following code snippet exports all columns from the fleet data table, where `LocalDateTime` is between `2021-11-01` and `2021-11-02`, and writes the results to a table named `fleet_data` in an Azure SQL Database instance. Make sure you replace `<AzureSQLServerName>` with the name of your Azure SQL Database instance:

```
.export to sql fleet_data
    h@"Server=tcp:<AzureSQLServerName>.database.windows.
net,1433;Database=droneanalyticssql;Authentication=Active
```

```
Directory Integrated;Connection Timeout=30;"
    with (createifnotexists="true")
    <| ['fleet data'] | where LocalDateTime between
(datetime(2021-11-01) .. datetime(2021-11-02))
```

This command produces the result seen in *Figure 9.5*:

Rows
17343

00:00:02 Query executed successfully.

Figure 9.5 – Confirmation of the data export task

This approach uses SQL Server's bulk copy API and does not provide any transactional guarantees, so it's up to you to ensure that the data export command worked as expected.

You can also use this command to export data to dedicated SQL pools in Azure Synapse but keep in mind that it may take a long time to complete the task, depending on your data volume. Even though Synapse SQL supports SQL Server's bulk copy API, the optimal way to load data into Synapse SQL is by exporting data to ADLS Gen2 first, and then using the `COPY` transact-SQL command to copy data from ADLS into SQL tables. To learn more about best practices to load data into Synapse SQL, visit `https://learn.microsoft.com/azure/synapse-analytics/sql/data-loading-best-practices`.

Exporting to external tables

Last but not least, Azure Synapse Data Explorer supports exporting data to Data Explorer external tables. Similarly to serverless SQL pools in Azure Synapse, external tables in Data Explorer pools allow you to query data directly on ADLS Gen2, without the need to ingest it into a Data Explorer pool database. This is useful, for example, when you have historical data that is not often accessed, yet it still needs to be available in case it is needed. Data that is frequently accessed should be kept in Data Explorer pools due to its performance benefits, while data that is less frequently used can be kept on ADLS Gen2 to help reduce costs.

The syntax to export data to external tables in Data Explorer pools is similar to the syntax for exporting to cloud storage or SQL tables, just a little simpler:

```
.export [async] to table ExternalTableName
[with (PropertyName = PropertyValue,...)] <| Query
```

Let's break down the syntax:

- `ExternalTableName`: The name of the external table that will receive the data
- `PropertyName` and `PropertyValue`: Lists any optional export parameters used by the export operation, such as the file size limit created for files created by the export operation

> **Note**
>
> To see all available optional parameters, refer to `https://learn.microsoft.com/azure/data-explorer/kusto/management/data-export/export-data-to-an-external-table`.

Exporting data to external tables does not support creating the destination table as part of the export task. So, to test this, we first need to create an external table. This can be achieved with the `.create external table` control command. The following command creates an external table with the name `TopPayloads` that contains three columns: `DeviceData` of type string, `LocalDateTime` of type datetime, and `PayloadWeight` of type long. This table is physically stored as CSV files. Make sure that you replace `<storageaccountname>` with the name of your storage account and `<SASKeyHere>` with the SAS key you created for this purpose:

```
.create external table TopPayloads (DeviceData:string,
LocalDateTime:datetime, PayloadWeight:long)
kind=storage
dataformat=csv
(
   h@"https://<storageaccountname>.blob.core.windows.net/data-
exports?<SASKeyHere>"
)
```

Now that the external table is created, you are ready to perform the data export operation. The following code snippet exports the top 10,000 rows by `PayloadWeight` in descending order, but only the `DeviceData`, `LocalDateTime`, and `PayloadWeight` columns (to align with the schema of our external table):

```
.export to table TopPayloads
<| ['fleet data'] | top 10000 by PayloadWeight desc | project
DeviceData, LocalDateTime, PayloadWeight
```

As the results in *Figure 9.6* illustrate, this command successfully exported 10,000 records to the `TopPayloads` external table.

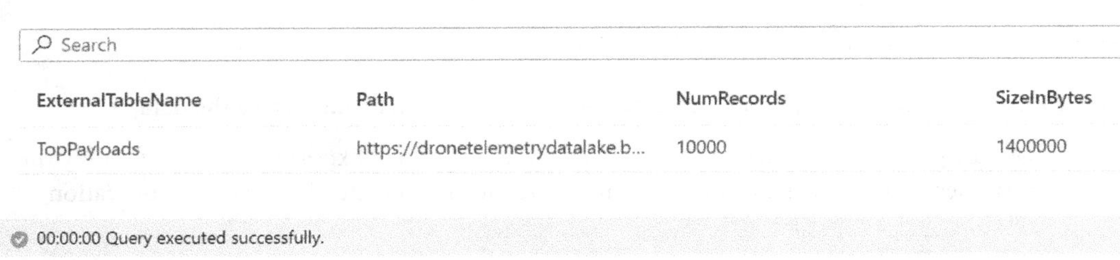

ExternalTableName	Path	NumRecords	SizeInBytes
TopPayloads	https://dronetelemetrydatalake.b...	10000	1400000

Figure 9.6 – Exporting data to external tables

If you'd like, you can verify that data was exported to the external table by querying it with the following code block:

```
external_table('TopPayloads')
| take 10
```

Your external table is now ready to use. You can query it normally using KQL as you would query any table, and the Data Explorer pool will provide fast results. To optimize queries to external tables and learn more about how Azure Data Explorer handles access to data in Azure Storage, refer to the article *Query Data in Azure Data Lake using Azure Data Explorer* at `https://learn.microsoft.com/azure/data-explorer/data-lake-query-data`.

Configuring continuous data export

All three data export mechanisms that we have explored so far with server-side data push are quite similar in syntax and in how they operate. They require someone to start the actual data export task, monitor it, and implement the logic to copy new data in the future. There is, however, one important advantage of exporting data to external tables: the possibility to use continuous data export to keep your external table updated as new records are inserted in the original table.

We'll continue to build on the table created in the *Exporting to external tables* section. We did the initial data load and now we would like to set up a process that updates the content of the external table with new records inserted in the `fleet data` table every two hours. Continuous data export allows you to configure this process, ensuring that each record in your table is processed exactly once.

Here's the syntax to create a new continuous data export job:

```
.create-or-alter continuous-export ContinuousExportName
[ over (T1, T2)]
to table ExternalTableName
[ with (PropertyName = PropertyValue,...)]
<| Query
```

Let's break down the syntax:

- `ContinuousExportName`: The name of the continuous data export job. Once the job gets created, you'll be able to disable it if needed, drop it, and monitor its execution. You'll need this name when you perform any of these tasks, so make sure you pick a name that helps you identify this job in the future.
- `over (T1, T2)`: An optional list of table names to pick from the query statement.
- `ExternalTableName`: The name of the external table that will receive data as a result of the export job.
- `PropertyName` and `PropertyValue`: Lists any optional export parameters used by the export operation, such as the time interval between runs.

> **Note**
> To see all available optional parameters, refer to `https://learn.microsoft.com/azure/data-explorer/kusto/management/data-export/create-alter-continuous`.

The following command creates a new continuous data export job named `ContinuousTelemetryData`, which runs every 2 hours and writes into the `TopPayloads` external table any new rows inserted into the `fleet data` table. Note that the query statement respects the schema of the external table:

```
.create-or-alter continuous-export ContinuousTelemetryData
to table TopPayloads
with
(intervalBetweenRuns=2h)
<|
['fleet data']
| project DeviceData, LocalDateTime, PayloadWeight
```

Upon creation, the continuous export job is active and ready to run on the interval determined by the `intervalBetweenRuns` property (set to 2 hours in our example). At any time, you can check the status of the job by using the `.show continuous export` command. Here's an example:

```
.show continuous-export ContinuousTelemetryData
| project ExternalTableName, IntervalBetweenRuns, IsDisabled,
LastRunResult, IsRunning
```

This command produces the output illustrated in *Figure 9.7*:

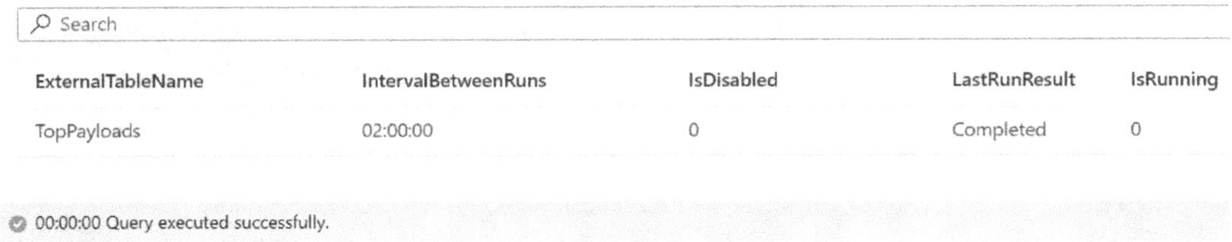

Search

ExternalTableName	IntervalBetweenRuns	IsDisabled	LastRunResult	IsRunning
TopPayloads	02:00:00	0	Completed	0

00:00:00 Query executed successfully.

Figure 9.7 – Checking the status and properties of the continuous export job

If you need to temporarily disable a continuous data export, you can use the `.disable continuous-export` command as follows:

```
.disable continuous-export ContinuousTelemetryData
```

By disabling the job, it will not run. However, it will keep its current state so that when you enable the job back again, it will resume from the last row it had exported. Keep in mind that if you've left the job disabled for a long period of time, enabling it will cause a data export that will process any data since it was disabled, which can be time-consuming depending on the data volume and how long it was disabled.

Finally, as you would expect, you can enable the continuous data export job back by using the `.enable continuous export` command, as exemplified in the following code snippet:

```
.enable continuous-export ContinuousTelemetryData
```

Note that creating the continuous data export job assumes you've already exported the initial copy of your table. It will not export the data in your table that was there before you created the job. You must manually export your table to the external table, and then configure the continuous data export job to copy the newly inserted records only.

Summary

In this chapter, you learned why you should care about data export scenarios and how they may be useful for you in real life. You've learned how to perform some basic data export work by using Azure Synapse Studio.

You then continued to learn about using server-side exports for a more robust data export experience. We looked at examples of how to export data using a pull approach, where we created a new table that pulls the result from a query to load data into it. Next, you exported data using server-side mechanisms that leverage the `.export` command to push data to cloud storage, SQL tables, and external tables in Data Explorer pools.

To close, you learned about configuring continuous data export jobs that, once configured, maintain an external table in sync with a source table by running recurrently at a pre-defined interval. We looked at how we can check the details of the job to see its status and how we can disable and enable the job again if needed.

This chapter closes the *Part 2*, *Working with Data*, section of the book. This part covered data ingestion, analysis, enrichment with machine learning, and data export. The next part of the book focuses on managing, monitoring, securing, and tuning Azure Synapse Data Explorer. So, let's get to it.

Part 3

Managing Azure Synapse Data Explorer

Once you learn the basics of Azure Synapse Data Explorer and learn to work with data, it's time to monitor, optimize, and tune your environment. In the last part of the book, you will learn how to stay on top of your implementation by using the web interface and KQL commands. You will learn how Azure Synapse Data Explorer implements resource management, and how you can configure your environment to prioritize certain users or workloads. Next, you will learn about all aspects of security and how to make sure that not only the right people get access to your data but also how to protect it at rest and in transit using encryption. To finalize the book, you will learn about advanced data management topics, and how to manage data while adhering to governmental regulations, such as the **General Data Protection Regulation** (**GDPR**).

The third and final part of the book includes the following chapters:

- *Chapter 10, System Monitoring and Diagnostics*
- *Chapter 11, Tuning and Resource Management*
- *Chapter 12, Securing Your Environment*
- *Chapter 13, Advanced Data Management*

10
System Monitoring and Diagnostics

Azure Synapse Data Explorer is a **platform-as-a-service** (**PaaS**) that brings together several components for an end-to-end analytics experience. As with any PaaS, you don't need to manage any of the underlying infrastructure of the service, such as the server operating system, storage, networking, or any hardware, which is different from running a service on your own premises, which requires you to manage the full stack of the service. *Figure 10.1* illustrates the common layers in each service hosting model, and how much you need to worry about management in each of them.

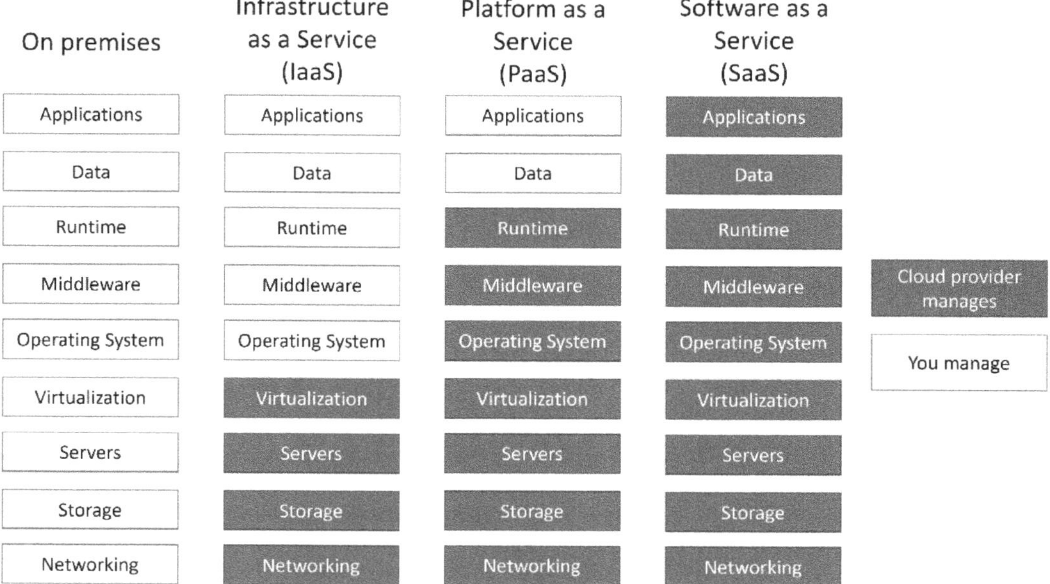

Figure 10.1 – Service management responsibilities per service hosting model

As you can see, for PaaS solutions, we'll typically only worry about managing the actual data and the applications that are used to connect to the platform. In this chapter, you'll look at ways to monitor your environment to manage your data, including metadata changes and query execution. As for the rest of the platform, Microsoft Azure offers an interface that allows you to observe what happens in the layers that you don't manage. You'll learn how to use Azure Monitor to help understand what's happening in these layers.

Here are the topics covered in this chapter:

- Monitoring your environment
- Setting up alerts

Before we dive into these topics, let's have a look at the technical requirements for this chapter, as usual, to make sure you are successful in your learning journey.

Technical requirements

In case you haven't yet, make sure you download the book materials from the repository at `https://github.com/PacktPublishing/Learn-Azure-Synapse-Data-Explorer`. You can download the full repository by selecting **Code**, and then **Download ZIP**, or by cloning the repository by using your Git client of choice.

This chapter uses the full `fleet data` table, which you ingested in the *Running your first query* section of *Chapter 3, Exploring Azure Synapse Studio*. If you did not perform the tasks in this section to load the data, take some time to go back and follow the instructions to ingest the full `fleet data` table.

The final example of this chapter sends a notification to your phone to let you know that a Data Explorer pool has started. This notification lands in the form of a cell phone **short message service** (**SMS**) message, and a notification in the Azure mobile app. Cell phone charges may apply, so make sure you understand any costs associated with this example and plan accordingly.

Last but not least, the installation of the Azure mobile app requires a device with an Android or iOS mobile operating system. To install the Azure mobile app on your phone or tablet, visit `http://aka.ms/azureapp`, or look for the Azure app in your phone's app store. Make sure you log on to the app with the same account that you use to log in to Azure.

Monitoring your environment

Azure Synapse Data Explorer offers a series of control commands that you can use to check for system status, ongoing queries, past operations, and more.

> **Note**
>
> All code examples in this chapter can be found in the `Chapter 10\System Monitoring.kql` file of the book repository.

The following examples will walk you through checking and adjusting the capacity of your Data Explorer pool, monitoring query execution, and reviewing the history of metadata changes in database objects.

Checking your Data Explorer pool capacity

If you are experiencing long execution times when executing queries or data ingestion jobs, the first step to take is to look at the overall system usage. The easiest way to check this is by running the `.show capacity` command. Simply open a new **Kusto Query Language** (**KQL**) script in Azure Synapse Studio and run the following command:

```
.show capacity
```

This command produces a result similar to what's illustrated in *Figure 10.2*. This table shows the estimated cluster capacity and usage for each resource type. For example, we can see that the current capacity policy on this example system allows a total of two concurrent ingestions, with zero being consumed at the moment.

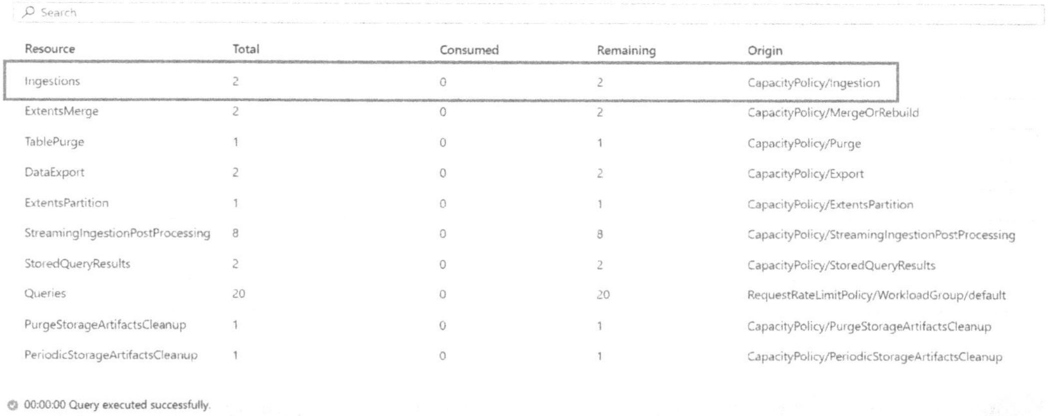

Search

Resource	Total	Consumed	Remaining	Origin
Ingestions	2	0	2	CapacityPolicy/Ingestion
ExtentsMerge	2	0	2	CapacityPolicy/MergeOrRebuild
TablePurge	1	0	1	CapacityPolicy/Purge
DataExport	2	0	2	CapacityPolicy/Export
ExtentsPartition	1	0	1	CapacityPolicy/ExtentsPartition
StreamingIngestionPostProcessing	8	0	8	CapacityPolicy/StreamingIngestionPostProcessing
StoredQueryResults	2	0	2	CapacityPolicy/StoredQueryResults
Queries	20	0	20	RequestRateLimitPolicy/WorkloadGroup/default
PurgeStorageArtifactsCleanup	1	0	1	CapacityPolicy/PurgeStorageArtifactsCleanup
PeriodicStorageArtifactsCleanup	1	0	1	CapacityPolicy/PeriodicStorageArtifactsCleanup

00:00:00 Query executed successfully.

Figure 10.2 – Showing the cluster's current capacity

If you are reaching your cluster's limits for a given resource and need to change the policy to allow more concurrent operations for that resource, you can change policy limits by using the `.alter cluster policy capacity` command, as follows:

```
.alter cluster policy capacity ```
{
```

```
  "IngestionCapacity": {
    "ClusterMaximumConcurrentOperations": 512
  }
}```
```

This command sets the maximum number of concurrent ingest operations allowed in a Data Explorer pool at any given time to `512`.

> **Note**
> For a detailed view of the capacity policy objects and their parameters, refer to the product documentation at `https://learn.microsoft.com/azure/data-explorer/kusto/management/capacitypolicy#ingestion-capacity`.

Be aware that when you raise the number of allowed operations for a certain resource type, you may unbalance your Data Explorer pool, as this resource will compete with other resources in the cluster for compute time.

Monitoring query execution

A common task in any database engine is to monitor ongoing queries for their execution time, resource usage, state, and, when they fail, the reasons why. Data Explorer pools expose the execution of control commands through the `.show commands` command and the execution of queries through the `.show queries` command. Better than that, it offers a `.show commands-and-queries` command, which joins control commands and queries together in a single result set.

As an example, to see queries and commands that failed, and sorted in descending order by the time they started (showing the most recent ones first), we can use the following command:

```
.show commands-and-queries
| where State == 'Failed'
| order by StartedOn desc
```

This command produces a result similar to the one in *Figure 10.3*. As you can see, it includes lots of relevant information, such as the client application that ran the command or query, the command or query text (limited to 64 KB), the duration of the execution attempt, and more.

ClientActivityId	CommandType	Text	Database	StartedOn	LastUpdatedOn	Duration	State
Studio;89b5752...	DataExportToFile	.export async c...	drone-telemetry	2023-01-06T23:...	2023-01-06T23:...	00:00:21.2744582	Failed
DM.IngestionEx...	DataIngestPull	.ingest-from-st...	drone-telemetry	2023-01-06T22:...	2023-01-06T22:...	00:00:00.1093904	Failed
AzureDataFacto...	Query	['fleet data pipe...	drone-telemetry	2023-01-06T22:...	2023-01-06T22:...	00:00:00.1874681	Failed
DM.IngestionEx...	DataIngestPull	.ingest-from-st...	drone-telemetry	2023-01-06T18:...	2023-01-06T18:...	00:00:00	Failed
AzureDataFacto...	Query	['fleet data pipe...	drone-telemetry	2023-01-06T18:...	2023-01-06T18:...	00:00:00	Failed

00:00:00 Query executed successfully.

Figure 10.3 – Showing recent queries and commands

It is worth explaining that the query or command text that you see here is only available to members of the database admin or database monitor roles. If a user who is not a member of these roles runs the same command, they will only see queries or commands that were invoked by themselves.

If you see long-running queries that you would like to cancel, or even cancel a query that you started yourself, you can use the `.cancel query` command. The command can even display a message in the user's query results to explain why the query was canceled. As an example, the following command cancels a query with the `Studio;7330d14b-f5de-46cb-acd9-6f30ffebf844` request ID and displays the `Canceled by the admin` message to the user who issued the query:

```
.cancel query 'Studio;7330d14b-f5de-46cb-acd9-6f30ffebf844'
with (reason='Canceled by the admin')
```

Members of the cluster-admin role can stop any ongoing queries in the Data Explorer pool, while members of the database admin role for a given database can stop queries only in such a database. As you would expect, users can cancel their own queries regardless of their role membership.

Another idea is to check for resource consumption per query to investigate queries that may have performance issues. For example, the following command shows you the queries that used the most CPU time in a given day. You can even select the columns you want to project to make your results easier to read:

```
.show commands-and-queries
| where StartedOn > ago(24h)
| order by TotalCpu desc
| project User, Application, TotalCpu, State, Text
```

This command produces a result similar to the one seen in *Figure 10.4*.

User	Application	TotalCpu	State	Text
peri@periclesrocha.com		00:00:06	Completed	.export async compressed to parquet ("https://sc0lddro...
peri@periclesrocha.com		00:00:05.8593750	Completed	.export async compressed to parquet ("https://sc0lddro...
peri@periclesrocha.com		00:00:05.7812500	Completed	.export async compressed to parquet ("https://48wlddr...
peri@periclesrocha.com		00:00:05.7343750	Completed	.export async compressed to parquet ("https://48wlddr...
peri@periclesrocha.com		00:00:05.7343750	Completed	.export async compressed to parquet ("https://48wlddr...

00:00:00 Query executed successfully.

Figure 10.4 – Showing top queries by CPU time

It's easy to see here which users are issuing the most CPU-intensive queries to your cluster, which applications they are using, and what the query text is.

> **Note**
> There are several aspects of queries that can be analyzed to make them run faster. For good practice with KQL queries, refer to the article *Query best practices* at `https://learn.microsoft.com/azure/data-explorer/kusto/query/best-practices`.

To check for queries with high memory usage, you can write the same query and replace `TotalCPU` with `MemoryPeak`. Queries that are high on CPU time and memory may be good candidates for optimization.

Reviewing object metadata and changes

Data Explorer pools support using control commands to investigate database object metadata (such as a table's schema and details) and the history of metadata changes. Let's look at some examples.

If you need to see details about how your table is stored in a Data Explorer pool and some of its properties, you can use the `.show table details` control command. As an example, the following command produces details about the `fleet data` table, such as the name of the database where the table resides, the total data size (in bytes), and the total row count and identifies principals who are authorized for this table (as JSON):

```
.show table ['fleet data'] details
| project DatabaseName, TableName, TotalOriginalSize,
TotalRowCount, AuthorizedPrincipals
```

This command produces a result similar to the one in *Figure 10.5*.

DatabaseName	TableName	TotalOriginalSize	TotalRowCount	AuthorizedPrincipals
drone-telemetry	fleet data	140546190	337500	[{ "Type": "AAD User", "DisplayName": "Peri Rocha (upn: peri@periclesr...

00:00:00 Query executed successfully.

Figure 10.5 – Reviewing table details

You can review any metadata changes that were made to objects using the `.show journal` command. In the following example, you can verify changes that were made to the `fleet data` table, including the date the change was made, the metadata event change, the original state of the table before the command, the command that was issued by the user to perform the metadata change, and the user who made the change:

```
.show journal
| where EntityName == 'fleet data'
| order by EventTimestamp desc
| project EventTimestamp, Event, OriginalEntityState,
ChangeCommand, User
```

This command produces a result similar to the one in *Figure 10.6*.

EventTimestamp	Event	OriginalEntityState	ChangeCommand	User
2022-12-13T05:37:59.8106643Z	SET-TABLE-ADMINS	Existing Principals: aaduser=0...	.set table ['fleet data'] ad...	peri@periclesro...
2022-12-03T05:36:55.3075008Z	ALTER-TABLE-CACHING-POLICY	DataHotSpan = '30.00:00:00', ...	.delete table ['drone-tele...	peri@periclesro...
2022-12-03T05:36:50.016267Z	ALTER-TABLE-CACHING-POLICY	DataHotSpan = '7.00:00:00', I...	.alter table ['drone-telem...	peri@periclesro...
2022-12-02T06:42:27.4361367Z	ALTER-TABLE-CACHING-POLICY		.alter table ['drone-telem...	peri@periclesro...
2022-10-26T05:20:34.5729832Z	ADJUST-TABLE-ROWSTORE-REF...	[]	.create table ['fleet data']...	peri@periclesro...

00:00:00 Query executed successfully.

Figure 10.6 – Reviewing table metadata changes

This makes it easy to understand, for instance, when a column was dropped, changed, or added into a table, a common operation that can cause queries to stop functioning or not work as expected. Reviewing these metadata changes is not only a way to audit database object modifications but also helps keep track of how these objects were updated after they were created.

Setting up alerts

A useful way to stay on top of what happens in your Azure Synapse workspace is to leverage its integration with the monitoring and observability capabilities available in the Azure platform. The service in the platform that offers these capabilities is called Azure Monitor. At a high level, Azure

Monitor allows you to detect and diagnose issues with your Azure services, look into monitoring data for troubleshooting, trigger automated actions depending on events, and much more. Feature availability for Azure Monitor varies from service to service.

Data Explorer pools offer native integration with Azure Monitor and allow you to leverage some of its capabilities to monitor your environment. You can find the capabilities available for your Data Explorer pool by navigating to your pool's page in the Azure portal using the following steps:

1. Navigate to the Azure portal at `https://portal.azure.com` and find your Azure Synapse workspace. The easiest way to find it is to simply type the workspace's name (`drone-analytics` is the name we are using in this book) in the global search box.
2. In your Azure Synapse workspace's page, select **Data Explorer pools (preview)** under **Analytics pools**.
3. You will see a list of the Data Explorer pools in your workspace. Select your Data Explorer pool to land on the resource page of your Data Explorer pool.
4. In the left-hand navigation menu of your Data Explorer pool's page, look for the **Monitoring** group, as highlighted in *Figure 10.7*.

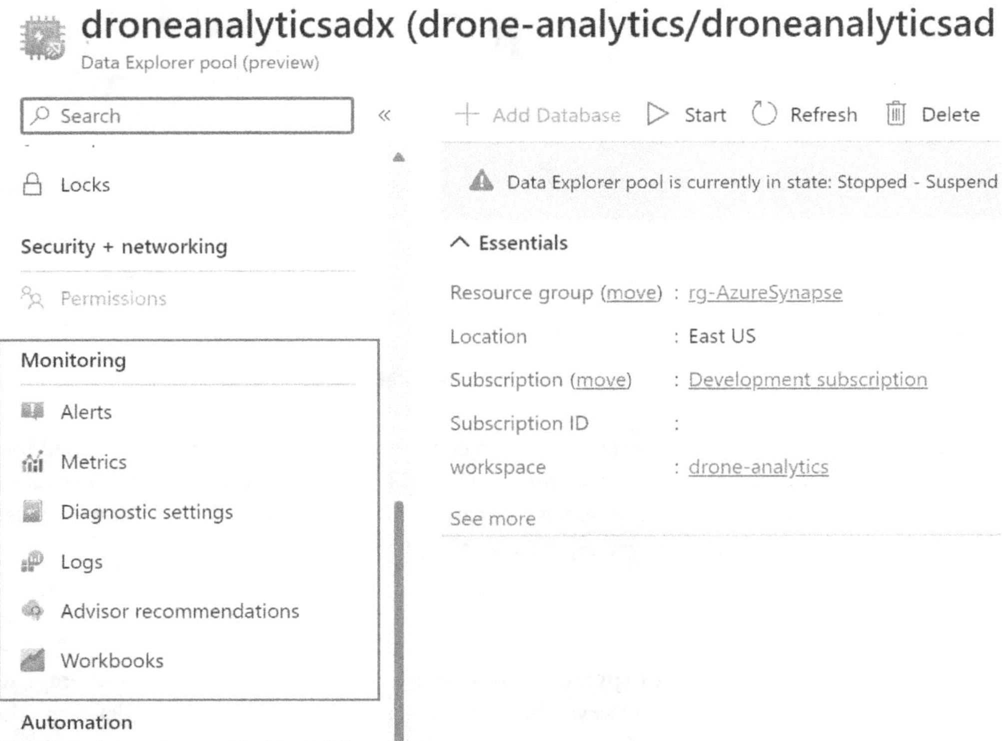

Figure 10.7 – Finding the Azure Monitor options in your Data Explorer pool

Azure Monitor allows you to monitor your Data Explorer pools in the following ways:

- **Alerts**: Azure Monitor Alerts help you become aware of issues that may indicate problems with your Data Explorer pool or problems caused by applications accessing it. You can set up notifications that monitor service metrics and log data to inform you of any issues with your Data Explorer pool.
- **Metrics**: Azure Monitor Metrics collects and saves data about service usage over a time series to help you understand aspects of the system at particular times. This allows you to easily visualize the data in charts, giving you insights into how resource usage has evolved over time, when queries are most issued into the system, and more.
- **Diagnostic settings**: This is a system that allows you to export metrics and logs to external data stores, such as **Azure Data Lake Storage** (**ADLS**) Gen2, or Azure Log Analytics. As an example, you can export metrics about table usage statistics or query execution to files on ADLS Gen2 and analyze them with your tool of choice.
- **Logs**: This allows you to run Azure Log Analytics queries over data captured with the **Diagnostic settings** section. Azure Log Analytics uses KQL for data analysis, so this should be home for you!
- **Advisor recommendations**: Azure Advisor analyses your implementation of Azure services and provides recommendations to keep your resources secure, reliable, and at a lower cost. The recommendations are service specific and include proposed actions you can take to address them.
- **Workbooks**: Azure Workbooks enables you to create rich analysis combining data from text, logs, and metrics, for freeform exploration.

> **Note**
>
> Azure Monitor is a robust service with multiple capabilities. To learn more about Azure Monitor, visit the product documentation at `https://learn.microsoft.com/azure/azure-monitor/`.

Let's look at an example of how you can use Azure Monitor Alerts to help save on Azure consumption costs. As you learned in *Chapter 3*, *Exploring Azure Synapse Studio*, you should pause your Data Explorer pools when not in use, as you are charged for the time they are running, and only start them when you know your users will need them. Azure Monitor Alerts can, among other things, notify you every time a Data Explorer pool is started or paused, which is helpful to know if you want to save costs and never forget to pause your pools if you don't need them.

Creating this alert requires two steps: the creation of an action group, and an alert rule. Let's look into the steps required in detail.

Creating action groups

Before we can configure the actual alert, we need to create an action group: a resource that invokes a set of notifications when an alert is triggered. To create the action group, run the following steps:

1. Select **Alerts**, under the **Monitoring** group.
2. Select + **Create**, and then pick **Action group**.
3. This opens the **Create action group** page, with a focus on the **Basics** tab. This tab requires the following parameters as input:

    - **Subscription**: This is the Azure subscription to where the resource will be deployed. Select the name of the subscription where you want this action group to reside.
    - **Resource group**: Select the resource group to where your action group will be deployed.
    - **Region**: This is the region where your action group will be stored and processed. If you select the **Global** option, the action groups service picks an Azure region to process and store the action group, whereas if you select a specific region, processing is guaranteed to happen in that Azure region. You should pick a specific region if you need to make sure your action group is processed in the region of your choice. This could include scenarios where you have to respect compliance and privacy rules for data processing. For this example, I've picked the **Global** option.
    - **Action group name**: As you would expect, this is the name of your action group resource. Make sure you pick a name that will help you identify this action group in the future. For our example, I've used the name `Kusto alerts`.
    - **Display name**: This is the name that will be displayed in the actual notification when you receive it via email or text message. This name can be different from the action group name, but for simplicity, I kept it the same.

4. When you are ready, click the **Next: Notifications** > button. This takes you to the **Notifications** tab, where you configure the actual notification methods that your action group will use.
5. Under **Notification type**, select **Email/SMS message/Push/Voice**. This opens the **Email/SMS message/Push/Voice** pane, seen in *Figure 10.8*. The options available are as follows:

    - **Email**: Check this box to receive an email when your notification triggers and type the email address that should receive this notification.
    - **SMS**: If you would like to be notified via an SMS message on your phone, you can check this option and provide your phone number.
    - **Azure mobile app notification**: The Azure mobile app is an app you can use on your phone or tablet to manage your Azure resources on the go. It allows you to see your Azure resources status and alerts, pause and resume some services, and even see your ongoing Azure costs. If you use the Azure mobile app and check this box, provide the email account that is logged into your app to receive a notification on your device when this action group is triggered.

- **Voice**: Check this box to receive an automated call when your action group is triggered. At the time of writing, this option was available only to phone numbers in the United States.
- **Enable the common alert schema**: In the past, Azure services used different formats to send messages in alerts. The common alert schema standardizes the message format across all services, making it easier to read them and build automation tasks for messages received.

For this example, I picked the **Email**, **SMS**, and **Azure mobile app** options. If you picked the same, your page should look similar to the one in *Figure 10.8*:

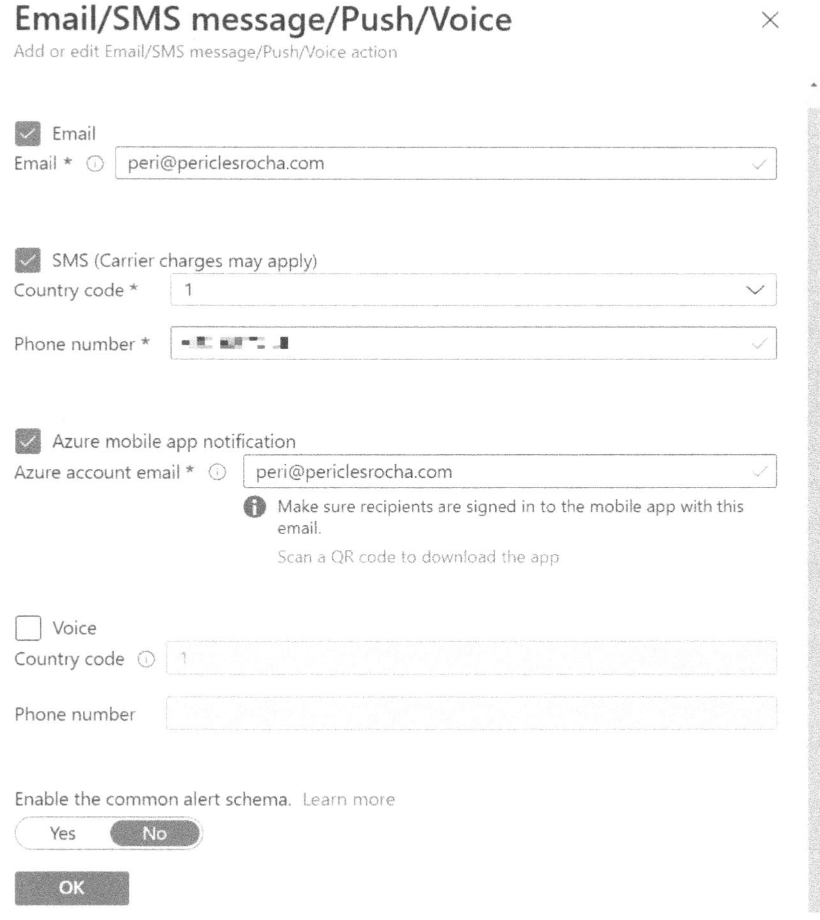

Figure 10.8 – Picking notification options for your action group

6. When you are ready, click **OK**. You'll be sent back to the **Notifications** tab. Provide a name for your notification and select **Next: Actions >** when you are ready.

7. The content in the **Actions** tab is optional. It allows you to select from a list of automated actions you can take when your action group is triggered, such as opening a ticket on an external IT service management tool, performing a web call, or others. For this example, we will skip this step.
8. When you are ready, select **Next: Tags** >. By now, you already know everything about tags in Azure. Feel free to provide any tags needed for this resource. If you need a refresher on tags, you can read more in the *Creating an Azure Synapse workspace* section of *Chapter 2, Creating Your First Data Explorer Pool.*
9. Finally, select the **Next: Review + create** > button, and then **Create**.

Your action group should get deployed within seconds. Once deployment is complete, you should move to the next step, which is to create the alert rule. We will look into this next.

Creating alert rules

Now that the action group that will process the notification is created, let's create the alert rule that will trigger it. To create this alert, follow these steps:

1. Select **Alerts** under **Monitoring**.
2. Select + **Create**, and then pick **Alert rule**.
3. This opens the **Create an alert rule** page with a focus on the **Scope** tab.
4. Note that your Data Explorer pool will already be listed under the **Resource** section, as shown in *Figure 10.9*. If your environment has more than one Data Explorer pool, you can add them here to be part of this alert processing rule so that you get notified when any of these pools trigger the rule. When you are ready, select the **Next: Condition** > button.

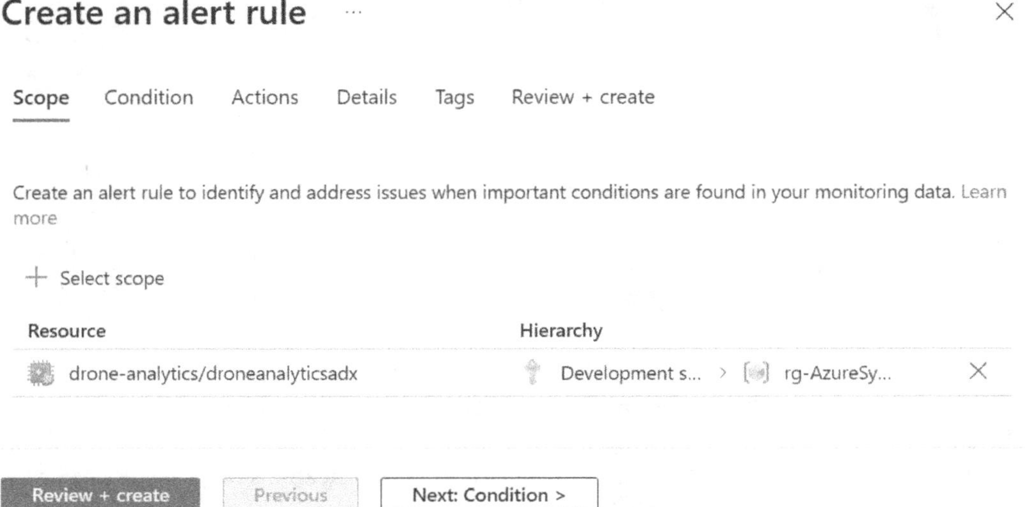

Figure 10.9 – Selecting the scope that applies to the alert rule

5. In the **Condition** settings tab, you'll pick the condition that, when triggered, will fire the notifications. A pane opens with the **Select a signal** title, allowing you to pick the event that you would like to monitor for changes to trigger the notification. As you can see in *Figure 10.10*, there are 11 signals you can pick from. For this example, you should pick **Cluster start action (Microsoft.Synapse/workspace/kustoPools)**. Select **Done** when you are ready.

Select a signal

Choose a signal below and configure the logic on the next screen to define the alert condition.

Signal type: All

Monitor service: All

Displaying 1 - 11 signals out of total 11 signals

Search by signal name

Signal name	Signal type	Monitor service
All Administrative operations	Activity log	Administrative
Write cluster resource (Microsoft.Synapse/workspaces/kustoPools)	Activity log	Administrative
Delete cluster resource (Microsoft.Synapse/workspaces/kustoPools)	Activity log	Administrative
Cluster start action (Microsoft.Synapse/workspaces/kustoPools)	Activity log	Administrative
Cluster stop action (Microsoft.Synapse/workspaces/kustoPools)	Activity log	Administrative
Cluster check name availability action (Microsoft.Synapse/workspaces/kustoPools)	Activity log	Administrative
Lists language extensions action (Microsoft.Synapse/workspaces/kustoPools)	Activity log	Administrative
Add language extensions action (Microsoft.Synapse/workspaces/kustoPools)	Activity log	Administrative
Remove language extensions action (Microsoft.Synapse/workspaces/kustoPools)	Activity log	Administrative
Detach follower databases action (Microsoft.Synapse/workspaces/kustoPools)	Activity log	Administrative
List follower's databases action (Microsoft.Synapse/workspaces/kustoPools)	Activity log	Administrative

Done

Figure 10.10 – Selecting the signal you want to use for your notification

6. Select **Next: Actions** >. This is where you select the action group that we created to send the actual notifications. Click + **Select action groups** and select the **Kusto alerts** action group that you created previously.
7. When you are ready, click **Next: Details** > to move to the next tab.
8. On the **Details** tab, you'll provide basic information about your new alert processing rule. It requires the following parameters:

   - **Subscription**: This is the Azure subscription to where the resource will be deployed. Select the name of the subscription where you want this resource to reside.

- **Resource group**: Select the resource group to where your alert processing rule will be deployed.
- **Alert rule name**: Type the name you would like to give to this resource. As always, make sure you pick a name that will help you identify this resource in the future. For this example, I picked the name `Data Explorer pool start alert`.
- **Description**: You can provide a description that will help you and others identify this resource in the future.
- **Enable rule upon creation**: Check this box to make sure your rule is enabled.

9. When you are ready, select **Next: Tags >** to land in the **Tags** tab. Once again, provide any tags that you would like for this resource. When you are done, click **Next: Review + create**.
10. Take a moment to review your choices and, when you are ready, click **Create**.

You are done! To test your Azure Monitor alert, go back to the **Overview** tab of your Data Explorer pool and select the **Start** button to resume. If it was already started, you'll need to pause it first, and then resume it. Once your Data Explorer pool starts, you'll start receiving notifications. *Figure 10.11* illustrates the SMS and the Azure mobile app notification produced by this alert received on an Android device.

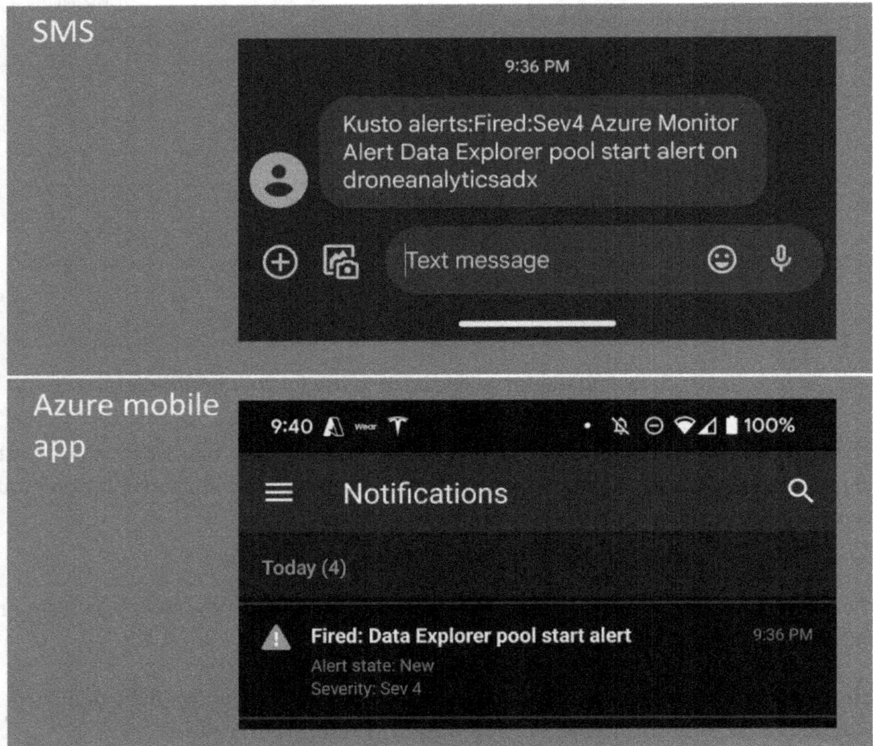

Figure 10.11 – The notifications received by the alert

You can configure a similar alert to get notified when your Data Explorer pool is paused, always making you aware of the status of your Data Explorer pool. Simply follow the same steps to create another alert rule using the same action group as before, but select **Cluster stop action (Microsoft.Synapse/workspace/kustoPools)** in the **Select a signal** window during the configuration of your new alert rule.

This is a useful and simple process you can follow to help you save the ongoing costs of your Data Explorer pool.

Summary

In this chapter, you learned ways to monitor your Azure Synapse Data Explorer environment.

You started by using control commands with KQL to check and update the capacity of your Data Explorer pools. You investigated how to monitor query execution, look for ongoing queries and commands, and stop a query's execution if needed. Still on the topic of system monitoring using KQL, you learned how to understand object metadata and the history of metadata changes in objects.

Next, you learned about the Azure Monitor integration with Azure Synapse Data Explorer. We explored ways Azure Monitor can help you with monitoring and observability of your Data Explorer pool. To close, you created an alert that sent notifications to your phone if your Data Explorer pool was started without you having to be in front of a computer.

In the next chapter, you will learn how to configure resource management in your Data Explorer pools to classify and prioritize user queries.

11
Tuning and Resource Management

As organizations build analytical environments and grow a data culture among their employees, more and more individuals will consume data to support their decision-making process, to carry out experiments, to analyze product data, or for any needs that can be resolved with a query to the source data. While the number of users in your analytical environment grows, remember that your compute resources are not unlimited. You have to find ways to govern how many compute resources each user gets to avoid exhausting them or losing control of your costs.

In this chapter, you will learn about ways to reserve resource usage based on the properties of the incoming request, such as who submitted the request, what application they are using, and more. This helps you prioritize how resources are used in the cluster and offer predictable performance for all users. You'll also learn how to queue requests for delayed execution when your compute cluster reaches the limit for concurrent requests, reducing user query errors caused by resource throttling.

Finally, we'll look at ways to decide which data should be pre-cached on SSDs or RAM for blazing-fast queries, avoiding having to read data from storage every time.

At a glance, here are the topics covered in this chapter:

- Implementing resource governance with workload groups
- Speeding up queries using cache policies

Before we dive into these topics, let's review the technical requirements that will help you make the most of the learning experience.

Technical requirements

In case you haven't yet, make sure you download the book materials from the repository at `https://github.com/PacktPublishing/Learn-Azure-Synapse-Data-Explorer`. You can download the full repository by selecting **Code**, and then **Download ZIP**, or by cloning the repository by using your Git client of choice.

This chapter uses the `fleet data` table, but it doesn't require any actual data loaded into it. If you did not perform the tasks in the *Running your first query* section of *Chapter 3, Exploring Azure Synapse Studio*, take some time to create the `drone-telemetry` database and the `fleet data` table, but you don't actually need any data loaded into it to complete the examples on this chapter. If you have the table you created in the previous chapters, then you are good to go.

Implementing resource governance with workload groups

Resource governance is an important feature of database management systems. It helps control how much compute resources incoming requests can use to avoid exhausting them with only a handful of very busy user requests. Your business may require that all users and applications get the same amount of resources and compete for them on a first-come, first-served basis, or it may require you to configure different rules depending on who the user is, the application they are using, or their department, for example. Azure Synapse Data Explorer offers a robust platform for resource governance by implementing resource policies through workload groups, and by offering mechanisms to classify user requests. There are a lot of concepts here, so let's get started.

Workload groups help define policies that allow us to limit compute resource usage by incoming requests. When a user request comes in, your Data Explorer pool determines which policy this incoming request should be assigned to. This policy assignment is implemented through **request classification policies**, which can classify incoming requests based on the user's account name, Active Directory group membership, application name, or other criteria. Once the request is classified to a policy, it runs under the resource boundaries specified for this policy.

Let's look in detail at the steps involved in using workload groups and how to classify requests using request classification policies.

Managing workload groups

Workload groups work in conjunction with workload group policies, which specify rules for resource usage by workload groups. A **workload group policy** can be one of the following types:

- **Request limits policy**: These policies allow limiting the compute resources that can be used by incoming requests, such as the amount of memory, the maximum number of rows returned, maximum query execution time, and others. For a complete list of configurable limits for request limits policies, refer to `https://learn.microsoft.com/azure/data-explorer/kusto/management/request-limits-policy`.
- **Request rate limit policy**: This allows you to configure the maximum number of concurrent requests that can be classified into a workload group. When these limits are reached, new incoming requests are throttled.
- **Request rate limits enforcement policy**: The enforcement policy helps control how request rate limit policies are enforced for incoming requests. They can be enforced at the cluster level, at the database level, or at the query head node level.

- **Request queueing policy**: This allows you to enable request queueing for throttled queries.
- **Query consistency policy**: Azure Synapse Data Explorer supports **strongly consistent queries**, which ensure any query issued after a command respects the result of the command, and **weakly consistent queries**, which may offer some latency between recent changes and the query results. This policy helps you enforce a consistency policy for user queries. To learn more about query consistency in Azure Synapse Data Explorer, refer to the product documentation at `https://learn.microsoft.com/en-us/azure/data-explorer/kusto/concepts/queryconsistency`.

Workload groups carry one or more of these workload group policies on them. To create or alter a new workload group, use the `.create-or-alter workload_group` command. This command allows you to specify which policy you are configuring, the policy limits, and whether the specified policy can be relaxed by the incoming request or not. As an example, the following command creates a new workload group named `Engineering Department WG` with a request limits policy that limits the maximum number of records in a result set to `1048576` (the row limit in Excel!), the size of the result set to 64 MB, and the execution time for any requests assigned to this policy to two and a half minutes:

```
.create-or alter workload_group ['Engineering Department
WG']  ```
{
  "RequestLimitsPolicy": {
    "MaxResultRecords": {
        "IsRelaxable": false,
        "Value": 1048576
    },
    "MaxResultBytes": {
        "IsRelaxable": true,
        "Value": 64000000
    },
    "MaxExecutiontime": {
        "IsRelaxable": false,
        "Value": "00:02:30"
    }
  }
}```
```

Note

All code examples in this section can be found in the `Chapter 11\Tuning and Resource Management.kql` file of the book's repository.

This produces a result similar to the one shown in *Figure 11.1*, confirming the successful creation of your workload group:

WorkloadGroupName	WorkloadGroup
Engineering Department WG	{ "RequestLimitsPolicy": { "MaxResultRecords": { "IsRelaxable": fals...

00:00:00 Query executed successfully.

Figure 11.1 – Confirmed creation of your new workload group

As mentioned, a workload group can contain one or more workload group policies. For example, in addition to specifying a request limits policy to limit cluster resource usage, you can also create a request rate limit policy to apply hard concurrency limits to requests. The following code snippet uses the `.alter-merge` control command to update your existing workload group by adding a request rate limit policy that limits the maximum number of concurrent requests for this workload group to `50`, and the maximum number of requests for any given user assigned to this workload group to `10` requests:

```
.alter-merge workload_group ['Engineering Department WG']  ```
{
  "RequestRateLimitPolicies": [
  {
      "IsEnabled": true,
      "Scope": "WorkloadGroup",
      "LimitKind": "ConcurrentRequests",
      "Properties": {
        "MaxConcurrentRequests": 50
      }
    },
    {
      "IsEnabled": true,
      "Scope": "Principal",
      "LimitKind": "ConcurrentRequests",
      "Properties": {
```

```
            "MaxConcurrentRequests": 10
          }
        }
      ]
    } ```
```

We can visualize the workload group with the merged changes by using the `.show workload groups` command. Running this command produces an output similar to the one shown in *Figure 11.2*. The results in the query window are truncated, but if you expand the `WorkloadGroup` column in this result set, you will see both your request limit policy and request rate limit policy combined in the serialized array of policy objects:

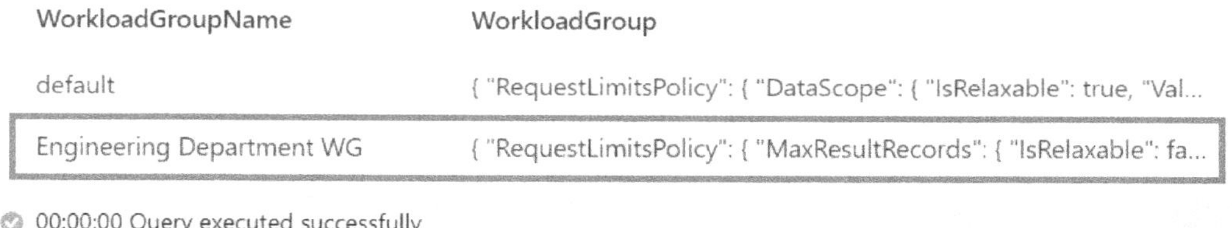

WorkloadGroupName	WorkloadGroup
default	{ "RequestLimitsPolicy": { "DataScope": { "IsRelaxable": true, "Val...
Engineering Department WG	{ "RequestLimitsPolicy": { "MaxResultRecords": { "IsRelaxable": fa...

00:00:00 Query executed successfully.

Figure 11.2 – Visualizing your workload groups

When you ran the `.show workload groups` command, you may have noticed that in addition to your `Engineering Department WG` workload group, there's also a `default` workload group. In fact, every Data Explorer pool has two built-in workload groups that are predefined: the `default` workload group and the `internal` workload group. User requests are classified to the `default` workload group when they are not classified to any other workload group, when you attempt to classify the request to a group that doesn't exist (for instance, one that was deleted), or when a classification error has occurred. The good news is that you can change the policies of the `default` workload group, so if requests are often classified there, you can have control over the resource limits that they will use.

The other predefined workload group is the `internal` workload group. This workload group is for internal use only, and you cannot alter its policies or classify user requests to this group.

In addition to the `default` and `internal` workload groups, you can create up to 10 additional ones. At first glance, this may not look like a lot, but keep in mind that maintaining several workload groups and ensuring there's no conflict in their policies can be a demanding, time-consuming task. Ideally, you should have a few workload policies created that help you govern resource usage somehow but avoid having too many specific rules. One way I like to think about workload groups is to reserve more resources for production environments while limiting resources needed for development and testing.

Now that you understand what workload groups are and how they implement resource policy limits, let's see how you can classify user requests to your workload groups.

Classifying user requests

The way users are classified to specific workload groups is through the use of request classification policies. Whenever new incoming requests reach the Data Explorer pools, they will be processed against a series of classification policies and, when matched, assigned to workload groups.

Classification is done through a user-defined function that helps understand the context of the incoming request. Each Data Explorer pool can have only one classification function, and this function processes every single incoming request.

> **Note**
>
> Since the request classification policy evaluates every incoming request, it's important to make sure that it runs quickly. If you have complex logic in the user-defined function that is used for the classification policy, you may introduce performance issues.

Request classification policies are a cluster-wide object. They have an `IsEnabled` Boolean property, which helps you control whether this policy is enabled, and a `ClassificationFunction` property, which is a string that stores the actual user-defined function that is used to classify incoming requests.

The user-defined function that is used by the request classification policy must return as a result the name of the workload group that the incoming request should be classified to. This return value needs to be a single scalar string value. If the user-defined function returns an invalid workgroup name, or if the function fails for some reason, the request will be classified to the default workload group. The function receives an object named `request_properties`, which helps you understand certain properties of the incoming request, such as the name of the current database, the name of the application submitting the request, the user's account name, the request type (command or query), and other useful information. You can then use any of these properties to evaluate the incoming request and determine which workload group it should be assigned to.

As an example, let's say I want to create a request classification policy with the following rules:

- If the user is a member of the `engineering@periclesrocha.com` Active Directory group, the connection will be assigned to the `Engineering Department WG` workload group.
- If the application submitting the incoming request is `Kusto.Explorer`, the connection will be assigned to a workload group named `General queries WG`.
- If the description submitted for the incoming request is equal to `Night shift` and the current hour is between 18 and 23, the incoming request will be classified to a workload group named `After hours WG`.
- If none of these conditions are satisfied, then classify the request to the `default` workload group.

This policy can be created with the following code block:

```
.alter cluster policy request_classification
'{"IsEnabled":true}' <|
    case(
      current_principal_is_member_of('aadgroup=engineering@
periclesrocha.com'), "Engineering Department WG",
      request_properties.current_application == "Kusto.
Explorer", "General queries WG",
      request_properties.request_description == "Night shift"
and hourofday(now()) between (18 .. 23), "After hours WG",
      "default"
    )
```

In this example, `engineering@periclesrocha.com` is the Azure AD group that I want the classification policy to apply to. To reproduce this example, you should replace this group name with a group within your own Azure AD organization. If you are running these examples outside of an Azure AD organization and using a Microsoft account to log in to Azure Synapse, you can replace this with your own user account to reproduce the example. Once you make any necessary adjustments, you should see a result similar to the one illustrated in *Figure 11.3*:

PolicyName	EntityName	Policy	ChildEntities
ClusterRequestClassificationPolicy		{ "ClassificationProperties": ["req...	

00:00:01 Query executed successfully.

Figure 11.3 – Your new request classification policy

From now on, incoming requests will be classified in accordance with this new policy. As you can see in this example, the user-defined function used by the classification policy can not only read properties about the incoming request but can also execute other evaluations, such as checking the current time to determine how the request should be classified.

If you need to disable your request classification policy temporarily for any reason, you can use the `.alter-merge cluster policy` control command to change the `IsEnabled` property, like so:

```
.alter-merge cluster policy request_classification
'{"IsEnabled":false}'
```

When the request classification policy is disabled, all incoming requests will be classified to the `default` workload group. To check the status of your request classification policy, you can use the `.show cluster policy` control command, like so:

```
.show cluster policy request_classification
```

This command outputs the details of your policy, which include the `IsEnabled` property. By resizing the column in the result set or selecting and copying its content to the clipboard and then pasting it into a text editor, you'll be able to see the details of your policy, as illustrated in *Figure 11.4*:

Policy
{ "ClassificationProperties": ["request_description", "current_application"], "IsEnabled": false, "Classificati...

Figure 11.4 – Checking the status of your request classification policy

That puts the two pieces of the puzzle together: workload groups define the rules that cap resource usage in your cluster to make sure the right people have the right resources available for them, and the request classification policy assigns an incoming request to that workload group policy. But what happens when a user reaches your Data Explorer pool, but the workload group policy determines that this request can't be satisfied as it violates your policy? This can happen, for example, if you set a maximum number of concurrent requests, and this limit has been reached. Should we drop the incoming connection and just return an error message to the user? Let's look at a way to mitigate that.

Queuing requests for delayed execution

If users submit queries to a Data Explorer pool and hit limits defined by a workload group named `Engineering Department WG`, their query will fail, and they will see an error message similar to the following:

```
The query was aborted due to throttling. Retrying
after some backoff might succeed. Capacity: 50, Origin:
'RequestRateLimitPolicy/WorkloadGroup/Engineering Department
WG'.
```

Azure Synapse Data Explorer offers a mechanism that allows throttled requests to be queued once the compute pool's thresholds have been met, avoiding the need for users to try submitting their query again. This is done by enabling a request queueing policy in your Data Explorer pool. By queueing queries, this policy helps reduce throttling errors when your pools are the busiest.

> **Note**
>
> At the time of writing, the request queueing policy feature was in preview, even for the standalone service Azure Data Explorer. There may be implications in the supportability and functionality of features in preview, so make sure you check its availability before you use this feature in production.

To enable the request queueing policy, all you need to do is switch its `IsEnabled` property to `true` in your workload group:

```
.alter-merge workload_group ['Engineering Department WG'] ```
{
  "RequestQueuingPolicy": {
      "IsEnabled": true
  }
} ```
```

Remember that the request queueing policy is set at the workload group level, so you can enable and disable it as needed per workload group. The `default` workload group also supports this policy, and you can enable it just like you would do for any other workload group. As an example, the following code enables the request queueing policy for the `default` workload group:

```
.alter-merge workload_group default ```
{
  "RequestQueuingPolicy": {
      "IsEnabled": true
  }
} ```
```

The request queueing policy can only be defined when your workgroup policy limits the maximum number of concurrent requests through a request rate limit policy. Still, this is a convenient feature for avoiding some common user errors caused by query throttling.

Speeding up queries using cache policies

As you have seen in *Chapter 1*, *Introducing Azure Synapse Data Explorer*, Data Explorer pools can manage very large amounts of data. They separate the compute layer from the storage layer, allowing you to scale massively in storage, regardless of how much compute you have allocated in your Data Explorer pool.

When dealing with large volumes of data, it's useful to understand what data you need readily available as needed, and what data can be stored as an archive, meaning that it is still available but maybe at a cheaper location that is slower to retrieve. This is the concept of hot data and cold data. Data accessed frequently is designated as *hot data* and should be quick to retrieve. Data that is less frequently accessed but is still needed is designated as *cold data*, and can typically be stored in cheaper storage that is still reliable but slower to retrieve. The implication here is not only on performance but also on cost: cold storage is much cheaper.

> **Note**
>
> To better understand data storage tiers in Azure, visit Azure Storage's documentation at `https://learn.microsoft.com/azure/storage/blobs/access-tiers-overview`.

Azure Synapse Data Explorer uses, mostly, Azure Blob storage to store data. It implements a robust mechanism for automatically determining which data should be cached to local **solid-state drives** (**SSDs**) and **random access memory** (**RAM**) for hot data, and which data should be stored as cold data. In some cases, however, you may want to force some data to always be treated as hot data by the compute engine as you know it is commonly used, or it needs to be retrieved quickly if needed. Similarly, maybe there's a table that stores historical log data that you would like to keep in cold storage as it is rarely used. Data Explorer pools allow you to define cache policies for hot and cold data.

Not all data can be cached. Azure Synapse Data Explorer uses 95% of the local SSD of its nodes to cache hot data. If the volume of your cached data needs more disk space than what is available, Data Explorer keeps the most recent data cached. It uses the date and time of ingestion to determine what data is most recent.

To view the cache policy for a table, you can use the `.show table policy caching` command. As an example, the following command shows the caching policy for the `fleet_data` table in the `drone telemetry` database:

```
.show table ['drone-telemetry'].['fleet data'] policy caching
```

This command produces a result similar to the one in *Figure 11.5*. Note that there's no caching policy specified, and Data Explorer will automatically define what data to cache:

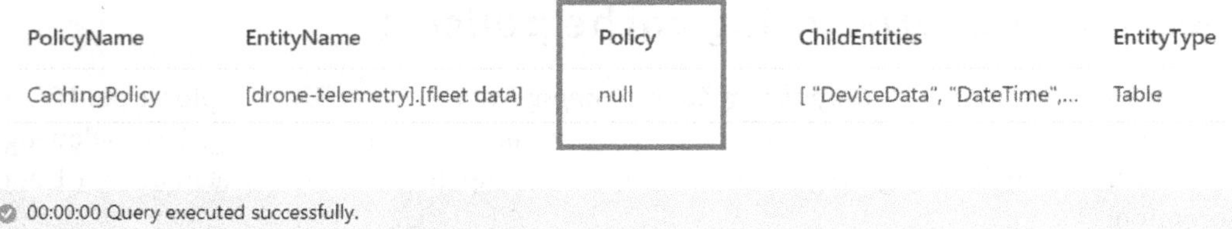

PolicyName	EntityName	Policy	ChildEntities	EntityType
CachingPolicy	[drone-telemetry].[fleet data]	null	["DeviceData", "DateTime",...	Table

00:00:00 Query executed successfully.

Figure 11.5 – Showing a table's caching policy

To define a caching policy for a table, you can use the `.alter table caching policy` command. For example, the following code specifies that data from the last 7 days in the `fleet_data` table should always be cached:

```
.alter table ['drone-telemetry'].['fleet data'] policy caching
hot = 7d
```

This command produces a result similar to the one in *Figure 11.6*. Note that we can now see the policy definition in the **Policy** column, serialized as a JSON value:

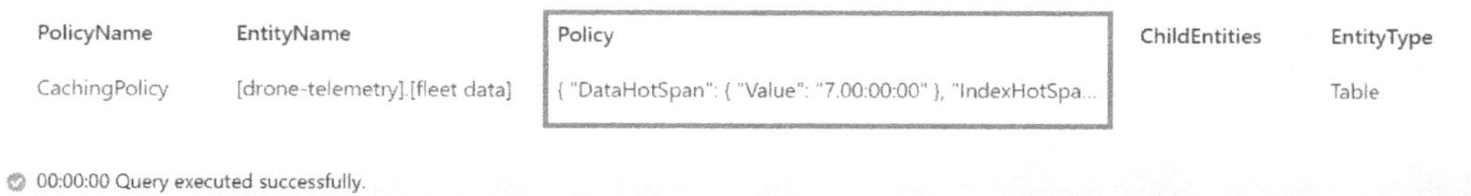

PolicyName	EntityName	Policy	ChildEntities	EntityType
CachingPolicy	[drone-telemetry].[fleet data]	{ "DataHotSpan": { "Value": "7.00:00:00" }, "IndexHotSpa...		Table

00:00:00 Query executed successfully.

Figure 11.6 – Defining the cache policy for a table

Besides defining the cache to work with recent data from the last few days, you can also set hot-cache windows from time ranges that you know are frequently used in queries. For example, if you want data from the holiday season to always be cached because it is used throughout the year, you can define your policy to cache not only the last 7 days, but also data from Black Friday (between November 24 and 28 for the year 2022), and all data for the month of December. The following example illustrates this scenario:

```
.alter table ['drone-telemetry'].['fleet data'] policy caching
        hot = 7d,
        hot_window = datetime(2022-11-24) ..
datetime(2022-11-28),
        hot_window = datetime(2022-12-01) ..
datetime(2022-12-31)
```

To delete a table's cache policy, use the `.delete table policy caching` command, like so:

```
.delete table ['drone-telemetry'].['fleet data'] policy caching
```

Cache policies can also be defined at the cluster level (that is, at the level of individual Data Explorer pools) or at the database level. In such cases, all data in the scope of the object to which the policy is applied is subject to hot caching. The command syntax to show, alter, and delete the caching policy for a database or a Data Explorer pool is the same as used for tables, but replacing the `table` keyword with `database` or `cluster`. For example, the following commands, which need to be run separately, one at a time, define the policy for a Data Explorer pool and delete the policy in the `drone-telemetry` database, respectively:

```
.alter cluster policy caching hot = 7d
.delete database ['drone-telemetry'] policy caching
```

Tables without a caching policy inherit any policies defined at the database level, which, in turn, inherit the policy defined at the cluster level. Any policies defined at the object level override that of the parent object.

Summary

As the usage of your Azure Synapse workspace grows, you may start to experience throttling errors at peak times due to extensive use of the compute resources in your Data Explorer pools. Having a plan to address resource usage through workgroup policies helps you better control the user experience and offer a balanced environment for all users. In addition to that, using techniques that help you speed up user queries helps you relieve stress on the platform, and resolve some requests more rapidly.

In this chapter, you learned how to implement resource governance to limit resource usage depending on who the user submitting the request is, the application that they are using, the description submitted with the request, and more. You learned how to create workload group policies, classify user requests to these policies, and enable the queueing of requests that are throttled for delayed execution.

You also learned how to pre-cache data to speed up queries. We explored the concepts of hot and cold data, and how to configure a caching policy for hot data through hot-cache windows, or for data ingested in the last few days only.

In the next chapter, you will learn about securing your Data Explorer pool to ensure that only the right people have access to your telemetry and log data.

12
Securing Your Environment

A long time ago, I attended a conference for the launch of Delphi 4, and someone in the audience asked a question that provoked me for years. When the Borland employee ran their demo for the Borland InterBase database management system and showed all of the marvelous features that it offered, one person asked *what about security?* The Borland employee returned a question: *what is database security?* Now that you know how old I am, I will further elaborate on why this question provoked me so much, for so long, and why you should think about it too.

Database security is a broad subject that involves not only keeping your data secure but also how you transmit it, how you store it, how you authenticate and authorize users to access it, and how you protect your data environment from malicious actors. Data is one of the most valuables assets for any organization, and keeping it secure is top of mind for executives, with 88% of board members classifying cybersecurity as a business risk, in accordance to Gartner's 2022 Board of Directors Survey (source: `https://www.gartner.com/en/articles/6-key-takeaways-from-the-gartner-board-of-directors-survey`).

The COVID-19 pandemic made the security landscape even more complicated with at least a third of households in the US reporting they were working from home more frequently than before, according to the US census (source: `https://www.census.gov/library/stories/2021/03/working-from-home-during-the-pandemic.html` – percentages may vary in different countries). These individuals moved from a managed network environment in corporate offices into their homes, where security standards are typically lower and unsupervised. This makes it easier for attackers to use different techniques and explore vulnerabilities for a wide variety of attacks.

With this context, allow me to go back to the question that has haunted me since the 90s: what is database security? This chapter focuses on different lenses that you need to have in mind for a complete and robust security posture in your Azure Synapse environment. Specifically, you will learn about the following topics in the context of Azure Synapse:

- Managing data encryption
- Authenticating users
- Configuring access to resources

- Implementing network security
- Protecting against external threats

Security is one of those things that takes a village to get done right. Depending on the size of your organization, you may need to work with your colleagues who are responsible for network infrastructure, identity, threat detection, and other areas. As you read through this chapter, make sure you identify who those individuals are and how you collaborate with them to stay on top of your security posture.

Technical requirements

In case you haven't yet, make sure you download the book materials from the repository at `https://github.com/PacktPublishing/Learn-Azure-Synapse-Data-Explorer`. You can download the full repository by selecting **Code**, and then **Download ZIP**, or by cloning the repository by using your Git client of choice.

This chapter uses the `fleet data` table, but it doesn't require any actual data loaded into it. If you did not perform the tasks in the *Running your first query* section of *Chapter 3*, *Exploring Azure Synapse Studio*, take some time to create the `drone-telemetry` database and the `fleet data` table, but you don't actually need any data loaded into it to complete the examples in this chapter. If you already have the table created from the previous chapters, then you are good to go.

Finally, this chapter covers several topics in the security spectrum. While it is not practical to describe every concept or technology in detail in this book, here are some resources that may help you navigate through the chapter more comfortably:

- *Azure encryption overview*: `https://learn.microsoft.com/azure/security/fundamentals/encryption-overview`
- *What is Azure Active Directory?*: `https://learn.microsoft.com/azure/active-directory/fundamentals/active-directory-whatis`
- *Network security groups*: `https://learn.microsoft.com/azure/virtual-network/network-security-groups-overview`
- *Data exfiltration*: `https://en.wikipedia.org/wiki/Data_exfiltration`
- *Azure VPN Gateway*: `https://azure.microsoft.com/products/vpn-gateway`
- *Azure ExpressRoute*: `https://azure.microsoft.com/products/expressroute`

Security overview

Security is a broad subject that covers several aspects of application, identity, and data protection. When thinking about security for cloud data environments, in general, there are five main pillars to consider:

- **Data encryption at rest and at transit**: Protecting your data as it is stored on disk and while it is transferred over the network

- **How users authenticate**: Providing an identity solution to authenticate users
- **How user access is determined**: Controlling who gets access to what data, and their access level
- **How to protect access to public endpoints**: Isolating your environment from other applications and protecting it from unwanted access over the internet
- **How to prevent online threats**: Monitoring and responding to security threats

We will cover each of those pillars in detail but in the context of Azure Synapse workspaces and Data Explorer pools. Remember that Azure Synapse is a unified analytics platform that brings together several cloud services as a single service offering, allowing you to deliver end-to-end analytics projects. As we have seen, it includes data integration pipelines, a SQL engine, an Apache Spark engine, the Data Explorer engine, as well as integrations with Azure Data Lake Storage Gen2, Power BI, Azure Machine Learning, and many other products. As a reminder, *Figure 12.1* illustrates the components contained within an Azure Synapse workspace:

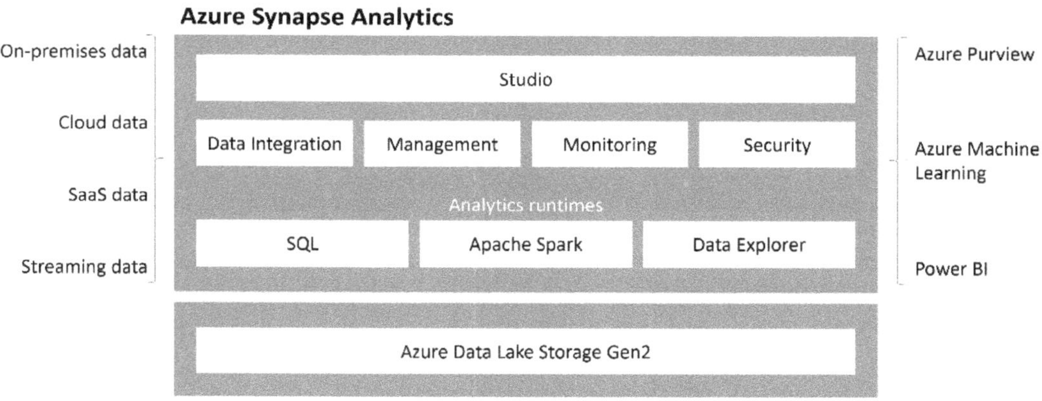

Figure 12.1 – Components of an Azure Synapse workspace

Each of these components offers its own set of security features to help you control usage data encryption at rest and at transit, access to data, authentication, network access, and more. For the purpose of this book, we will focus on the security features of Azure Synapse workspaces and Data Explorer pools. Let's explore how these five security pillars protect your Azure Synapse workspace.

Managing data encryption

For Azure Synapse workspaces, all data is stored with at least one layer of encryption, ensuring that data is never persisted on disk in its original, clear form. You don't have the option to not encrypt data at rest. This applies to any of the analytical engines in Azure Synapse as well as to data persisted, even temporarily, by integration pipelines.

In addition to this first layer of encryption, Azure Synapse workspaces offer an optional, additional layer of encryption, named double encryption. This feature helps protect data and keeps it encrypted even if one of the encryption layers gets compromised. **Double encryption** uses a customer-managed key, which is stored in Azure Key Vault, giving you full responsibility for key management. The first layer of encryption, on the other hand, uses platform-managed keys, which you don't have access to. Both layers use 256-bit **Advanced Encryption Standard** encryption, known simply as **AES 256**.

Configuring data encryption at rest

As you saw in the *Creating an Azure Synapse workspace* section of *Chapter 2, Creating Your First Data Explorer Pool*, double encryption can only be enabled during the creation of your Azure Synapse workspace. To enable it, you must select the **Enable** option next to **Double encryption using a customer-managed key**, in the **Security** tab of the **Create Synapse workspace** wizard. Upon selecting this option, as illustrated in *Figure 12.2*, you are required to specify the Azure key vault and the key that will be used to apply double encryption:

Workspace encryption
Double encryption configuration cannot be changed after opting into using a customer-managed key at the time of workspace creation.
Choose to encrypt all data at rest in the workspace with a key managed by you (customer-managed key). This will provide double encryption with encryption at the infrastructure layer that uses platform-managed keys. Learn more
Double encryption using a customer-managed key
Enable
Disable
Encryption key *
Select a key
Enter a key identifier
Key vault and key *
Key vault: dronetelemetryvault
Key: DataEncryptionKey
Select key vault and key
Azure Key Vaults in the same region as the workspace will be listed.
Managed identity *
User assigned
System assigned
The workspace will not be activated until the purge protection is enabled and system-assigned managed identity is granted required permissions (Get, Wrap Key, Unwrap Key) on the selected key vault. Learn more

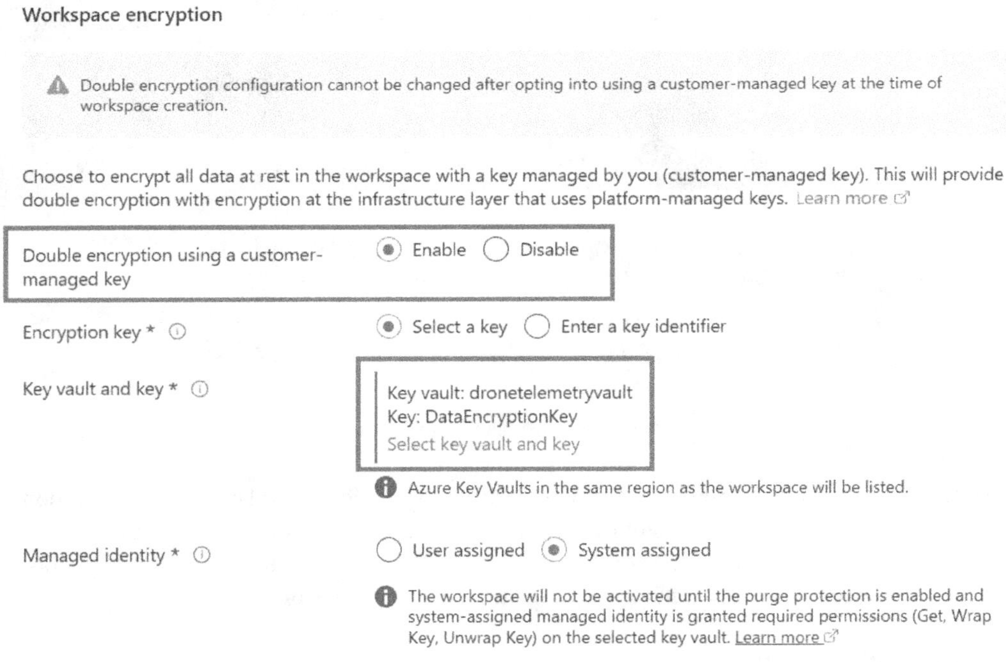

Figure 12.2 – Enabling double encryption

Once you create your Azure Synapse workspace with double encryption enabled, you cannot disable it. You can, however, change the key used for encryption as needed, simply by going to the **Encryption** page of your workspace resource in the Azure portal, as shown in *Figure 12.3*:

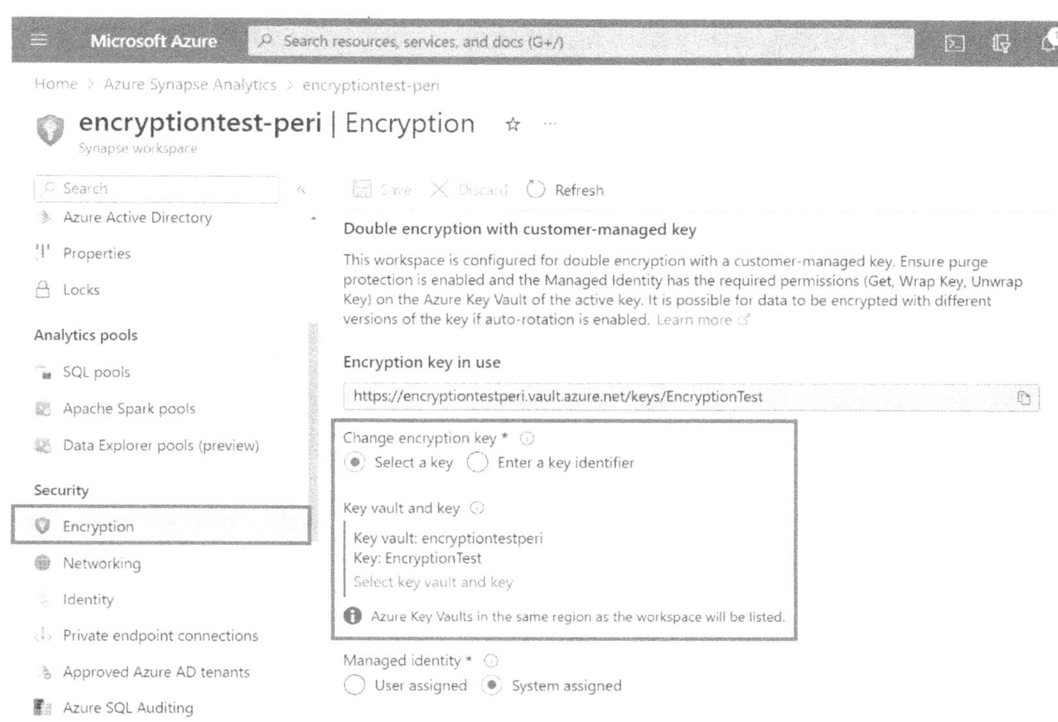

Figure 12.3 – Changing the key used in double encryption

Changing the encryption key causes the data to be decrypted and encrypted again, so make sure to keep the original key before you replace it with a new one, as it will be needed for decryption. Also note that if you are using key rotation to create new versions of a key periodically, the workspace will not re-encrypt data with the new version of your key. You must forcibly change the workspace key to a temporary key, wait for encryption with the temporary key to finish, and then go back to your original key (which will have a new version) to encrypt it again.

Using a key also requires you to specify the managed identity that will be used to access your key in Azure Key Vault. A managed identity is a type of Azure Active Directory identity that is used by applications to connect to other Azure resources without requiring a username and password. Your managed identity needs to have the **WrapKey**, **UnwrapKey**, and **Get** permissions on the key to be used for double encryption.

Now that you understand how Azure Synapse workspaces use encryption for data at rest, let's see how it deals with data in transit.

Understanding data encryption in transit

The analytical engines in Azure Synapse, namely Synapse SQL, Apache Spark, and Data Explorer, all use the **Tabular Data Stream** (**TDS**) protocol for communication between clients performing requests and the analytical engine endpoint. TDS works on top of **Transport Layer Security** (**TLS**), which encrypts all data sent between endpoints and clients with AES 256 encryption.

> **Note**
>
> To learn more about the TDS protocol, visit `https://learn.microsoft.com/openspecs/windows_protocols/ms-tds/893fcc7e-8a39-4b3c-815a-773b7b982c50`.

This encryption process uses a signed server certificate that is managed by Microsoft, so there is nothing you need to do to enable encryption in transit. In fact, you can't disable data encryption in transit. This work is done transparently for you and there's no way to configure it, but it's important to understand that your data is secured while sent over the internet.

Authenticating users

As discussed in the *Security overview* section of this chapter, each component of Azure Synapse may offer different security features or characteristics. Just as an example, Data Explorer pools only support **Azure Active Directory** (**Azure AD**) authentication, but dedicated SQL pools and serverless SQL pools support Azure AD authentication as well as SQL authentication, a mechanism to authenticate identities using a SQL Server-managed credential.

Azure AD is Microsoft's cloud-based identity and access management service for enterprises. It authenticates users and allows organizations to authorize access to multiple applications and network resources using one single user credential. It also simplifies access to cloud resources, such as the Azure portal, Azure Synapse Studio, and even Microsoft Office 365. Azure AD offers the following added benefits:

- Unification of user and service authentication on a single location, eliminating the need for companies to worry about managing credentials
- **Single sign-on** (**SSO**) enables you to use the same credential to authenticate to any service that offers Azure AD integration, and to avoid constant sign-in prompts
- Credential hardening and advanced security through the application of encryption in user credentials, strong password policies, **multi-factor authentication** (**MFA**), protection from brute-force attacks, and other security features
- Audit of sign-in events and updates applies to Azure AD accounts

Azure Synapse workspaces rely exclusively on Azure AD for authentication. Any user or application that needs to connect to an Azure Synapse workspace must use an Azure AD credential. Data Explorer pools, and Apache Spark pools in Azure Synapse, both currently only support Azure AD authentication as well.

As previously mentioned, Synapse SQL pools (serverless and dedicated) support SQL authentication as an option to Azure AD. SQL authentication can be enabled or disabled for each SQL pool, or you can disable SQL authentication at the workspace level and disallow any SQL pool from using it. This can be done on the **Azure Active Directory** page of your workspace's configuration page in the Azure portal, under the **Settings** options, as shown in *Figure 12.4*:

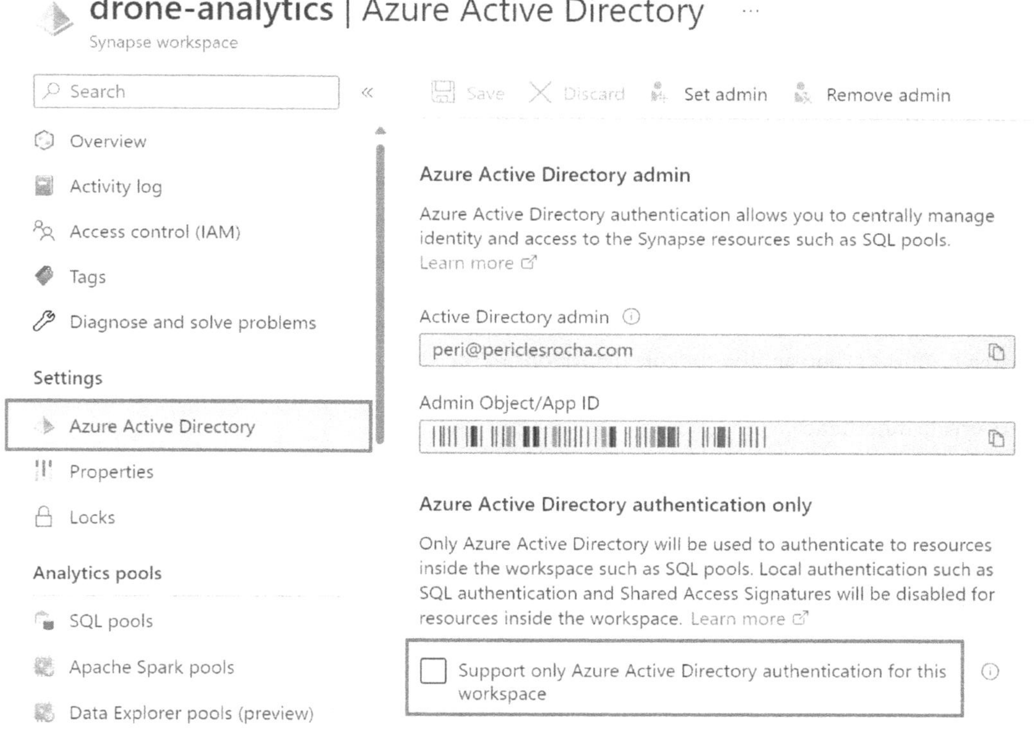

Figure 12.4 – Configuring a workspace to use Azure AD authentication only

Microsoft recommends using Azure AD whenever possible, and enabling SQL authentication only if you have users or applications that need to connect to a SQL pool but are outside of an Azure AD environment – a situation that can happen for legacy applications that do not support Azure AD or external environments without a trust relationship with your Azure AD tenant.

Configuring access to resources

Once users are authenticated into an Azure Synapse endpoint, such as a Data Explorer pool, a series of access rules will determine whether the user has access to the resource they are requesting, and what level of access they have. Azure Synapse offers a robust and granular mechanism to control access to resources.

Azure Synapse leverages Azure's **Role-Based Access Control** (**RBAC**), the authorization system that is used in Microsoft Azure to provide fine-grained access to Azure resources. RBAC roles consist of three elements:

- **Security principals**: The leaf object that represents a user account, a group account, a service principal, or a managed identity. This is the object that will be granted access to the desired resource through RBAC.
- **Role definitions**: The definition of the permission, or group of permissions for the actions that we want to allow with this role. A role definition can contain a single action, such as allowing reading data, or an array of actions, such as allowing reading, deleting, and inserting data.
- **Scope**: The list of resources that this role allows access to. The scope allows you to specify a role assignment to the levels of the management group directly, to the subscription level, to the resource group level, or to a specific resource. When you assign a role with the scope to a certain level, it cascades down to child resources. For example, specifying the scope of a role to a resource group applies the role definitions to the security principals of every resource in that Azure resource group. You can still deny access to individual resources as needed and override inheritance.

> **Note**
>
> If you are a seasoned Azure user, you should already be familiar with Azure RBAC. If that's not the case, you can learn more about it in the article *What is Azure role-based access control (Azure RBAC)?* at `https://learn.microsoft.com/azure/role-based-access-control/overview`.

Azure Synapse extends Azure RBAC by implementing its own RBAC roles, with role definitions that are specific to Azure Synapse workspaces. These roles can be used to control who can publish, access, and execute code artifacts, execute Azure Synapse notebooks with Apache Spark pools, access linked services, monitor job execution, and more. Data Explorer pools, on the other hand, implement their own role-based authentication, unrelated to the Azure RBAC model. Let's explore next how Azure Synapse RBAC roles work, and how to implement access control in Data Explorer pools.

Synapse RBAC roles

Azure Synapse RBAC help provide fine-grained permissions to users in an Azure Synapse workspace. Here is the list of all roles available and their definitions:

- **Synapse Administrator**: This is the most permissive role and should be used only for a limited set of users. It provides full access to compute pools and integration runtimes in Azure Synapse. It allows create, read, update, and delete actions on workspace code artifacts, such as KQL scripts or notebooks. It also allows the management of user permissions and RBAC role assignments.
- **Synapse Apache Spark Administrator**: This role should be used for users who administer Apache Spark pools. It provides full access to Apache Spark pools, including permissions to monitor and stop Spark jobs submitted from any user, and read access to notebooks. It does not allow, however, granting permissions to other users.
- **Synapse SQL Administrator**: This role is the Synapse SQL equivalent to the Synapse Apache Spark Administrator role.
- **Synapse Contributor**: This role gives users permission to read and write code artifacts, view notebooks and integration pipeline outputs. This role also allows performing all actions on Apache Spark activities and viewing Apache Spark pool logs.
- **Synapse Artifact Publisher**: This role allows users to create, read, update, and delete any code artifacts. It does not allow running code artifacts.
- **Synapse Artifact User**: This role is similar to the Synapse Artifact Publisher role, but it is more restrictive. It allows only reading code artifacts and their outputs.
- **Synapse Compute Operator**: This role allows the submission of Apache Spark jobs and the viewing of their outputs. It also allows canceling any Spark jobs, regardless of who submitted them.
- **Synapse Monitoring Operator**: This role allows viewing of code artifacts and their outputs, as well as reading details from Synapse SQL pools, Data Explorer pools, or Apache Spark pools.
- **Synapse Credential User**: This role is needed for jobs that need to use credentials for unattended tasks that run without user interaction, such as an integration pipeline run that uses credentials with linked services.
- **Synapse Linked Data Manager**: This role gives a user the create and update permissions to credentials, linked services, and managed private endpoints.
- **Synapse User**: This role gives a user access to an Azure Synapse workspace. As a member of the Synapse User role, the user will be able to connect to the workspace and view workspace assets, such as analytical pools, credentials, and linked services.

> **Note**
>
> For an extensive description of these Azure Synapse RBAC roles, the actions they permit, and the scope allowed for each role, visit the product documentation at `https://learn.microsoft.com/azure/synapse-analytics/security/synapse-workspace-synapse-rbac-roles`.

You always should provide users with the least possible permissions they need to perform their jobs. One easy tactic you can use to govern user access is to start by providing a user with the **Synapse User** role, which gives them access to the workspace, and then adding other roles as needed. When users need to perform a certain task and they don't have enough permissions for it, Azure Synapse Studio will disable that function on-screen and provide the user with messages to help them understand what roles they may need to perform that action. For example, *Figure 12.5* shows the pop-up message that a user sees when they do not have the right to create Data Explorer pools and hovers the mouse over the + **New** button. In addition to that, there's an informative message at the top of the page that provides hints as to what roles may be needed to manage Data Explorer pools:

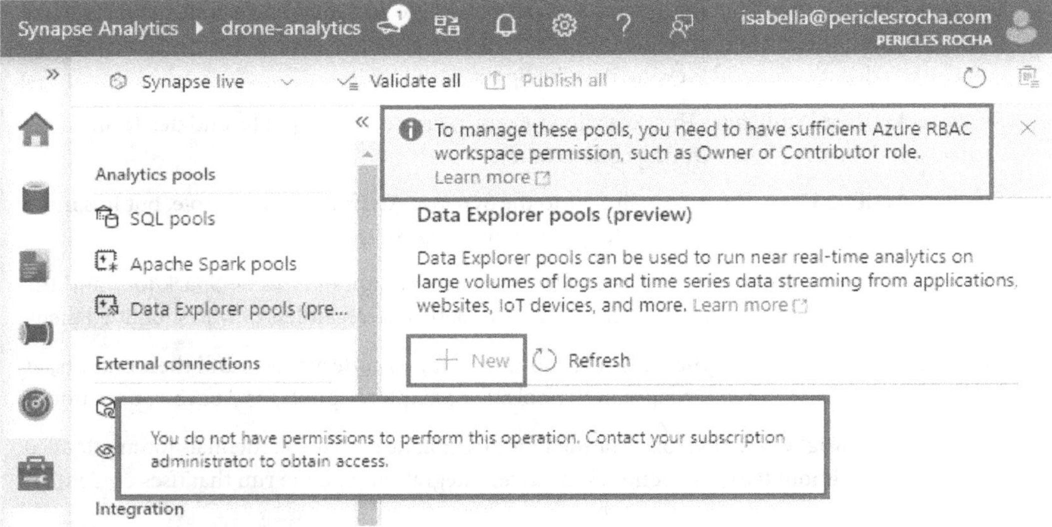

Figure 12.5 – On-screen feedback to users lacking permissions

One thing that none of the Azure Synapse RBAC roles covered was the creation of compute resources, such as Data Explorer pools. To create and manage Synapse SQL pools, Data Explorer pools, Apache Spark pools, and integration runtimes, you need to be a member of the **Azure Owner** or **Azure Contributor** Azure RBAC roles (and not Azure Synapse RBAC roles) *in the scope of the resource group* where the compute resource will be created.

Azure Synapse Studio offers you the management interface to review and manage role assignments. Now that you know what the Azure Synapse RBAC roles are, let's see how we can review existing role assignments and assign roles to users.

Reviewing role assignments

As a member of any Azure Synapse RBAC role, a user can view all role assignments for any user through Azure Synapse Studio. To review role assignments, follow these steps:

1. Navigate to Azure Synapse Studio at `https://web.azuresynapse.net` and log in to your workspace.
2. In the hubs region, select the **Manage** hub.
3. Under **Security**, select **Access control**. You will land on the **Access control** page.

As illustrated in *Figure 12.6*, you can see all the identities (users and service principals) that are members of Azure Synapse RBAC roles, including the scope at which the role assignment was made:

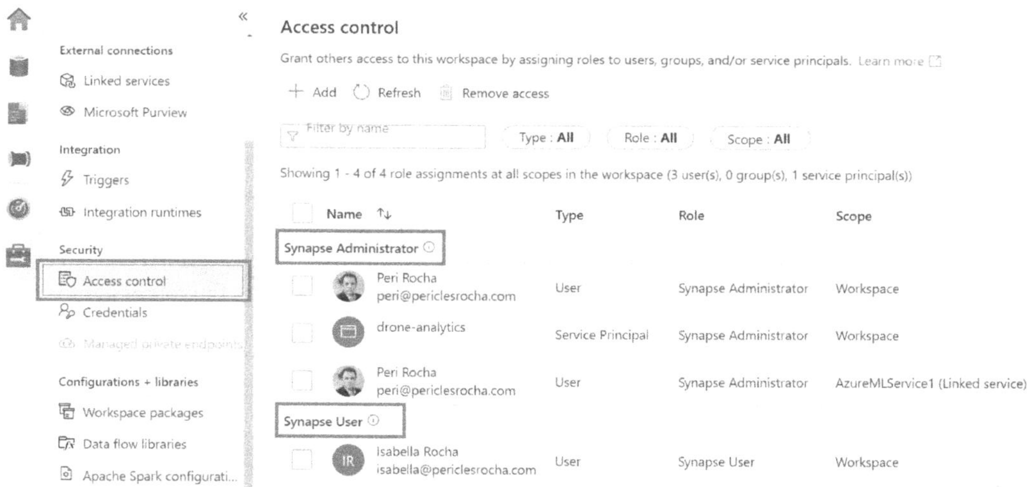

Figure 12.6 – Reviewing role assignments

You can also see in *Figure 12.6* that some users may be listed twice in the **Synapse Administrator** role. This happens because the user, `Peri Rocha` in this case, was added to the role at different scopes: once at the **Workspace** level and once at the resource level – in this case, a linked service named `AzureMLService1`.

Assigning RBAC roles

Users who are members of the Synapse Administrator role at the workspace scope can add or remove any Synapse RBAC role assignments. For lower-level scopes, members of the Synapse Administrator level for a given scope can perform Synapse RBAC role assignments for objects at the given scope, or any objects under the given scope.

Synapse RBAC role assignments can also be done using Azure Synapse Studio. To perform new role assignments, use the following steps:

1. Navigate to Azure Synapse Studio at `https://web.azuresynapse.net` and log in to your workspace.
2. In the hubs region, select the **Manage** hub.
3. Under **Security**, select **Access control**. You will land on the **Access control** page.
4. Select the + **Add** button. This opens the **Add role assignment** pane, as seen in *Figure 12.7*. This pane allows you to perform a role assignment at the workspace scope, or at the scope of a workspace item in specific. Let's explore the difference between these options:

   - **Workspace scope**: As discussed, when you perform a role assignment at the **Workspace** scope, your role assignment will apply to every object under your workspace, including code assets, analytical pools, credentials, and anything else. This option requires you to provide the Synapse RBAC role that you want to assign, and the user who will get the role assignment. Typing just a few letters of the user's name in the **Select user** textbox performs an Azure AD search and shows similar results for selection, as shown in *Figure 12.7*. You can select multiple Azure AD users or groups here to apply the role assignment to several users in one batch. Select the desired users and click **Apply** to complete the role assignment:

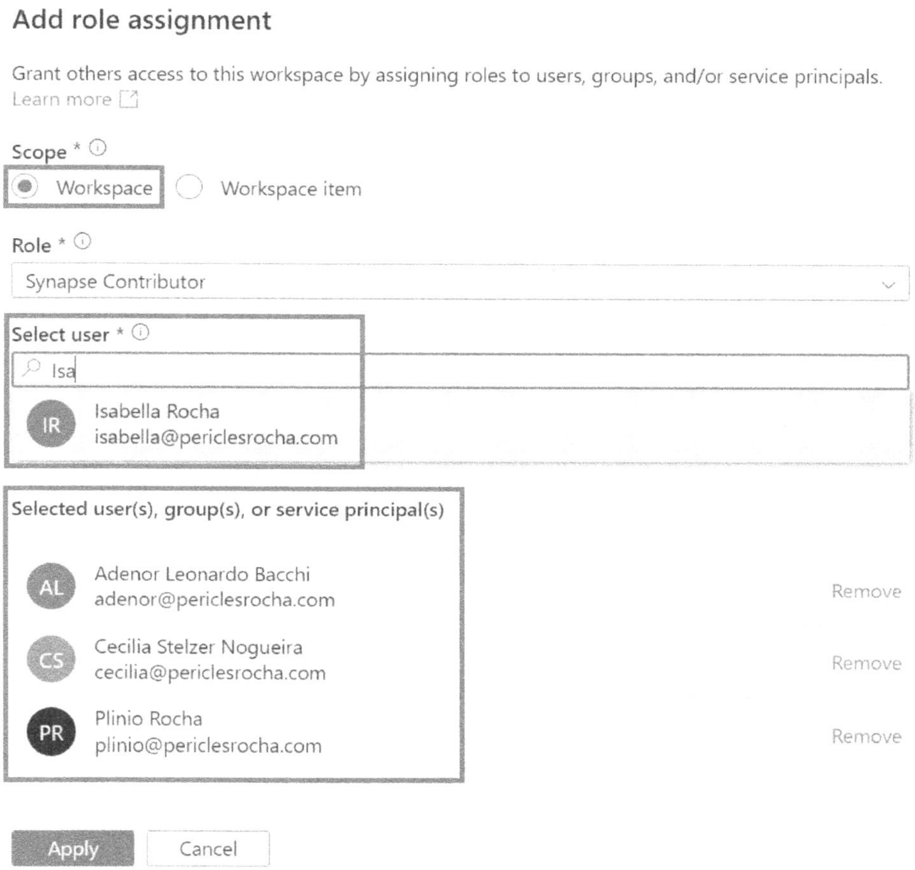

Figure 12.7 – Performing a role assignment at the Workspace scope

- **Workspace item scope**: Assignments at the **Workspace item** scope will apply to the workspace item that you choose, as well as any objects under the selected scope. Selecting this option requires you to specify the type of item you want to give rights to and the item of such type. For example, picking the **Linked service** item type allows you to pick any external service that is connected to your workspace through the Linked Services feature of Azure Synapse. Data Explorer pools are such an example, so if you want to perform role assignments for your Data Explorer pool, you should pick the **Linked service** option in the **Item type** box, and then pick your Data Explorer pool name in the **Item** box. *Figure 12.8* illustrates this example, with the `droneanalyticsadx` Data Explorer pool selected. You also need to pick the desired role and the user that will receive the role assignment, the same way you would do for the workspace scope.

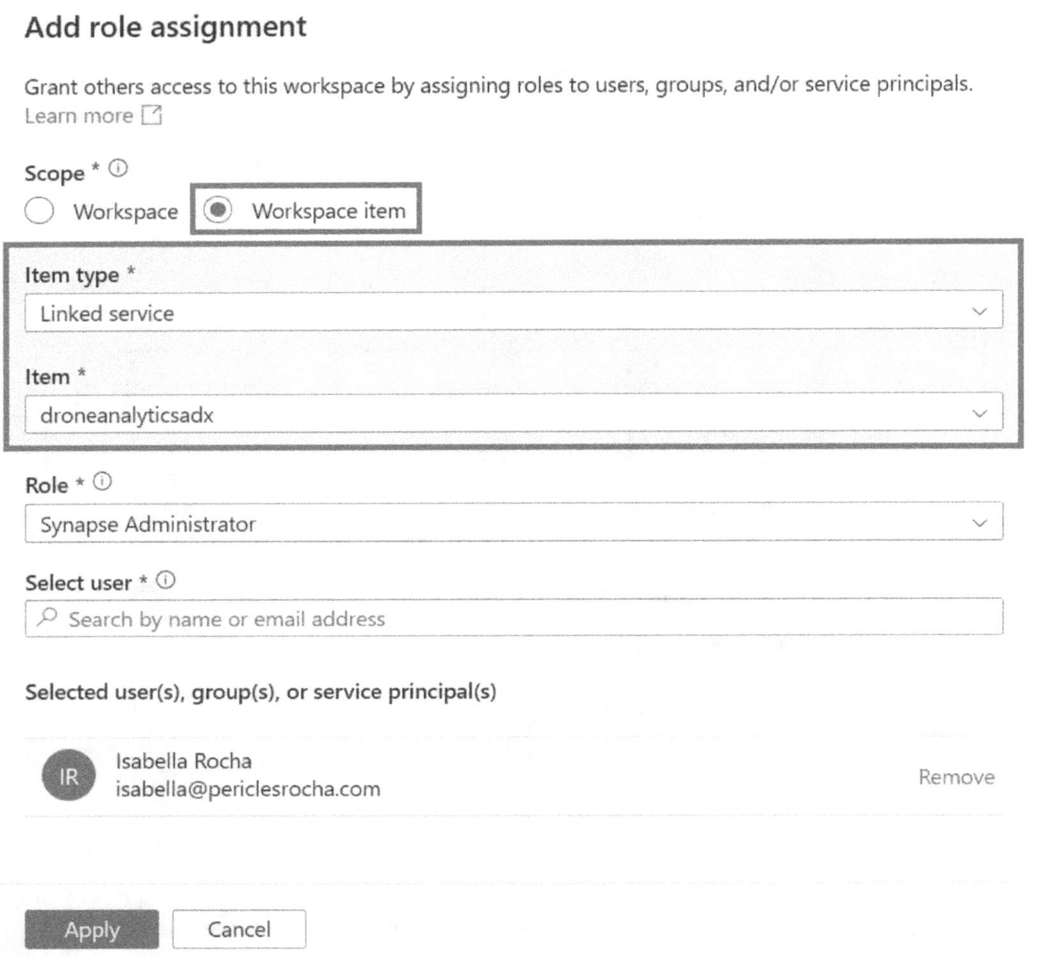

Figure 12.8 – Performing a role assignment at a Workspace item scope

Azure Synapse RBAC roles are useful to maintain access to Azure Synapse workspace assets and provide fine-grained access control to individuals who will use the workspace for development and data analysis. They are not needed, however, if you intend to allow access to Data Explorer pools directly through its endpoints, without using a Synapse workspace. In fact, the only Synapse RBAC roles that can be applied to the scope of Data Explorer pools are the **Synapse Administrator** or the **Synapse Credential User** roles. For regular usage of Data Explorer pools for querying and maintaining data, you cannot use Synapse RBAC roles. Instead, Data Explorer pools offer their own role-based authorization scheme, which we will explore next.

Data Explorer database roles

While Synapse RBAC roles can be helpful to provide fine-grained access control to workspace assets, Data Explorer pools use their own database roles to control who gets access to data. This authorization model is helpful as it allows you to govern access to data served by Data Explorer pools regardless of the application that is being used to query data that has access to Azure Synapse. You don't need to be an Azure Synapse user and have access to Azure Synapse workspace resources to be able to access data in Data Explorer pools.

Data Explorer offers the following database roles:

- **All Databases admin**: Users with this role can perform any actions in the scope of any database in a Data Explorer pool.
- **Database admin**: Users with this role can perform any actions in the scope of the database where the role was assigned.
- **Database user**: Use this role to allow users to read all data and metadata for a given database. This role also allows the user to create tables and functions in the database.
- **All Databases viewer**: This role allows reading data and metadata in all databases of a Data Explorer pool.
- **Database viewer**: This role allows reading data and metadata of a database.
- **Database ingestor**: This role allows users to ingest data into any table of a database.
- **Database unrestricted viewer**: This role allows users to query tables with the `RestrictedViewAccess` policy enabled on them. This policy, when enabled, allows only members of this role to query data on tables of a given database. Even cluster admins who are not members of this role can't access data if this policy is enabled.
- **All Databases monitor**: This role allows the execution of the `.show` control command in all databases in a Data Explorer pool to see database metadata, query execution status, and more.
- **Database monitor**: This role allows the execution of the `.show` control command in a given database.
- **Function admin**: This role allows users to create, alter, and manage permissions in functions.
- **Table admin**: Users with this role can perform any actions on the scope of a database table.
- **Table ingestor**: This role allows users to ingest data into a given table.

To review role assignments for a given database object, you can use the `.show` control command. As an example, the following command shows the security principals with access to the `fleet data` table:

```
.show table ['fleet data'] principals
```

> **Note**
> All code examples in this section can be found in the `Chapter 12\Data Explorer Roles.kql` file of the book's repository.

This command produces a result similar to the one in *Figure 12.9*. As you can see, the Azure AD user (**AAD User**, as shown in the results) `Peri Rocha` is a member of the `AllDatabasesAdmin` role at the **Cluster** scope, a member of the `Database admin` role at the scope of the `drone-telemetry` database, and a member of the `Table admin` role at the scope of the `fleet data` table. The results are truncated in this figure due to space, but the most relevant columns are shown:

Role	PrincipalType	PrincipalDisplayName
Cluster AllDatabasesAdmin	AAD User	Peri Rocha (upn: peri@periclesrocha.com)
Cluster AllDatabasesAdmin	AAD Application	drone-analytics (app id: d47a6ac7-88d7-4912-a39...
Database drone-telemetry Admin	AAD User	Peri Rocha (upn: peri@periclesrocha.com)
Table fleet data Admin	AAD User	Peri Rocha (upn: peri@periclesrocha.com)

00:00:00 Query executed successfully.

Figure 12.9 – Showing role membership by security principal

You can use the `.add` command to add new users to a role without affecting the existing role assignments. For example, the following code block adds three new users to the `Database admin` role in the scope of the `drone-telemetry` database:

```
.add database ['drone-telemetry'] admins
    ('aaduser=adenor@periclesrocha.com',
     'aaduser=cecilia@periclesrocha.com',
     'aaduser=plinio@periclesrocha.com')
     'Adding new users'
```

This produces a result similar to the one in *Figure 12.10*:

Role	PrincipalType	PrincipalDisplayName
Cluster AllDatabasesAdmin	AAD User	Peri Rocha (upn: peri@periclesrocha.com)
Cluster AllDatabasesAdmin	AAD Application	drone-analytics (app id: d47a6ac7-88d7-4912-a39...
Database drone-telemetry Admin	AAD User	Peri Rocha (upn: peri@periclesrocha.com)
Database drone-telemetry Admin	AAD User	Cecilia Stelzer Nogueira (upn: cecilia@periclesroch...
Database drone-telemetry Admin	AAD User	Plinio Rocha (upn: plinio@periclesrocha.com)
Database drone-telemetry Admin	AAD User	Adenor Leonardo Bacchi (upn: adenor@periclesro...

00:00:00 Query executed successfully.

Figure 12.10 – New users added to a role

To remove individual role assignments, you can use the `.drop` control command. For example, the following code removes the user `adenor@periclesrocha.com` from the `Database admin` role at the scope of the `drone-telemetry` database:

```
.drop database ['drone-telemetry'] admins ('aaduser=adenor@
periclesrocha.com')
```

As you can see in *Figure 12.11*, the user `adenor@periclesrocha.com` is no longer listed in the role:

Role	PrincipalType	PrincipalDisplayName
Cluster AllDatabasesAdmin	AAD User	Peri Rocha (upn: peri@periclesrocha.com)
Cluster AllDatabasesAdmin	AAD Application	drone-analytics (app id: d47a6ac7-88d7-4912-a39...
Database drone-telemetry Admin	AAD User	Peri Rocha (upn: peri@periclesrocha.com)
Database drone-telemetry Admin	AAD User	Cecilia Stelzer Nogueira (upn: cecilia@periclesroch...
Database drone-telemetry Admin	AAD User	Plinio Rocha (upn: plinio@periclesrocha.com)

00:00:00 Query executed successfully.

Figure 12.11 – Removing a user from a role

The `.drop` command also supports removing more than one user at a time, so you could provide an array of users instead of one user alone if you needed to do that.

Finally, you can use the `.set` control command to replace any existing users from a given role with a new list of users. Here's an example:

```
.set database ['drone-telemetry'] admins ('aaduser=peri@
periclesrocha.com') 'Replacing all roles'
```

This produces a result similar to the one shown in *Figure 12.12*. Note that all principals we added in the previous step were removed, and only the new role assignment is present:

Role	PrincipalType	PrincipalDisplayName
Cluster AllDatabasesAdmin	AAD User	Peri Rocha (upn: peri@periclesrocha.com)
Cluster AllDatabasesAdmin	AAD Application	drone-analytics (app id: d47a6ac7-88d7-4912-a39...
Database drone-telemetry Admin	AAD User	Peri Rocha (upn: peri@periclesrocha.com)

00:00:00 Query executed successfully.

Figure 12.12 – Resetting all role assignments for an object

Just as with RBAC roles, you can also use security groups defined in Azure AD to authorize users based on their Azure AD group membership. In that case, you would need only to replace the **principal** parameter used in the code samples with the **user principal name** (**UPN**) of the desired security group.

Implementing network security

In the previous section, *Configuring access to resources*, we discussed the fact that Data Explorer pools can be accessed directly through their public endpoints, without requiring users to connect to them via the Azure Synapse workspace. The implementation of a successful security strategy to protect Azure Synapse resources involves thinking about not only controlling access to your workspace but also how to protect access to each of these public endpoints. Thankfully, the platform offers some important features to help us achieve this objective.

Let's take this chance to establish some key terminology that will help us understand how network isolation is implemented:

- **Azure Synapse workspaces**: Workspaces are containers that offer one or more analytical services that you can leverage to deliver an end-to-end analytical experience. By combining different services in your workspace, you can combine the capabilities of these services to maximize the potential of your data projects. For example, you can use an Apache Spark pool to analyze data from a Data Explorer pool, combining these services together in one Azure Synapse workspace, as we did in *Chapter 6*, *Data Analysis and Exploration with KQL and Python*. The workspace also works as a security boundary and central point of management for the security of workspace components.

- **Public endpoints**: Each analytical engine of an Azure Synapse workspace offers public endpoints that allow users to connect to them directly, without the need to use Azure Synapse Studio and log in to the workspace. When accessed through these endpoints, the analytical engines behave the same way they would as standalone services. For example, endpoints exposed for Data Explorer pools allow users to connect to your compute pools directly, via the internet, without users even knowing that they are part of an Azure Synapse workspace. Authorization is controlled behind the curtains by using Data Explorer database roles, and, as we have seen, all data transmitted over the network is encrypted. The public query endpoint of Data Explorer pools uses the `https://<dataexplorerpoolname>.<workspacename>.kusto.azuresynapse.net` format, where `<dataexplorerpoolname>` is the name of the Data Explorer pool, and `<workspacename>` is the name of the Azure Synapse workspace where the compute pool belongs. There are ways to protect this public endpoint from undesired access or to disable public access altogether, and we will discuss these options in the following pages.
- **Azure Synapse Studio**: This is the development and management interface for Azure Synapse, where you can provision new compute pools, connect external services, perform data analysis, configure access control, and much more. Azure Synapse Studio can be reached at `https://web.azuresynapse.net`.

The way to think about network security in the context of these three items is that users access Azure Synapse workspace components through public endpoints, or through Azure Synapse Studio. In the *Synapse RBAC roles* section, you learned how to control access to Azure Synapse workspaces using RBAC, including who can monitor and manage the service. But how do we protect a workspace's public endpoints?

The first thing to know about public endpoints is that, like any Azure service, they are protected by Azure **Distributed Denial of Service** (**DDoS**), a service that actively monitors and protects endpoints from DDoS attacks, provides attack mitigations, and more. The second thing to know is that all network traffic to and from workspace endpoints is done using the TLS protocol, which encrypts all data at transit, so your data is always protected, even when transmitted over the internet. Azure Synapse workspaces require that any client connecting to the service uses TLS 1.2 or higher.

> **Note**
>
> To learn more about Azure DDoS protection, visit `https://learn.microsoft.com/azure/ddos-protection/ddos-protection-overview`.

Even with these levels of protection, you may not be comfortable with knowing that the services hosting your data are available publicly on the internet. Thankfully, there are ways to provide further protection to these workspaces. The first step to take is to fully isolate your workspace on its own Azure Virtual Network. We'll discuss this next.

Using a managed virtual network

When you are implementing services of any kind in Microsoft Azure, one of the considerations for you to take into account is the use of an Azure **Virtual Network** (**VNet**). You can create Azure VNets for the same reasons you would create virtual networks on your local network: they allow you to isolate network traffic, secure communication in a set of servers or services, filter and route network traffic, and more.

When you create an Azure Synapse workspace, you have the option to enable what's called a **managed virtual network**. This feature, which, as you may have guessed, is called **Azure Synapse Managed Virtual Network**, creates an Azure VNet associated with your Azure Synapse workspace but removes all of the management overhead from you. You won't need to configure IP ranges, routing, or anything at all. Azure Synapse implements the VNet for you and manages it behind the curtains, but your workspace still gets all the benefits of an Azure VNet, such as achieving full network isolation from any other Azure resources, including other Azure Synapse workspaces.

Enabling a managed virtual network for Azure Synapse workspaces is done at the time of the workspace's creation. When you are creating your new workspace in the Azure portal, navigate to the **Networking** tab and select **Enable** next to **Managed virtual network**, as illustrated in *Figure 12.13*:

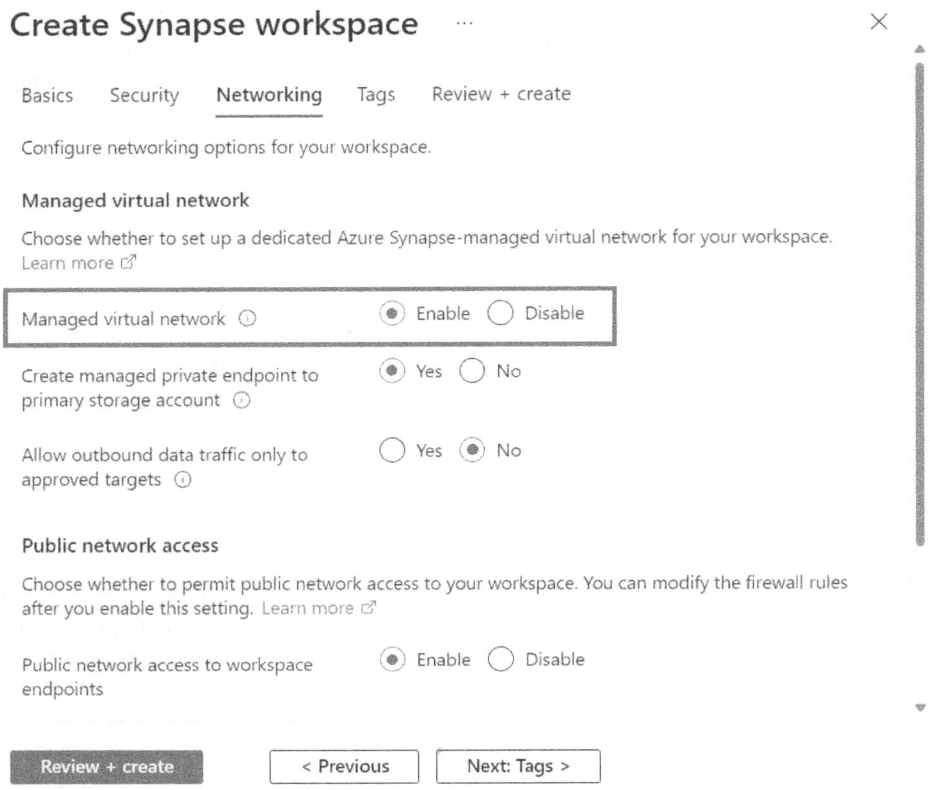

Figure 12.13 – Enabling the Managed virtual network at workspace creation time

> **Note**
>
> Once the workspace is created, you cannot enable or disable the managed virtual network. This can only be done at the time you create your workspace.

Enabling a managed virtual network for your workspace unlocks other workspace capabilities, such as the use of managed private endpoint connections to workspace resources, and the ability to disable public access altogether. Let's understand better what these are and how you can use them.

Managed private endpoint connection

Another helpful service offered by Microsoft Azure to help you protect your cloud network environment is Azure Private Link, a service that enables the communication between Azure services using your own virtual network. The main benefit of this service is that all network traffic between Azure services that you are using within a virtual network communicates using Microsoft's backbone network, which is isolated from the internet. As a result, data that is transmitted within a virtual network that uses Azure Private Link transits only in this isolated network, without exposure to the internet.

Azure Private Link implements communication through network interfaces that use private IP addresses in your virtual network, which are called **private endpoints**. Similar to what it does with Azure VNets, Azure Synapse offers **managed private endpoints** – Azure Private Link private endpoints that are managed by Azure Synapse for you. When enabled, all traffic between resources in your Azure Synapse workspace and other Azure services happens through Microsoft's backbone network. This applies to ingesting data from ADLS Gen2 into a Data Explorer pool, using an Apache Spark pool to read data from an Azure SQL database, or any other scenario between your Azure Synapse workspace and other Azure services.

Using managed private endpoints requires you to configure the endpoints in between Azure services. When you are creating your Azure Synapse workspace, however, you have an opportunity to create a private endpoint with the primary ADLS Gen 2 account used by the workspace (it is a separate Azure service, after all). To create this during your workspace's creation, on the **Networking** tab, you will find the **Create managed private endpoint to primary storage account** option, as shown in *Figure 12.14*. If you select this option, which is only available if you enabled the **Managed virtual network** option, Azure Synapse adds a managed private endpoint for your workspace's primary storage account.

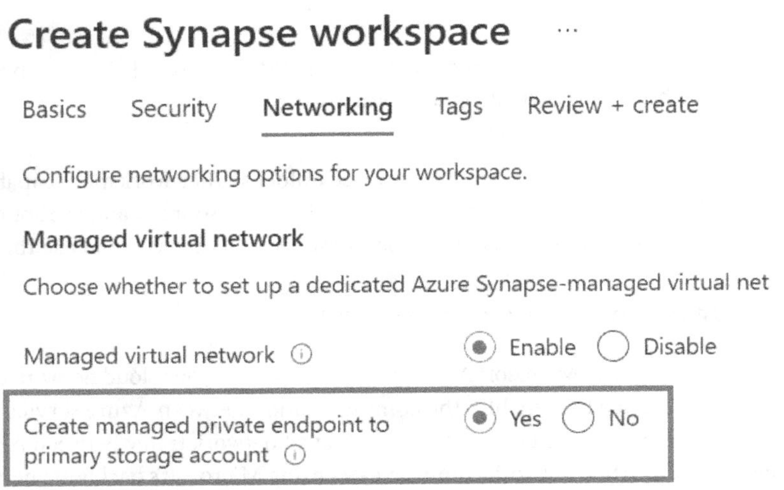

Create Synapse workspace ...

Basics Security **Networking** Tags Review + create

Configure networking options for your workspace.

Managed virtual network

Choose whether to set up a dedicated Azure Synapse-managed virtual net

Managed virtual network ⓘ (●) Enable ( ) Disable

Create managed private endpoint to primary storage account ⓘ (●) Yes ( ) No

Figure 12.14 – Creating a managed private endpoint for your workspace storage account

In addition to the workspace's primary storage account, Azure Synapse automatically adds two managed endpoints for you when you create your workspace: one for dedicated SQL pools, and one for serverless SQL pools.

You can review and create new managed private endpoints in Azure Synapse Studio. Simply go to the **Manage** hub and select **Managed private endpoints**, under **Security**, as shown in *Figure 12.15*:

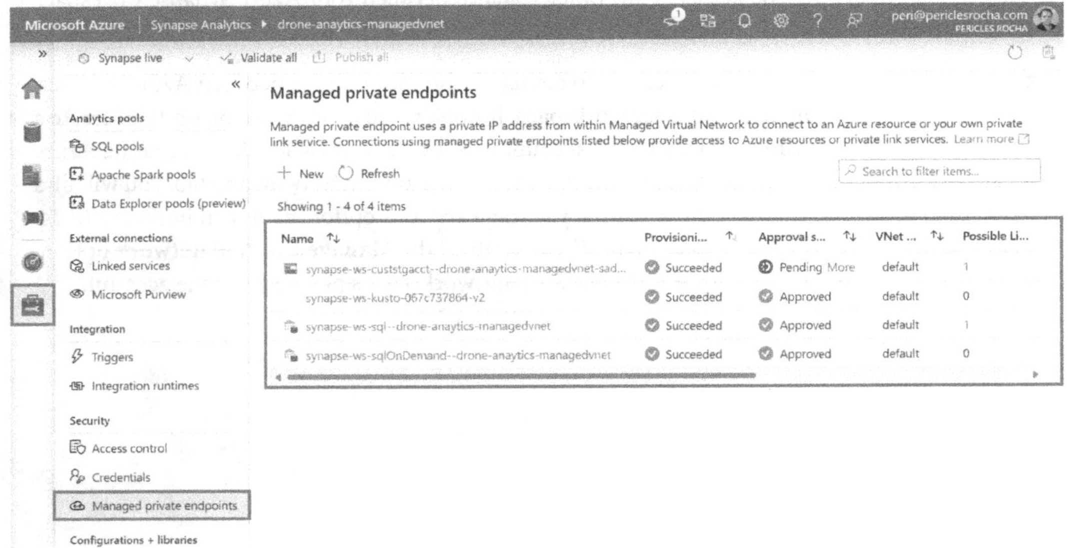

Figure 12.15 – Reviewing Managed private endpoints

You can create additional managed private endpoints by selecting the **+ New** button, which launches the **New managed private endpoint** page. As an example, to create a new managed private endpoint to another storage account, use the following steps:

1. Click the **+ New** button on the **Managed private endpoints** page, as shown in *Figure 12.15*.
2. In the list of possible destinations, select **Azure Data Lake Storage Gen2**. You can find it easily if you start typing its name in the search box, as shown in *Figure 12.16*. Select **Continue** once you have picked the right option.

New managed private endpoint

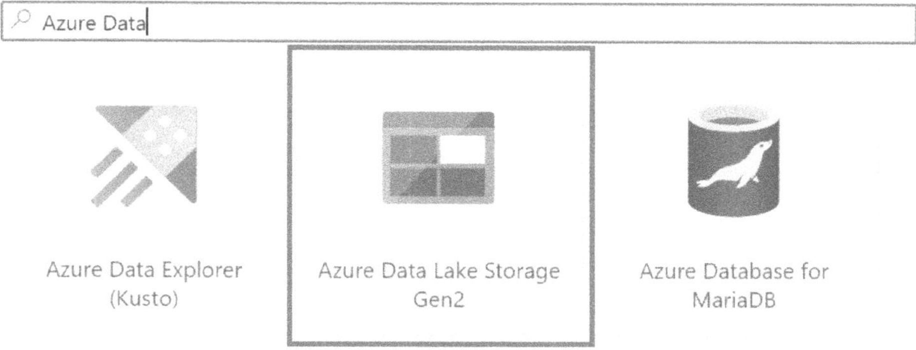

Figure 12.16 – Finding Azure Data Lake Storage Gen2

3. On the **New managed private endpoint (Azure Data Lake Storage Gen2)** page, illustrated in *Figure 12.17*, provide input for the following parameters:
   - **Name**: This is the name of your managed private endpoint. As usual, provide a name that will help you recognize it in the future.
   - **Description**: Provide a quick description of your resource. This field is optional.
   - **Account selection method**: Select **From Azure subscription** here to easily navigate through your subscription and find the desired storage account.
   - **Azure subscription**: Select here the Azure subscription where the storage account resides to filter the storage account selections, or optionally leave the default **Select all** option to see all storage accounts.
   - **Storage account name**: Finally, select here the storage account to which you will create the managed private endpoint.

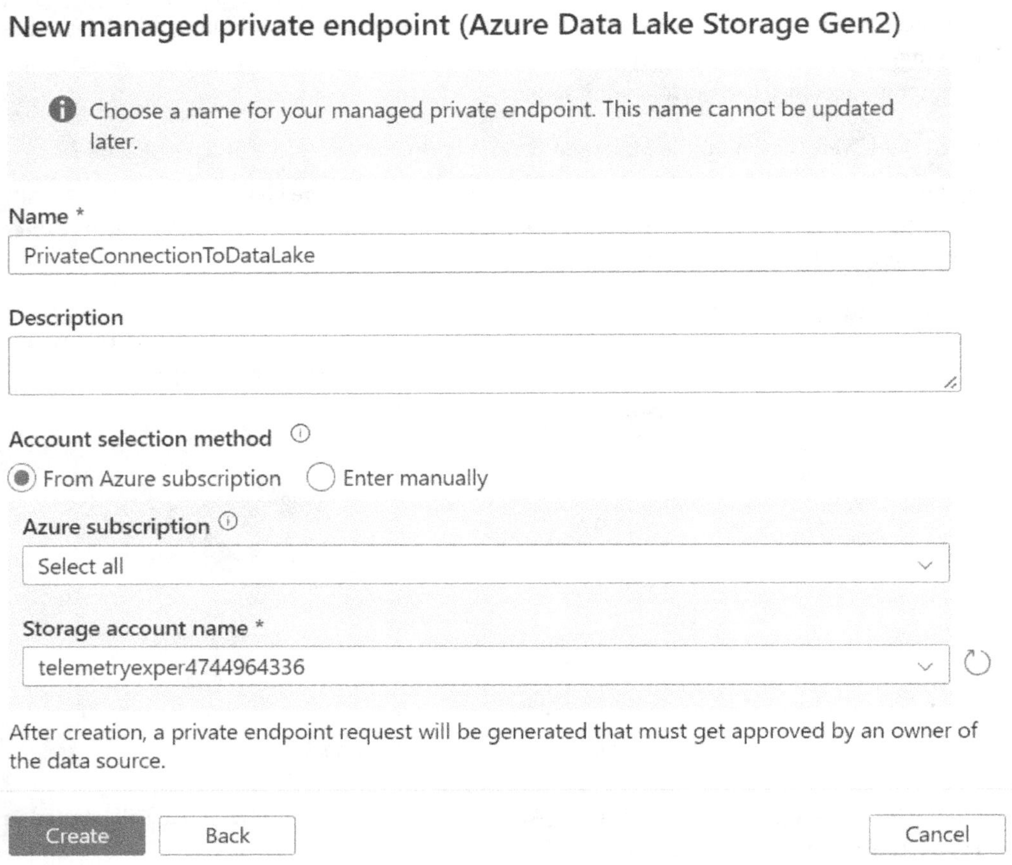

New managed private endpoint (Azure Data Lake Storage Gen2)

Choose a name for your managed private endpoint. This name cannot be updated later.

Name *

PrivateConnectionToDataLake

Description

Account selection method

From Azure subscription Enter manually

Azure subscription

Select all

Storage account name *

telemetryexper4744964336

After creation, a private endpoint request will be generated that must get approved by an owner of the data source.

Create Back Cancel

Figure 12.17 – Creating a new managed private endpoint for an ADLS Gen2 account

4. When you are done, click **Create**.

Once you create the managed private endpoint connection, an owner of the storage account will need to manually approve the connection. This can be done by navigating to the storage account in the Azure portal, selecting **Networking** under **Security + networking**, and then **Private endpoint connections**, as shown in *Figure 12.18*. Here, an account owner will be able to approve the connection.

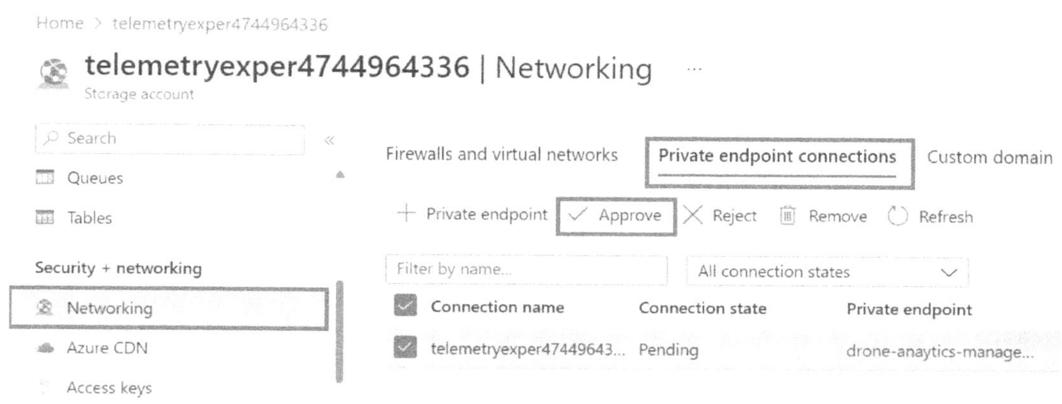

Figure 12.18 – Approving requests for a new private endpoint connection

Just like managed virtual networks, managed private endpoints do not require further configuration or maintenance. Azure Synapse will manage these connections without requiring any management effort from you, aside from having to approve the connection.

> **Note**
>
> To connect on-premises servers to the private endpoint connections, you need to connect these servers to your virtual network using Azure VPN Gateway or Azure ExpressRoute.

The use of a managed VNet with managed private endpoint connections enables an additional security measure that we can take to protect Azure Synapse workspaces with secure data egress. Let's explore this next.

Enabling data exfiltration protection

Now that you have taken the steps to enable secure data transfer between Azure services using managed VNet and private endpoint connection, let's understand how you can protect your Azure Synapse workspace data from being unwillingly moved outside of your organization.

Azure Synapse workspaces offer the option to enable data exfiltration protection during workspace creation. By enabling this setting, outbound data traffic will only be authorized to targets on approved Azure AD tenants, who also need to connect using managed private endpoint connections. Anyone trying to access your workspace resources who are not on the list of authorized targets for data transfer will be denied access, including any attempts by malicious attackers trying to move data from your environment – even if they manage to connect to your virtual network somehow.

Note

Azure AD tenants are a security boundary that IT organizations use to manage identity and authorization to an organization's resources. To learn more about Azure AD tenants, visit `https://learn.microsoft.com/microsoft-365/education/deploy/intro-azure-active-directory`.

Data exfiltration protection is enabled during workspace creation. During the creation of your Azure Synapse workspace, under the **Networking** tab, you will find the **Allow outbound traffic only to approved targets** option, as shown in *Figure 12.19*. By default, your Azure AD tenant will be approved and users in this tenant with the right to connect to the resources will be allowed to access data in the workspace.

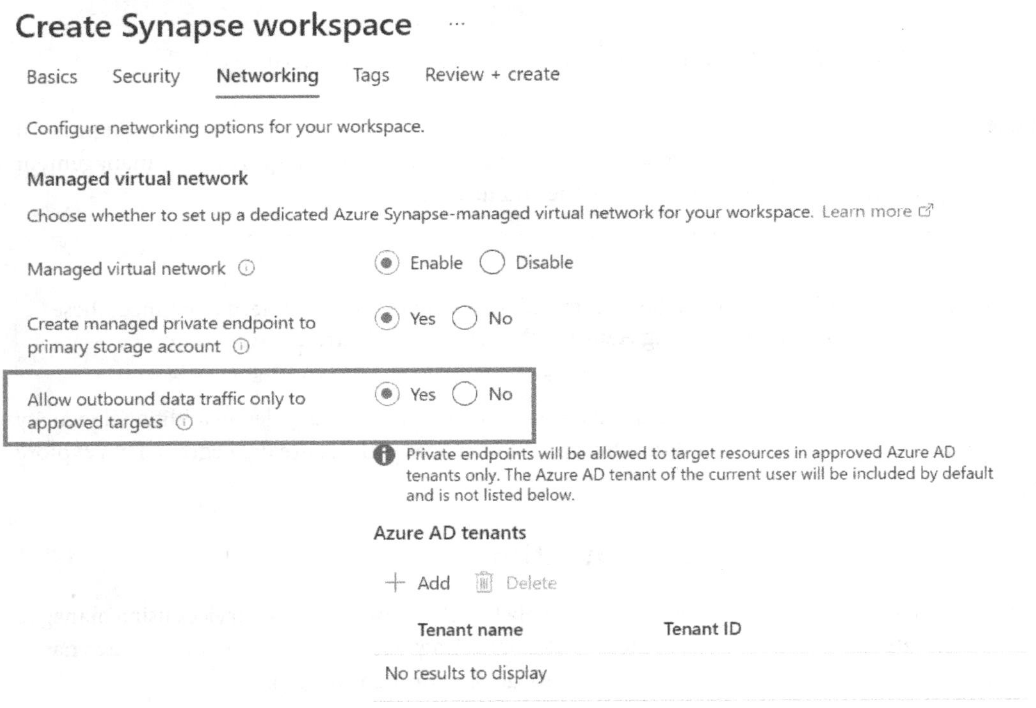

Figure 12.19 – Enabling data exfiltration protection

You can add additional Azure AD tenants here by selecting the + **Add** button.

Enabling data exfiltration protection has some important consequences for your Azure Synapse experience. For example, if you are using a notebook in Apache Spark that requires a package that relies on public repositories, this package will not work. You must explicitly upload workspace packages into your Apache Spark pool. Also, Microsoft states that ingesting data from Event Hub into a Data Explorer pool is not supported and does not work at this time.

Controlling public network access

The ultimate feature you can use to secure your data is to disable public access to your Azure Synapse workspace endpoints altogether. When you disable public access, all access to the public endpoints is disabled, and users can only connect to the resources through managed private endpoint connections. Microsoft recommends disabling public access to workspace resources to protect sensitive data.

> **Note**
>
> The option to remove public network access is only available if you created your workspace with a managed virtual network.

Public access can be disabled for an Azure Synapse workspace using the Azure portal. To disable it, you can follow these steps:

1. Navigate to the Azure portal at `https://portal.azure.com` and find your Azure Synapse workspace.
2. On your workspace's page, navigate to **Networking** under **Security**, as shown in *Figure 12.20*.
3. You will find the **Public network access to workspace endpoints** option. Select **Disabled**, and then click **Save**.

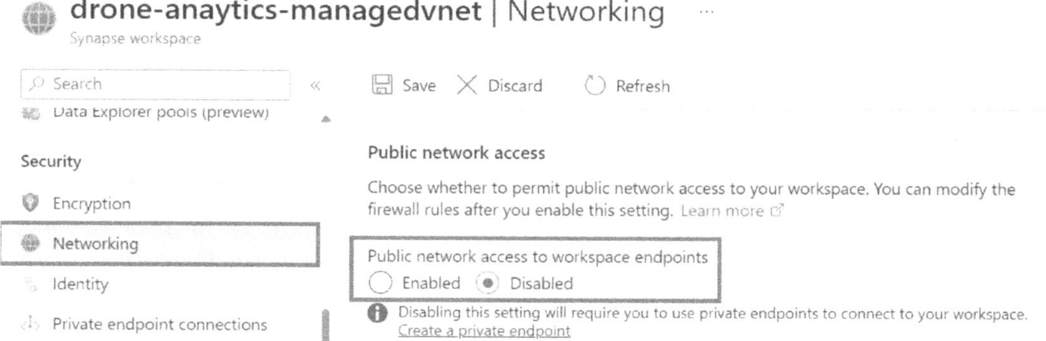

Figure 12.20 – Disabling network access to workspace endpoints

Once you disable public access, you will not be able to access resources in your Azure Synapse workspace using your client PC. You will need to create an Azure **Virtual Private Network** (**VPN**) gateway associated with your Azure Synapse workspace's managed virtual network, and connect to the gateway using a point-to-site VPN connection. For detailed instructions on how to create this VPN connection, refer to `https://learn.microsoft.com/azure/vpn-gateway/vpn-gateway-howto-point-to-site-resource-manager-portal`.

Using firewall rules

If you have external applications or users that need public access to your workspace, one way to further protect it is by using firewall rules to restrict access to known client IP addresses only. By default, the workspace has a firewall rule to allow all IP addresses to access the workspace, but you can remove this rule and add only IP addresses from clients that need access, as shown in *Figure 12.21*:

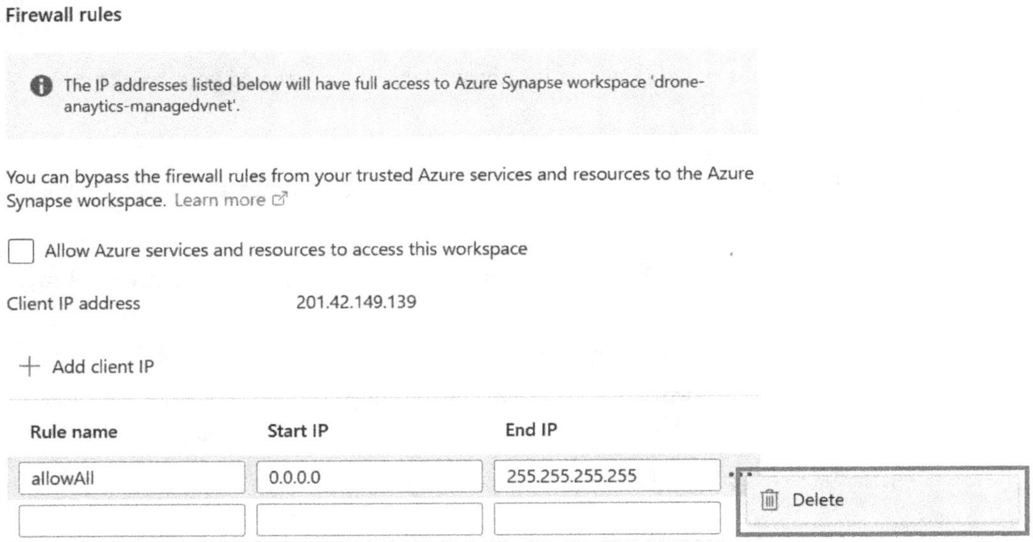

Firewall rules

The IP addresses listed below will have full access to Azure Synapse workspace 'drone-anaytics-managedvnet'.

You can bypass the firewall rules from your trusted Azure services and resources to the Azure Synapse workspace. Learn more

Allow Azure services and resources to access this workspace

Client IP address 201.42.149.139

+ Add client IP

Rule name	Start IP	End IP
allowAll	0.0.0.0	255.255.255.255

Delete

Figure 12.21 – Deleting the default firewall rule

To add new IP addresses, simply type a name to the new rule in the textbox under **Rule name**, and the allowed IP address range in the **Start IP** and **End IP** textboxes. You can add as many rules as you need. These firewall rules apply to access to the workspace via Azure Synapse Studio as well as direct access to any of the public endpoints in the workspace.

> **Note**
>
> If you need to configure firewall rules on your own network to access Azure Synapse Studio, you should allow outbound traffic to TCP ports 80, 443, and 1443.

You can also select **Allow Azure services and resources to access this workspace for Azure services** to allow other Azure services to bypass these firewall rules and communicate to your workspace normally, which is something quite common in Azure Synapse as you work with external databases, storage accounts, and other services.

Protecting against external threats

Azure Synapse integrates with the Microsoft Defender for Cloud platform to assess, secure, and defend your cloud resources. The platform actively monitors your cloud environments on Microsoft Azure, **Amazon Web Services** (**AWS**), and **Google Cloud Platform** (**GCP**) to track your security efforts, provide security recommendations, and defend and alert you of security incidents in real time.

The Microsoft Defender for Cloud integration with Azure Synapse offers support for Microsoft Defender for SQL only, allowing you to monitor your dedicated SQL pools for SQL injection attacks, uncommon access attempts for database queries, and suspicious activity that could be malicious. This integration does not currently include serverless SQL pools, Apache Spark pools, or Data Explorer pools. For that reason, we will not cover this capability in detail. You can learn more about protecting your SQL environment at `https://learn.microsoft.com/azure/azure-sql/database/azure-defender-for-sql`.

Summary

Security is paramount for any project dealing with data. Keeping your environment safe from unwanted access and ensuring your data is always encrypted at rest and at transit should be considered fundamental problems that you need to address when you are working with data. Remember, data is arguably the most important intangible asset of any organization.

In this chapter, you learned about ways to keep your Azure Synapse Data Explorer environment secure. We started with an overview of the security challenges, and then we drilled into each one of them. You learned about how Azure Synapse handles data encryption, and how to enable double encryption on your workspaces for an added layer of security. Once the data is encrypted, you learned about how Azure Synapse handles user authentication through Azure AD integration.

Once you had learned about authentication, we discussed ways to control access to resources using Azure Synapse RBAC and Data Explorer database roles. These roles help you give the right permissions to individuals depending on why they need access to data or the management environment. You learned how to review role assignments and perform new ones, and how to use KQL control commands to assign users to Data Explorer database roles.

Next, we covered all the options you have at hand to protect the network layer of your workspace. We discussed managed VNets, and you learned that they are the cornerstone when it comes to the security approach for the network layer. After you enable the managed VNet in your Azure Synapse workspace, it enables you to use other important features such as private endpoint connections, and the ability to disable public access to your workspace and its public endpoints. We also discussed how these features combined help you enable data exfiltration protection, allowing only known (and desired) actors to access your data.

Finally, the last section of this chapter briefly discussed Azure Synapse's support for threat detection using Microsoft Defender for Cloud and how it helps protect dedicated SQL pools.

In the next chapter, you will learn about advanced data management aspects of Data Explorer pools, such as managing shards, and tools to help you satisfy governmental regulations.

13
Advanced Data Management

In this final chapter, we will discuss advanced data management aspects of Data Explorer pools. You will learn what extents are and how to manage them for your benefit. We will also discuss moving large chunks of data for archival, dropping large volumes of data, and tagging extents so you can easily manage them. Dealing with very large data volumes presents significant challenges for data management, and Data Explorer deals with these challenges elegantly.

You will also learn about purging personal data in your databases, including why you should care, how it happens, and how to do it at scale. These data purge operations are an essential part of handling personal data and ensuring you respect an individual's right to govern their data. These operations may take several days to complete, so it is important to stay on top of this important topic.

Here are the topics presented in this chapter:

- Managing extents
- Purging personal data

As your data volume grows in Data Explorer pools, these advanced data management topics will become more and more important. Without any further ado, let's get to it.

Technical requirements

In case you haven't yet, make sure you download this book's material from the GitHub repository at `https://github.com/PacktPublishing/Learn-Azure-Synapse-Data-Explorer`. You can download the full repository by selecting **Code**, and then **Download ZIP**, or by cloning the repository by using your git client of choice. This chapter uses a version of the `fleet data` table, but we will create it and ingest some data for this chapter, so don't worry about creating it at this time.

In this chapter, we will go back to data ingestion for a quick example of how we can tag extents at the time of data ingestion. Later, we will use this data to move extents around and perform a data purge operation. This ingestion uses the simplified drone telemetry dataset, which needs to be copied to a container on your ADLS Gen2 storage account before we can use it. If you don't have this file on your storage account yet, make sure you follow the steps provided in the *Technical requirements* section of *Chapter 5, Ingesting Data into Data Explorer Pools*, to copy the simplified dataset to your storage account.

Managing extents

In *Chapter 1, Introducing Azure Synapse Data Explorer*, you learned about the architecture of Data Explorer pools and how it uses database sharding techniques to split tables and physically store them on cluster nodes. Database sharding is broadly used by database management systems to split large tables (or databases) into smaller, more manageable files, offering performance that scales linearly as you add more compute and removing limitations that would arise from managing very large individual files to persist data on disk.

In Data Explorer, each of these data shards is called an extent. When you create a new table and ingest data, the Data Explorer engine creates and distributes your table into a series of extents across the cluster. These extents contain not only your data, but also some associated metadata that indicates the creation time of the extent, some optional tags (which we will discuss shortly), and information that may help the Data Explorer engine process queries more efficiently, such as indexes for columns stored on this extent.

To decide how to split data into extents, Data Explorer uses a data sharding policy. This policy controls three properties:

- `MaxRowCount`: The maximum row count each extent can have. The default is 750,000 rows.
- `MaxExtentSizeInMb`: The maximum data size each extent can have, after compression. The default is 1 GB. This parameter is used for extent merge operations only.
- `MaxOriginalSizeInMb`: The maximum size of the original data. The default is 2 GB. This parameter is used for extent rebuild operations only.

Every new Data Explorer pool database contains a data sharding policy. Tables on a database inherit this policy from the database by default, but the policy can be overridden at the table level.

> **Note**
>
> Microsoft offers control commands that you can use to change the data sharding policy. However, you should not alter these policies without an explicit recommendation from Microsoft. For more information, consult `https://learn.microsoft.com/azure/data-explorer/kusto/management/shardingpolicy`.

Once an extent has been created, it can never be modified. Data Explorer can, however, execute merge operations to extents for better efficiency. As you ingest data on Data Explorer pools, your extents may have inefficient sizes or row distributions, and performing merge operations can help increase efficiency. These merge operations can be of two types: a merge operation that rebuilds indexes, or a rebuild operation that ingests data from existing extents into a new extent, and drops the old extents.

To decide how to merge extents, Data Explorer uses a merge policy. This policy uses a series of properties:

- `RowCountUpperBoundForMerge`: This parameter controls the maximum row count for an extent that is merged. The default is 16,000,000 rows. It does not apply to rebuild operations.
- `OriginalSizeMBUpperBoundForMerge`: Defines the maximum size of the original extent that will be merged. The default is 30 GB. It does not apply to rebuild operations.
- `MaxExtentsToMerge`: Defines the maximum number of extents to be merged in a single operation. The default is 100 extents. It does not apply to rebuild operations.
- `LoopPeriod`: The maximum time the Data Management service should wait between two consecutive merge or rebuild operations. The default is 1 hour and applies to merge and rebuild operations.
- `AllowRebuild`: This property defines if rebuild operations are allowed. The default is true. Rebuild operations are always preferred over merge operations unless this property is set to false.
- `AllowMerge`: Similarly, this defines if merge operations are allowed. The default is true.
- `MaxRangeInHours`: The maximum time difference between the creation times of two extents that will be merged. The default is 24 hours. This applies to merge and rebuild operations.
- `Lookback`: Controls the timespan used to consider if extents should be merged or rebuilt. The supported values are:
    - `Default`: Sets the timespan to 14 days.
    - `All`: All extents are included.
    - `HotCache`: Only extents in the hot cache are included. We discussed caching policies in the *Speeding up queries using cache policies* section of *Chapter 11, Tuning and Resource Management*.
    - `Custom`: Allows you to set a custom timespan, in days, to determine the extent's age.

Just like the data sharding policy, databases contain a default merge policy, which implements the default values for the policy properties. Tables on a database inherit this policy from the database by default, but the policy can be overridden at the table level.

To see your database merge policy, you can use the following control command:

```
.show database ['drone-telemetry'] policy merge
```

> **Note**
>
> All the code examples in this section can be found in the `Chapter 13\Managing Extents.kql` file of this book's GitHub repository.

This command produces an output similar to the one shown in *Figure 13.1*:

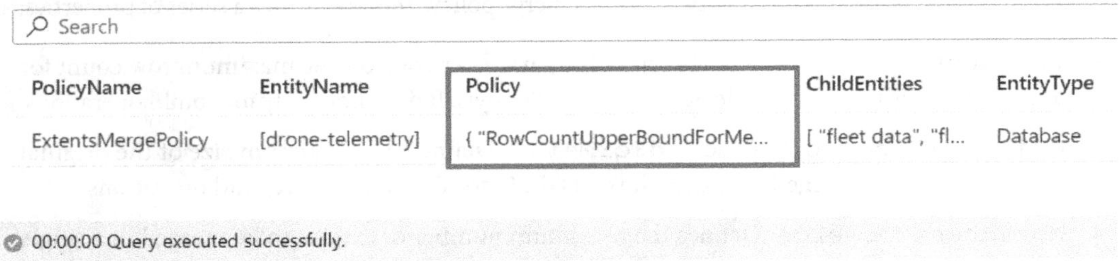

PolicyName	EntityName	Policy	ChildEntities	EntityType
ExtentsMergePolicy	[drone-telemetry]	{ "RowCountUpperBoundForMe...	["fleet data", "fl...	Database

00:00:00 Query executed successfully.

Figure 13.1 – Reviewing a database's merge policy

If you expand the **Policy** column from the results, you will see the JSON representation of your policy:

```
{
      "RowCountUpperBoundForMerge": 16000000,
"OriginalSizeMBUpperBoundForMerge": 30000,
      "MaxExtentsToMerge": 100,
      "LoopPeriod": "01:00:00",
      "MaxRangeInHours": 24,
      "AllowRebuild": true,
      "AllowMerge": true,
      "Lookback":
      {
         "Kind": "Default",
         "CustomPeriod": null
      }
}
```

> **Note**
>
> Microsoft offers control commands that you can use to change the merge policy. However, you should not alter this policy without an explicit recommendation from Microsoft. For more information, consult `https://learn.microsoft.com/azure/data-explorer/kusto/management/mergepolicy`.

One thing that can affect merge operations is extent tagging. If an extent is tagged with multiple `drop-by` tags, it will not be merged, regardless of the merge policy. But what is extent tagging?

Extent tagging

Data Explorer allows you to add metadata to extents using tags. These tags are string properties and can contain any information you think may be useful in the future, such as the source of the data, a contact person, an observation about the data, or anything you'd like. Tags can be added at ingestion time, or at any time after ingestion using control commands.

When extents are merged, the new extent that is created as a result of the merge operation will keep the tags from all the extents that were merged, so your tags are always kept, even when moved to new extents. Also, when you add tags to an extent, this extent gets dropped and a new one is created. Therefore, adding tags at ingestion time is a more optimal approach.

To add tags at ingestion time, you can use the `tags` property with the `.ingest into` command. Let's look at an example. First, we will create a new table named `fleet data with tags` to keep the original `fleet data` table intact for your experiments. This table will use the schema of the simplified drone telemetry dataset:

```
.create table ['fleet data with tags'] (
    DeviceData: string,
    LocalDateTime: string,
    DeviceState: string,
    CoreTemp: long,
    CoreStatus: string,
    Compass: long,
    Altitude: long,
    Speed: long,
    PayloadWeight: long,
    EventData: string
)
```

Now that we have the table, let's go ahead and run the `.ingest into` command, providing the `tags` property as an option. To run this command, you'll need to generate a **shared access signature** (**SAS**) token to the container where the `drone-telemetry-simplified.csv` file is stored and use it in the ingestion task. For a reminder of how to generate the SAS token, refer to the *Performing data ingestion* section of *Chapter 5, Ingesting Data into Data Explorer Pools*. Once you have your SAS key, make sure you replace it in this code example:

```
.ingest into table ['fleet data with tags'] 'https://
dronetelemetrydatalake.blob.core.windows.net/raw-data/drone-
```

```
telemetry-simplified.csv?sp=r&st=2022-12-23T12:35:01Z&se=2022-
12-30T20:35:01Z&spr=https&sv=2021-06-08&sr=c&sig=8O
FOuVnfha0%2BqhDF2eIqfNJlm5xl33BGDTmnxXjxTGc%3D' with
(ignoreFirstRecord=true, tags='["drop-by:2023-03-23"]')
```

This adds a tag of `drop-by:2023-03-23` to all new extents that are created on this table as a result of this ingestion task.

> **Note**
>
> As discussed in *Chapter 5, Ingesting Data into Data Explorer Pools*, in real life, the best practice is to create a data mapping for your table before you start ingesting data into a table. In this example, we skipped creating a data mapping for simplicity.

You can use the `.alter extent tags` control command to change the tags of an extent with a new one, or the `.alter-merge extent tags` command to add a new tag to whichever tags are already present in the extent. For example, the following code block replaces tags on an extent with a new tag that says `Drone telemetry data.`:

```
.alter extent tags ('Drone telemetry data.') <| .show table
['fleet data with tags'] extents
```

This produces a result similar to the one shown in *Figure 13.2*:

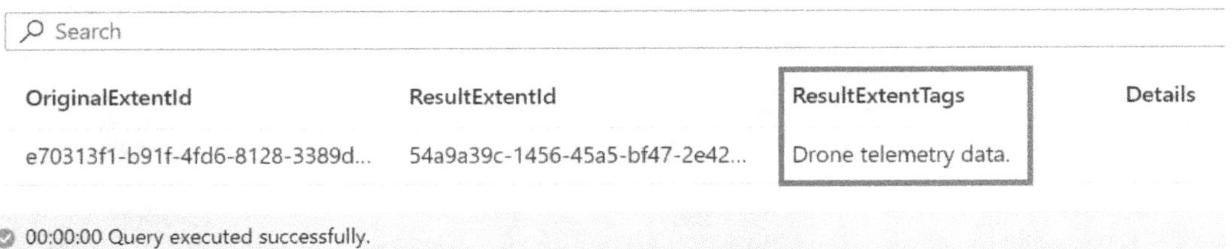

Figure 13.2 – Altering the tags on all the extents of a table

As another example, the following code adds a tag to existing tags in extents in the `fleet data with tags` table:

```
.alter-merge extent tags ('Data source: telemetry data lake.')
<| .show table ['fleet data with tags'] extents
```

As shown in *Figure 13.3*, this command appended the new tag to the existing one:

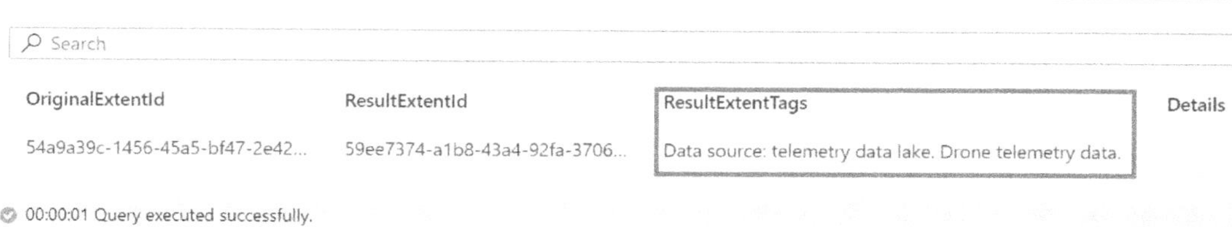

Figure 13.3 – Appending a tag to the existing tags

In the syntax of these commands, note that the control command is followed by a query, which specifies the scope at which the tags will be added. In this case, I used `.show table ['flee data with tags'] extents`, which includes all extents for the fleet data table. If you replace this query to select only specific extents, then only the extents that you specify will be tagged.

At any time, you can review the tags on extents using the `.show table … extents` command. For example, the following command shows the tags for each extent on a table (since our `fleet data with tags` table is not that large, it has one extent only):

```
.show table ['fleet data with tags'] extents
| project ExtentId, Tags
```

You should see a result similar to the one shown in *Figure 13.4*:

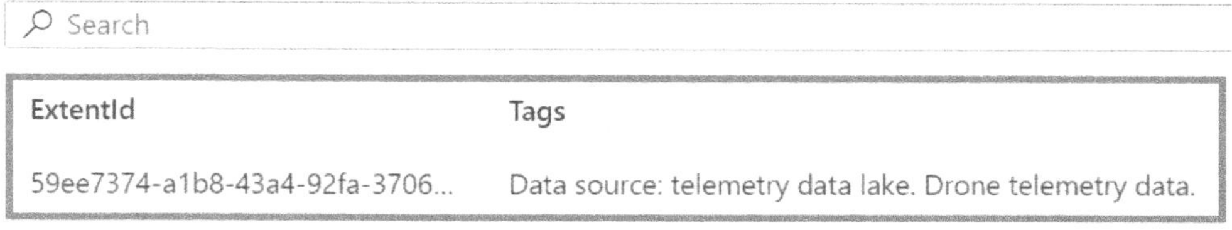

Figure 13.4 – Reviewing the tags on an extent

You can use tags freely to help you manage extents, but Data Explorer also uses tags for some special purposes. There are two special tag prefixes that, when used, help Data Explorer determine how extents are merged and help control data ingestion:

- `drop-by`: When you set extent tags that start with the `drop-by` prefix, in the event of an extent merge operation, only extents with the same value for the `drop-by` prefix will be merged. When you drop extents after a certain time, which is a common scenario regarding log and telemetry data, these extents can be dropped based on the value of your `drop-by` prefix. Since these extents were likely merged before, this makes an extent drop operation (which we will discuss shortly) more effective.

- `ingest-by`: Continuously ingesting data from a data lake can be an error-prone process that may cause data to be ingested more than once. This tag prefix helps you ensure data is only ingested once, or that data is only ingested if it does not exist. You can combine this with the `ingestIfNotExists` property to prevent data from being ingested if extents with the same value for the `ingest-by` tag already exist. For example, the `.ingest into … with (ingestIfNotExists = '["2023-03-23"], tags = '["drop-by:2023-03-23"]'')` command would insert data using the `["drop-by:2023-03-23"]` tag, but only if the value of `["2023-03-23"]` is not present on any `drop-by` tags of any of the extents in this table. If such a tag is present, the ingestion task will do nothing.

These tag prefixes are useful to help control extent merge operations and ingestion task behavior, but Microsoft recommends not overusing them. You should remove them if they are not needed, as they may have an impact on performance.

Another use of tags is to help perform other management tasks on extents. We'll explore this next.

Moving extents

Imagine having to deal with billions of records that are becoming stale and need to be archived. In traditional databases, you would need to write routines that copy data from one table to an archival table or cold storage of some sort, and then delete the records from the source table. This whole task would be part of a single transaction and could take a long time to complete, depending on your data volume.

Data Explorer helps you easily move large chunks of data simply by moving extents in between tables. To help illustrate this with examples, let's create an empty archive table that is similar to our fleet data table in the simplified version:

```
.create table ['fleet data with tags - archive'] (
    DeviceData: string,
    LocalDateTime: string,
    DeviceState: string,
    CoreTemp: long,
    CoreStatus: string,
    Compass: long,
    Altitude: long,
    Speed: long,
    PayloadWeight: long,
    EventData: string
)
```

Now that we have our table, we'll look at some examples. In the first one, we will move all extents from one table to the other:

```
.move extents all from table ['fleet data with tags'] to table
['fleet data with tags - archive']
```

This produces a result similar to the one seen in *Figure 13.5*, which shows the extent identifier (ID) of the original extent on the source table, and the ID of the extent in the destination (result) table. Since our table is quite small, it only contains one extent, so the result is only one line. For very large tables, you will see a row in these results for each extent that was moved. The **Details** column will only include any information if the operation fails:

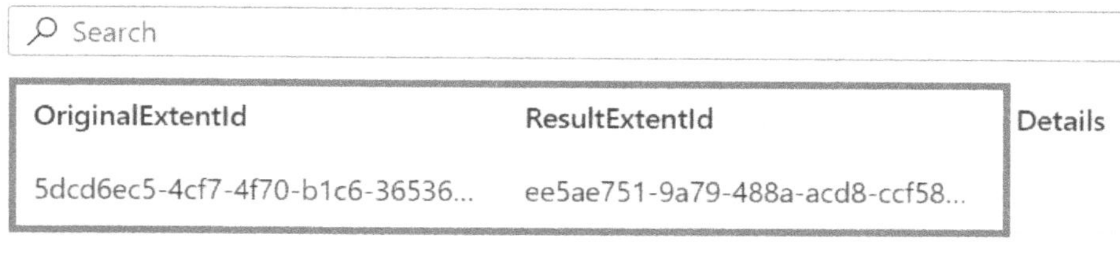

Figure 13.5 – The result of an extent move operation

You can also move specific extents in between tables by supplying their extent IDs. In the following example, we are moving the extent with an ID of `ee5ae751-9a79-488a-acd8-ccf58b315479` back to the original table:

```
.move extents from table ['fleet data with tags - archive']
to table ['fleet data with tags'] (ee5ae751-9a79-488a-acd8-
ccf58b315479)
```

You could also supply a comma-separated list of extent IDs to select which extents to move.

Finally, as you would expect, you can move extents based on their tags. The following example moves all extents with a `drop-by:2023-03-23` tag from the `fleet data with tags` table back to the archive table:

```
.move extents to table ['fleet data with tags - archive'] <|
.show table ['fleet data with tags'] extents | where Tags ==
'drop-by:2023-03-23'
```

These operations are much faster than filtering data and moving it around, so extent tagging and moving make a lot of sense when you need to implement a data archival strategy. Note that these operations can only be performed within the same database and you can't move extents in between tables on different databases.

Dropping extents

Dropping extents is just as simple. Since extents have unique IDs across the database, you can run the `.drop extent` command and provide the ID of the extent you want to drop, or a list of IDs separated by commas, and Data Explorer will drop the extents for you. Let's look at an example:

```
.drop extent 49aa2250-d8b8-4127-9049-739399e2a89f
```

A more useful way is to delete extents a certain time after they were created. For instance, you may choose to keep only the last 90 days of telemetry data in your Data Explorer pool to reduce storage costs. If that were the case, you could use the following command to drop extents older than 90 days:

```
.drop extents <| .show database ['drone-telemetry'] extents |
where MaxCreatedOn < now() - time(90d)
```

Dropping extents permanently deletes all rows within it, so make sure you plan these actions accordingly. One way to make sure you understand the impact of what you are doing and which extents would be affected by a `.drop extents` command is to use the convenient `whatif` command option. This option shows you what would have happened if you ran a command, but without actually running it. For example, the following command shows you the list of extents that would have been affected if you tried to drop extents that were created in the last 90 days:

```
.drop extents whatif  <| .show database ['drone-telemetry']
extents | where MaxCreatedOn < now() - time(90d)
```

This command produces a result similar to the one seen in *Figure 13.6*:

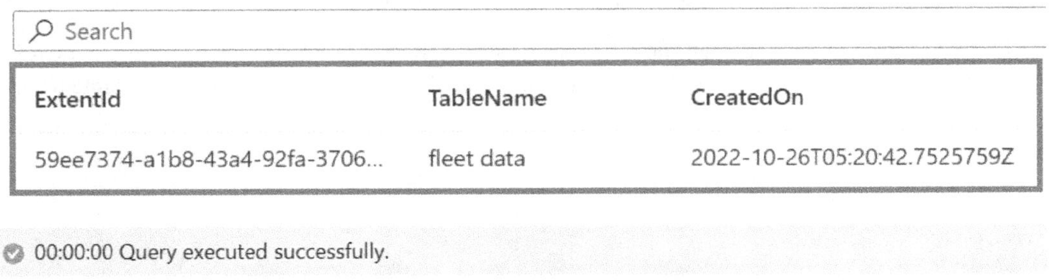

Search

ExtentId	TableName	CreatedOn
59ee7374-a1b8-43a4-92fa-3706...	fleet data	2022-10-26T05:20:42.7525759Z

00:00:00 Query executed successfully.

Figure 13.6 – Reviewing extents that would be dropped without the whatif command option

In this case, we can see that if we were to run this command without the `whatif` option, it would delete an extent in the fleet data table that was created on `2022-10-26`. You should review this output and determine whether your query predicate is including only the extents that you want to drop.

Moving and dropping extents helps you manage data moves and deletions at the scale of Data Explorer. In some cases, however, you may need to delete data more granularly for different needs, such as to satisfy regulatory requirements from laws that govern how and when you should delete data. We'll look at an approach to addressing such needs next.

Purging personal data

Data protection and privacy are important issues that we need to consider when handling personal data. Since May 2018, the **European Union** (**EU**) has enforced the **General Data Protection Regulation** (**GDPR**) to govern how entities gather, use, and manage personal data. GDPR aims to enhance the individual's control and rights of their data. Even though GDPR is a European regulation that is binding and applicable to EU member countries only, it has influenced legislation in many countries outside of the EU, such as Argentina, Brazil, Turkey, Japan, and South Africa, to name a few. In the United States, the **California Consumer Privacy Act** (**CCPA**) offers similar policies to the ones seen in GDPR.

> **Note**
> To learn more about GDPR, visit `https://gdpr.eu/what-is-gdpr/`.

Depending on the type of data you are working with, you may be dealing with log and telemetry data, which may include personal information. For example, you may have telemetry data that describes actions an individual is taking with your product, or log data that reveals an individual's web search habits or purchasing history. As a data professional, you must comply with your local regulations, and with the local regulations of wherever your product is serving users, to ensure you are handling data as per the law.

When you are working with personal data on Azure Synapse Data Explorer, Microsoft recommends the following data purge guidelines:

1. Implement a retention policy. Ideally, your retention policy should automatically satisfy regulatory requirements and delete data automatically. We discussed retention policies in the *Defining a retention policy* section of *Chapter 5, Ingesting Data into Data Explorer Pools.*
2. If your default retention policy does not satisfy regulatory requirements, or you need to keep the rest of your data for longer than you would keep personal data, then you should isolate personal data on one or more tables, apart from the rest of your data. Design your database schema with this consideration in mind.

3. When a data purge is needed to delete personal data, you should batch these requests and run them up to twice a day, per table, at the most. As we will see next, the `.purge` command scans tables, looking for extents that contain data that needs to be purged. This is an inefficient process for constant data deletion that has a significant performance impact on your Data Explorer pool. Also, only one purge request can run on a Data Explorer pool at any given time, so purge requests that are sent to the cluster when there is already one running will be queued.

Before you run any purge operations, you need to enable this capability on your Data Explorer pool. We'll explore how to enable this feature next.

Enabling purge on Data Explorer pools

Running purge operations requires you to enable purge at the Data Explorer pool level. You can enable purging in your Data Explorer pool by following these steps:

1. Navigate to the Azure portal (`https://portal.azure.com`) and find your Azure Synapse workspace.
2. Select **Data Explorer pools (preview)** under **Analytics pools** and select your Data Explorer pool.
3. On your Data Explorer pool's page, select **Configurations** under **Settings**.
4. Select the **On** toggle next to **Enable purge**, as seen in *Figure 13.7*:

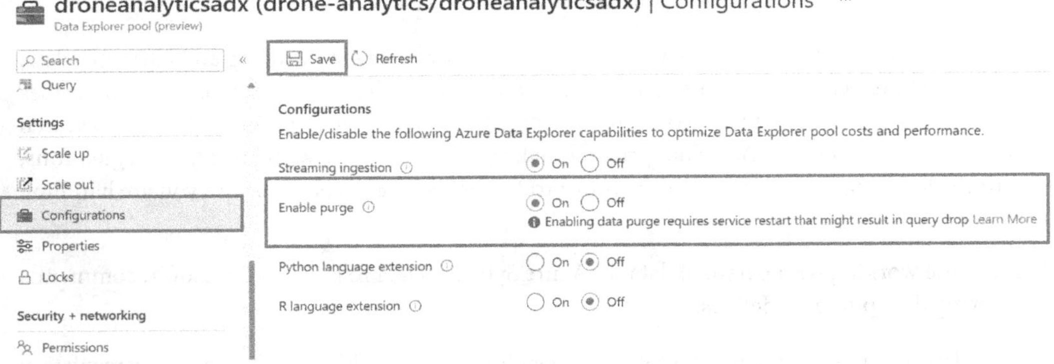

Figure 13.7 – Enabling data purges on a Data Explorer pool

> **Note**
> Changing this setting will cause your Data Explorer pool to restart and remain unavailable for a few minutes. Any ongoing queries will be canceled and users will be disconnected. You should plan this change accordingly to avoid user impact.

5. When you are ready, select **Save**.

After your Data Explorer pool restarts, you will be ready to perform purge operations. There are two ways you can invoke a purge operation:

- **Programmatic invocation**: This option runs as a single step and is meant to be used by automated processes only, without human intervention. This option is enforced by using the `noregrets` option of the `.purge` command.
- **Human invocation**: This option runs as a two-step process, requiring explicit confirmation of the purge action on the second step. This is the equivalent of when you are asked a question such as "Are you sure you want to delete all data?" on applications or operating systems. The second step requires a verification token to proceed, which is obtained as a result of running the first step of the process.

Next, let's look at how to use the human invocation approach to perform a data purge operation.

Executing data purge operations

To run data purge tasks, you need to connect to the Data Management service of the Data Explorer pool. You can use the data ingestion endpoint of your Data Explorer pool to connect to the Data Management service. The data ingestion endpoint for Azure Synapse Data Explorer is in the format of `https://ingest-<DataExplorerPoolName>.<WorkspaceName>.kusto.azuresynapse.net`. The data ingestion endpoint can be found on the **Overview** page of your Data Explorer pool in the Azure portal or Azure Synapse Studio by selecting your Data Explorer pool under the **Manage** hub, as seen in *Figure 13.8*.

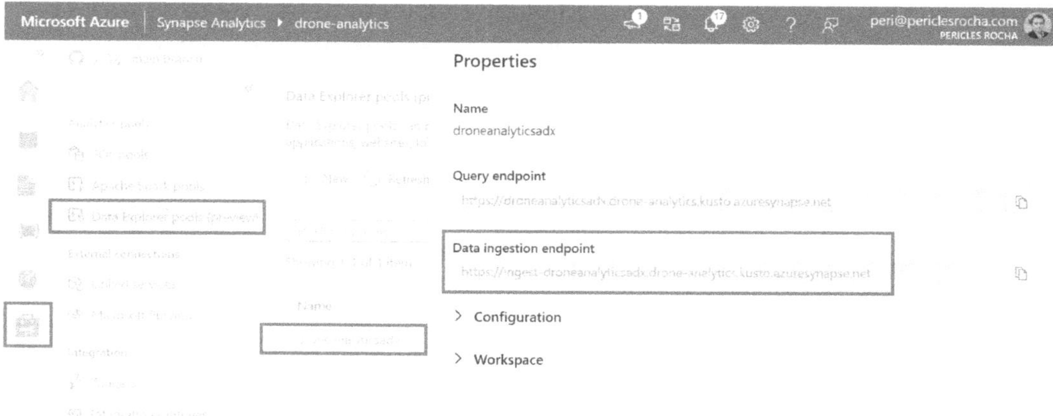

Figure 13.8 – Finding your data ingestion endpoint

At the time of writing this book, it was not possible to connect to the data ingestion endpoint directly from Azure Synapse Studio, so we will use Azure Data Explorer's web interface to do that. To connect to the data ingestion endpoint, perform the following steps:

1. Navigate to `https://dataexplorer.azure.com/` and log in with your Azure credentials.
2. On the **Query** hub, select the **Add cluster** button, as seen in *Figure 13.9*:

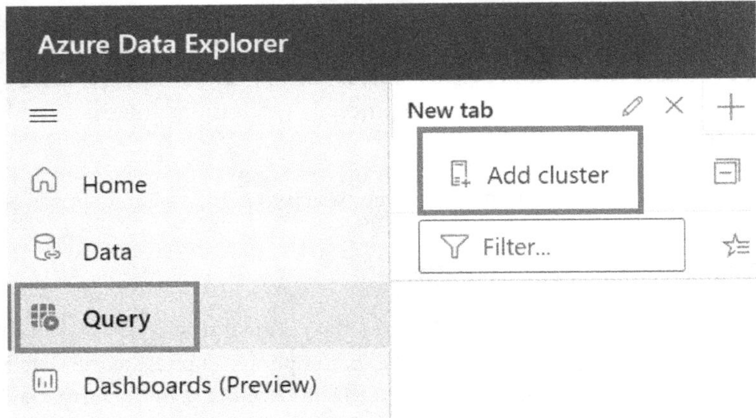

Figure 13.9 – Adding a new cluster connection

3. In the **Add cluster** dialog, provide the URI of your data ingestion endpoint in the **Connection URI** text box. As an example, the URI of my endpoint was `https://ingest-droneanalyticsadx.drone-analytics.kusto.azuresynapse.net`.
4. Click **Add** when you are ready.

To run the two-step purge process, first, we must run the `.purge table` command, providing the name of the table and database where data will be purged with the query predicate that selects the desired rows. For example, the following code block deletes all records of the `fleet data with tags - archive` table in the `drone telemetry` database that are older than 90 days:

```
.purge table ['fleet data with tags - archive'] records in
database ['drone-telemetry'] <| where todatetime(LocalDateTime)
> now() - time(90d)
```

> **Note**
>
> All the code examples in this section can be found in the `Chapter 13\Purging Personal Data.kql` file of this book's GitHub repository.

As you run this command, as shown in *Figure 13.10*, as a result, you get the number of records that will be purged, the estimated time needed to run the purge process, and the verification token (truncated due to space):

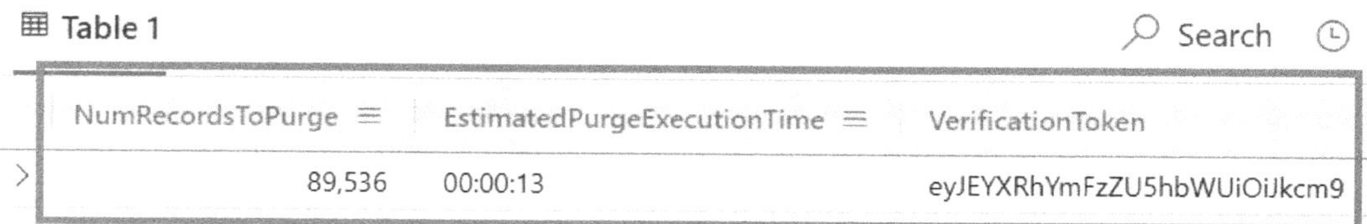

Table 1
Search
NumRecordsToPurge
EstimatedPurgeExecutionTime
VerificationToken
89,536
00:00:13
eyJEYXRhYmFzZU5hbWUiOiJkcm9

Figure 13.10 – Retrieving a verification token

Make sure you select the value in the **VerificationToken** column of the command output and copy it to the clipboard; you will need it next.

> **Note**
> This command purges data from the last 90 days, based on the values of the `LocalDateTime` column, at the time of writing this book. Depending on when you read this book, you may need to adjust the date range to select more or fewer rows accordingly, based on your current date and the values stored in the `LocalDateTime` column.

Now that you have the verification token, you can copy it and apply it to the following command to confirm the data purge operation:

```
.purge table ['fleet data with tags - archive'] records in
database ['drone-telemetry'] with (verificationtoken=h'ey-
JEYXRhYmFzZU5hbWUiOiJkcm9uZS10ZWxlbWV0cnkiLCJUYWJsZU5hbWUi-
OiJmbGVldCBkYXRhIHdpdGggdGFncyAtIGFyY2hpdmUiLCJQcmVkaWNhdGUi-
OiJ3aGVyZSB0b2RhdGV0aW1lKExvY2FsRGF0ZVRpbWUpID4gbm93KCkgLS-
B0aW1lKDQwNWQpIn0=') <| where todatetime(LocalDateTime) > now()
- time(90d)
```

This command schedules the data purge operation and produces an output similar to what is seen in *Figure 13.11*. Among all the information that it outputs, you will receive an **OperationId**, which you can use to track the status of the purge operation later:

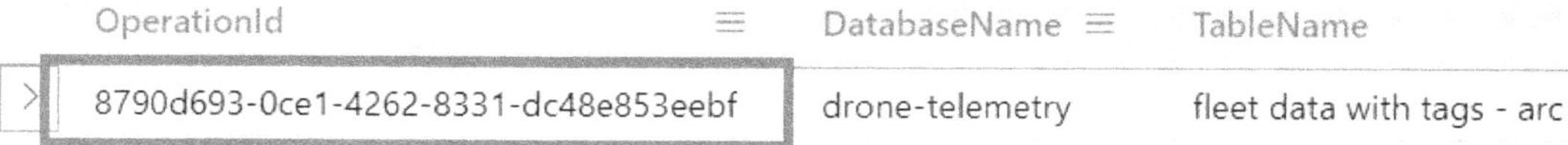

OperationId
DatabaseName
TableName
8790d693-0ce1-4262-8331-dc48e853eebf
drone-telemetry
fleet data with tags - arc

Figure 13.11 – OperationId obtained from starting the data purge process

The operation runs asynchronously, and you can monitor the status of the purge operation to see its status. We'll look at this next.

Monitoring data purge operations

You can track the status of the purge operation by using the `.show purges` command. This command accepts `OperationId` as a parameter, though you can use this command to show the status of all purge operations on a database, or in a date range. For example, the following command shows the status of a purge operation with a specific ID:

```
.show purges 8790d693-0ce1-4262-8331-dc48e853eebf
```

As an alternative, if you would like to see all purge operations in a given database, you can use the following command:

```
.show purges in database ['drone-telemetry']
```

To see purge operations on a date range, you can use the following command:

```
.show purges from '2022-12-19 00:00' to '2022-12-24 23:59' in
database ['drone-telemetry']
```

In the date range example, you can also omit the end date (the `to` clause) to see all data purges since the date provided in the `from` clause:

```
.show purges from '2022-12-19 00:00' in database ['drone-
telemetry']
```

All these options produce a similar output, which provides extensive details about the purge operation, including its status, similar to what you can see in *Figure 13.12*:

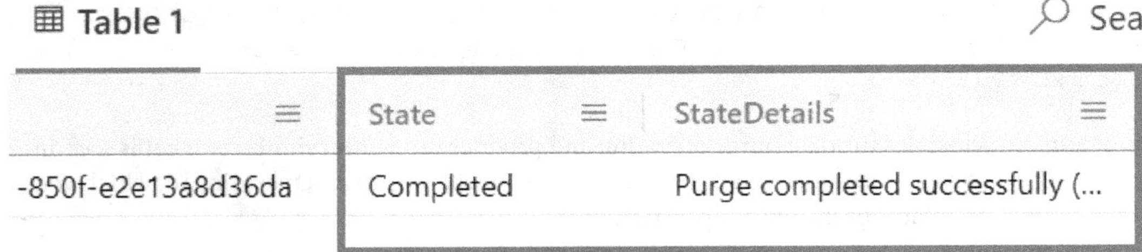

Figure 13.12 – Tracking the status of data purge commands

The purge process is quite complex behind the scenes, and hard deletion of all data can take days. First, the Data Explorer engine scans a table and identifies extents that contain rows that need to be purged. Any extent with one or more rows is selected at this stage. Next, Data Explorer performs a soft deletion of data: it writes new extents that replace the extents selected in the first step, but only inserts rows that should not be purged into the new extents. This stage can last anything from a few seconds to several hours, depending on the number of records to be purged, the size of your Data Explorer pool, how records are distributed across the cluster, and other factors. In the third and final stage, Data Explorer performs hard deletion: it deletes storage artifacts that contain the purged data. This stage takes anything between 5 and 30 days to complete after the purge operation started. There is no way to recover purged data, so make sure you plan these actions accordingly.

One final consideration for data purges is that you pay attention to data that may persist in materialized views. A materialized view is a database object that contains the results of a query persisted on disk for faster results of subsequent requests. Even though materialized views are refreshed frequently, they may still contain data that you would like to have purged from your data environment. You should always make sure you purge data that needs to be removed from materialized views to ensure you are satisfying regulatory requirements.

> **Note**
>
> Materialized views are an important resource to provide fast results to frequent queries. You can learn more about it at `https://learn.microsoft.com/azure/data-explorer/kusto/management/materialized-views/materialized-view-overview`.

Purging materialized views involves using the same syntax that was demonstrated for tables, except that you should replace `.purge table` with `.purge materialized-view`. The rest of the syntax is the same. For example, the following command purges data from a materialized view using the single-step approach (note the `noregrets` parameter):

```
.purge materialized-view ['fleet data with tags - archive
- mv'] records in database ['drone-telemetry'] with
(noregrets='true') <| where todatetime(LocalDateTime) > now() -
time(90d)
```

The rest of the syntax for tracking purge operations also applies to materialized views.

Summary

In this chapter, you learned advanced aspects of data management regarding Data Explorer pools. We started this journey by discussing extents, and how Azure Synapse Data Explorer uses data sharding techniques to split very large tables into smaller, more manageable files across the cluster. You learned about merge policies and how the Data Explorer pool engine decides which extents should be merged when needed. You also learned about tagging extents and how to move them to a different table for archival, as well as how to drop extents.

Next, you learned about how to implement data purges in Data Explorer pool databases to ensure you adhere to regulatory requirements, such as the GDPR. You learned how to enable a data purge in your Data Explorer pool, and how to use the two-step process to start and eventually trigger the data purge process. We also covered how you could track data purges to see their statuses, as they run asynchronously, and how the data purge process takes several days to complete behind the scenes.

This was the last chapter of this book. I sincerely hope the topics presented and the examples provided help inspire your journey of Azure Synapse Data Explorer. The possibilities are truly endless and some of these chapters could be a book of their own. I expect that you learn the topics with enough detail that they trigger your curiosity and make you want to dig even deeper into the topics proposed.

From one data person to another, thank you for being a part of this journey.

Index

Symbols

A

B

C

D

I

J

K

L

M

R

S

T

U

V

W

packtpub.com

Subscribe to our online digital library for full access to over 7,000 books and videos, as well as industry leading tools to help you plan your personal development and advance your career. For more information, please visit our website.

Why subscribe?

- Spend less time learning and more time coding with practical eBooks and Videos from over 4,000 industry professionals
- Improve your learning with Skill Plans built especially for you
- Get a free eBook or video every month
- Fully searchable for easy access to vital information
- Copy and paste, print, and bookmark content

Did you know that Packt offers eBook versions of every book published, with PDF and ePub files available? You can upgrade to the eBook version at packtpub.com and as a print book customer, you are entitled to a discount on the eBook copy. Get in touch with us at customercare@packtpub.com for more details.

At www.packtpub.com, you can also read a collection of free technical articles, sign up for a range of free newsletters, and receive exclusive discounts and offers on Packt books and eBooks.

Other Books You May Enjoy

If you enjoyed this book, you may be interested in these other books by Packt:

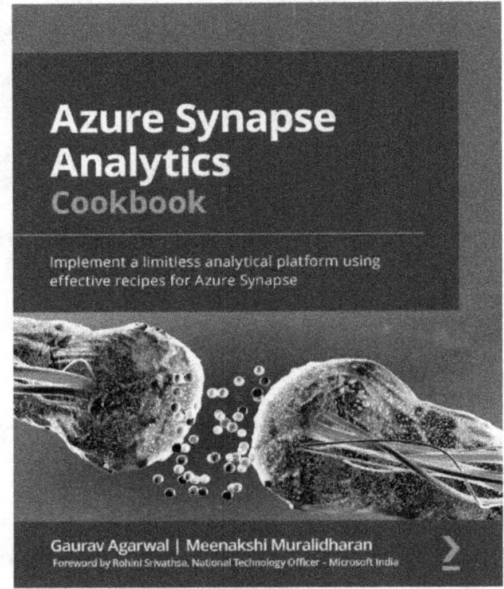

Azure Synapse Analytics Cookbook

Gaurav Agarwal, Meenakshi Muralidharan

ISBN: 978-1-80323-150-1

- Discover the optimal approach for loading and managing data
- Work with notebooks for various tasks, including ML
- Run real-time analytics using Azure Synapse Link for Cosmos DB
- Perform exploratory data analytics using Apache Spark
- Read and write DataFrames into Parquet files using PySpark
- Create reports on various metrics for monitoring key KPIs
- Combine Power BI and Serverless for distributed analysis
- Enhance your Synapse analysis with data visualizations

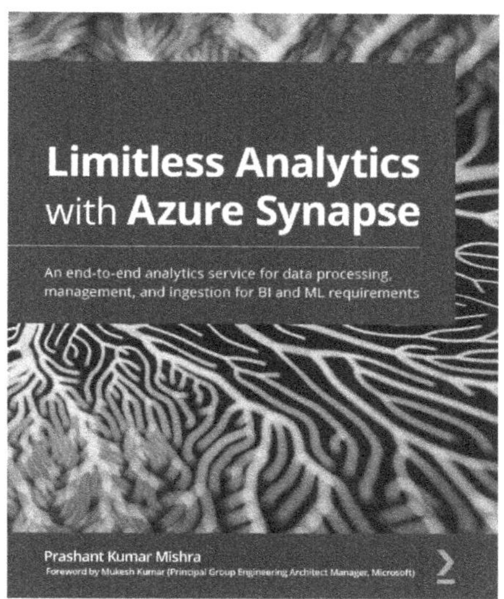

Limitless Analytics with Azure Synapse

Prashant Kumar Mishra

ISBN: 978-1-80020-565-9

- Explore the necessary considerations for data ingestion and orchestration while building analytical pipelines
- Understand pipelines and activities in Synapse pipelines and use them to construct end-to-end data-driven workflows
- Query data using various coding languages on Azure Synapse
- Focus on Synapse SQL and Synapse Spark
- Manage and monitor resource utilization and query activity in Azure Synapse
- Connect Power BI workspaces with Azure Synapse and create or modify reports directly from Synapse Studio
- Create and manage IP firewall rules in Azure Synapse

Packt is searching for authors like you

If you're interested in becoming an author for Packt, please visit `authors.packtpub.com` and apply today. We have worked with thousands of developers and tech professionals, just like you, to help them share their insight with the global tech community. You can make a general application, apply for a specific hot topic that we are recruiting an author for, or submit your own idea.

Share Your Thoughts

Now you've finished *Learn Azure Synapse Data Explorer*, we'd love to hear your thoughts! Scan the QR code below to go straight to the Amazon review page for this book and share your feedback or leave a review on the site that you purchased it from.

`https://packt.link/r/1-803-23395-8`

Your review is important to us and the tech community and will help us make sure we're delivering excellent quality content.

Download a free PDF copy of this book

Thanks for purchasing this book!

Do you like to read on the go but are unable to carry your print books everywhere? Is your eBook purchase not compatible with the device of your choice?

Don't worry, now with every Packt book you get a DRM-free PDF version of that book at no cost.

Read anywhere, any place, on any device. Search, copy, and paste code from your favorite technical books directly into your application.

The perks don't stop there, you can get exclusive access to discounts, newsletters, and great free content in your inbox daily

Follow these simple steps to get the benefits:

1. Scan the QR code or visit the link below

`https://packt.link/free-ebook/9781803233956`

2. Submit your proof of purchase
3. That's it! We'll send your free PDF and other benefits to your email directly

www.ingramcontent.com/pod-product-compliance
Lightning Source LLC
Chambersburg PA
CBHW062056050326
40690CB00016B/3117

* 9 7 8 1 8 0 3 2 3 3 9 5 6 *